To Go Where Nature Leads: Focus on Palmitoylethanolamide and Related ALIAmides as Innovative Approach to Neuroinflammatory and Pain-Related Disease States in Honor of Doctor Francesco Della Valle

To Go Where Nature Leads: Focus on Palmitoylethanolamide and Related ALIAmides as Innovative Approach to Neuroinflammatory and Pain-Related Disease States in Honor of Doctor Francesco Della Valle

Editors

Salvatore Cuzzocrea
Rosalia Crupi

MDPI

Basel • Beijing • Wuhan • Barcelona • Belgrade • Novi Sad • Cluj • Manchester

Editors

Salvatore Cuzzocrea
Department of Chemical,
Biological, Pharmaceutical
and Environmental Sciences,
University of Messina
Messina, Italy

Rosalia Crupi
Department of Chemical,
Biological, Pharmaceutical
and Environmental Sciences,
University of Messina
Messina, Italy

Editorial Office
MDPI
St. Alban-Anlage 66
4052 Basel, Switzerland

This is a reprint of articles from the Special Issue published online in the open access journal *Biomolecules* (ISSN 2218-273X) (available at: https://www.mdpi.com/journal/biomolecules/special_issues/Palmitoylethanolamide).

For citation purposes, cite each article independently as indicated on the article page online and as indicated below:

Lastname, A.A.; Lastname, B.B. Article Title. *Journal Name* **Year**, *Volume Number*, Page Range.

ISBN 978-3-0365-9474-3 (Hbk)
ISBN 978-3-0365-9475-0 (PDF)
doi.org/10.3390/books978-3-0365-9475-0

Cover image courtesy of Epitech Group

Contents

biomolecules

MDPI

Editorial

To Go Where Nature Leads: Focus on Palmitoylethanolamide and Related ALIAmides as Innovative Approach to Neuroinflammatory and Pain-Related Disease States in Honor of Doctor Francesco Della Valle

Salvatore Cuzzocrea [1,*] and Rosalia Crupi [2,*]

1 Department of Chemical, Biological, Pharmaceutical and Environmental Sciences, University of Messina, 98166 Messina, Italy
2 Department of Veterinary Science, University of Messina, 98166 Messina, Italy
* Correspondence: salvatore.cuzzocrea@unime.it (S.C.); rosalia.crupi@unime.it (R.C.)

Citation: Cuzzocrea, S.; Crupi, R. To Go Where Nature Leads: Focus on Palmitoylethanolamide and Related ALIAmides as Innovative Approach to Neuroinflammatory and Pain-Related Disease States in Honor of Doctor Francesco Della Valle. *Biomolecules* **2023**, *13*, 1583. https://doi.org/10.3390/biom13111583

Received: 23 October 2023
Accepted: 24 October 2023
Published: 26 October 2023

Dr. Francesco Della Valle made a significant impact in both the business and scientific domains. Reading for a degree in chemistry, with an emphasis on organic and biological disciplines, he quickly distinguished himself in the pharmaceutical industry. This distinction was largely due to the industrial ventures he passionately and adeptly led. His fervor for science resulted in him receiving three honorary degrees and co-inventing numerous pharmaceutical patents, including more than 100 in Europe, with broad recognition in major countries, and as many active patents in the United States.

A quintessential man of science, Dr. Della Valle was simultaneously a visionary entrepreneur always in search of innovation and a thorough, meticulous manager who steered his company with determination. Recognizing the importance of open innovation, he valued the synergistic relationships between academia and industry [1].

A devoted scholar of neuroscience, Dr. Della Valle collaborated with Nobel laureate Rita Levi Montalcini for many years. He deeply grasped the significance of biological breakthroughs like the nerve growth factor (NGF), drawing him closer to Professor Montalcini [2]. "Try to understand the strategy that Nature would use to protect itself from harm", was the counsel given to him by the Nobel laureate in 1991 [3]. Inspired by this advice, Della Valle pursued the discovery and development of substances that could emulate nature's protective strategies through the ALIA mechanism [4]. He was profoundly invested in studying neuroinflammation as the foundation for various ailments, ranging from neuropathic pain to neurological diseases [5]. His scientific journey was marked by a relentless "quest for life", gleaning insights and motivation from nature.

Dr. Francesco Della Valle leaves behind a legacy of pharmaceutical patents, scientific publications, pharmacological innovations, and concepts of neuroinflammation and neuropathic pain. Yet, his most invaluable legacy is his passion for science, truth, dialogue, and life itself.

The research presented in this Issue underscores the importance of continuing studies on PEA and related ALIAmides, using insights from nature to tackle new and increasingly complex challenges. We would like to express a heartfelt thank you to all the authors, reviewers, and contributors who have made the publication of this Special Issue possible. Our hope is that the information shared will inspire further research and exploration.

Here is a List of contributions contained in the Special Issue.

Conflicts of Interest: The authors declare no conflict of interest.

List of Contributions:

1. Palenca, I.; Seguella, L.; Del Re, A.; Franzin, S.B.; Corpetti, C.; Pesce, M.; Rurgo, S.; Steardo, L.; Sarnelli, G.; Esposito, G. N-Palmitoyl-D-Glucosamine Inhibits TLR-4/NLRP3 and Improves DNBS-Induced Colon Inflammation through a PPAR-α-Dependent Mechanism. *Biomolecules* **2022**, *12*, 1163. https://doi.org/10.3390/biom12081163.
2. Cristiano, C.; Avagliano, C.; Cuozzo, M.; Liguori, F.M.; Calignano, A.; Russo, R. The Beneficial Effects of Ultramicronized Palmitoylethanolamide in the Management of Neuropathic Pain and Associated Mood Disorders Induced by Paclitaxel in Mice. *Biomolecules* **2022**, *12*, 1155. https://doi.org/10.3390/biom12081155.
3. Guida, F.; Rocco, M.; Luongo, L.; Persiani, P.; Vulpiani, M.C.; Nusca, S.M.; Maione, S.; Coluzzi, F. Targeting Neuroinflammation in Osteoarthritis with Intra-Articular Adelmidrol. *Biomolecules* **2022**, *12*, 1453. https://doi.org/10.3390/biom12101453.
4. Valenza, M.; Facchinetti, R.; Steardo, L.; Scuderi, C. Palmitoylethanolamide and White Matter Lesions: Evidence for Therapeutic Implications. *Biomolecules* **2022**, *12*, 1191. https://doi.org/10.3390/biom12091191.
5. della Rocca, G.; Re, G. Palmitoylethanolamide and Related ALIAmides for Small Animal Health: State of the Art. *Biomolecules* **2022**, *12*, 1186. https://doi.org/10.3390/biom12091186.
6. Assogna, M.; Di Lorenzo, F.; Martorana, A.; Koch, G. Synaptic Effects of Palmitoylethanolamide in Neurodegenerative Disorders. *Biomolecules* **2022**, *12*, 1161. https://doi.org/10.3390/biom12081161.
7. Cifelli, P.; Ruffolo, G.; Ceccanti, M.; Cambieri, C.; Libonati, L.; Palma, E.; Inghilleri, M. Classical and Unexpected Effects of Ultra-Micronized PEA in Neuromuscular Function. *Biomolecules* **2022**, *12*, 758. https://doi.org/10.3390/biom12060758.
8. Landolfo, E.; Cutuli, D.; Petrosini, L.; Caltagirone, C. Effects of Palmitoylethanolamide on Neurodegenerative Diseases: A Review from Rodents to Humans. *Biomolecules* **2022**, *12*, 667. https://doi.org/10.3390/biom12050667.
9. Crupi, R.; Cuzzocrea, S. Role of EPA in Inflammation: Mechanisms, Effects, and Clinical Relevance. *Biomolecules* **2022**, *12*, 242. https://doi.org/10.3390/biom12020242.
10. Toti, A.; Micheli, L.; Lucarini, E.; Ferrara, V.; Ciampi, C.; Margiotta, F.; Failli, P.; Gomiero, C.; Pallecchi, M.; Bartolucci, G.; et al. Ultramicronized N-Palmitoylethanolamine Regulates Mast Cell-Astrocyte Crosstalk: A New Potential Mechanism Underlying the Inhibition of Morphine Tolerance. *Biomolecules* **2023**, *13*, 233. https://doi.org/10.3390/biom13020233.
11. Schiano Moriello, A.; Roviezzo, F.; Iannotti, F.A.; Rea, G.; Allarà, M.; Camerlingo, R.; Verde, R.; Di Marzo, V.; Petrosino, S. First Evidence of the Protective Effects of 2-Pentadecyl-2-Oxazoline (PEA-OXA) in In Vitro Models of Acute Lung Injury. *Biomolecules* **2023**, *13*, 33. https://doi.org/10.3390/biom13010033.
12. Elfarnawany, A.; Dehghani, F. Palmitoylethanolamide Mitigates Paclitaxel Toxicity in Primary Dorsal Root Ganglion Neurons. *Biomolecules* **2022**, *12*, 1873. https://doi.org/10.3390/biom12121873.
13. Interdonato, L.; D'amico, R.; Cordaro, M.; Siracusa, R.; Fusco, R.; Peritore, A.F.; Gugliandolo, E.; Crupi, R.; Coaccioli, S.; Genovese, T.; et al. Aerosol-Administered Adelmidrol Attenuates Lung Inflammation in a Murine Model of Acute Lung Injury. *Biomolecules* **2022**, *12*, 1308. https://doi.org/10.3390/biom12091308.

References

1. Della Valle, F.; Gambardella, A. 'Biological' Revolution and Strategies for Innovation in Pharmaceutical Companies. *R&D Manag.* **1993**, *23*, 287–302.
2. Aloe, L. Rita Levi-Montalcini: The Discovery of Nerve Growth Factor and Modern Neurobiology. *Trends Cell Biol.* **2004**, *14*, 395–399. [CrossRef] [PubMed]
3. Aloe, L.; Leon, A.; Levi-Montalcini, R. A Proposed Autacoid Mechanism Controlling Mastocyte Behaviour. *Agents Actions* **1993**, *39*, C145–C147. [CrossRef] [PubMed]
4. Levi-Montalcini, R.; Skaper, S.D.; Dal Toso, R.; Petrelli, L.; Leon, A. Nerve Growth Factor: From Neurotrophin to Neurokine. *Trends Neurosci.* **1996**, *19*, 514–520. [CrossRef] [PubMed]
5. Petrosino, S.; Schiano Moriello, A. Palmitoylethanolamide: A Nutritional Approach to Keep Neuroinflammation within Physiological Boundaries—A Systematic Review. *Int. J. Mol. Sci.* **2020**, *21*, 9526. [CrossRef] [PubMed]

biomolecules

MDPI

Article

Ultramicronized *N*-Palmitoylethanolamine Regulates Mast Cell-Astrocyte Crosstalk: A New Potential Mechanism Underlying the Inhibition of Morphine Tolerance

Alessandra Toti [1], Laura Micheli [1], Elena Lucarini [1,*], Valentina Ferrara [1], Clara Ciampi [1], Francesco Margiotta [1], Paola Failli [1], Chiara Gomiero [2], Marco Pallecchi [3], Gianluca Bartolucci [3], Carla Ghelardini [1] and Lorenzo Di Cesare Mannelli [1]

1 Department of Neuroscience, Psychology, Drug Research and Child Health—NEUROFARBA—Pharmacology and Toxicology Section, University of Florence, Viale Pieraccini 6, 50139 Florence, Italy
2 Epitech Group SpA, Via Luigi Einaudi 13, 35030 Padua, Italy
3 Department of Chemistry, University of Florence, Via Ugo Schiff 6, 50019 Florence, Italy
* Correspondence: elena.lucarini@unifi.it; Tel.: +39-055-2758395

Abstract: Persistent pain can be managed with opioids, but their use is limited by the onset of tolerance. Ultramicronized *N*-palmitoylethanolamine (PEA) in vivo delays morphine tolerance with mechanisms that are still unclear. Since glial cells are involved in opioid tolerance and mast cells (MCs) are pivotal targets of PEA, we hypothesized that a potential mechanism by which PEA delays opioid tolerance might depend on the control of the crosstalk between these cells. Morphine treatment (30 μM, 30 min) significantly increased MC degranulation of RBL-2H3 cells, which was prevented by pre-treatment with PEA (100 μM, 18 h), as evaluated by β-hexosaminidase assay and histamine quantification. The impact of RBL-2H3 secretome on glial cells was studied. Six-hour incubation of astrocytes with control RBL-2H3-conditioned medium, and even more so co-incubation with morphine, enhanced CCL2, IL-1β, IL-6, Serpina3n, EAAT2 and GFAP mRNA levels. The response was significantly prevented by the secretome from PEA pre-treated RBL-2H3, except for GFAP, which was further upregulated, suggesting a selective modulation of glial signaling. In conclusion, ultramicronized PEA down-modulated both morphine-induced MC degranulation and the expression of inflammatory and pain-related genes from astrocytes challenged with RBL-2H3 medium, suggesting that PEA may delay morphine tolerance, regulating MC-astrocyte crosstalk.

Keywords: morphine; tolerance; *N*-palmitoylethanolamine; mast cell; glial cells; cross-talk

Citation: Toti, A.; Micheli, L.; Lucarini, E.; Ferrara, V.; Ciampi, C.; Margiotta, F.; Failli, P.; Gomiero, C.; Pallecchi, M.; Bartolucci, G.; et al. Ultramicronized *N*-Palmitoylethanolamine Regulates Mast Cell-Astrocyte Crosstalk: A New Potential Mechanism Underlying the Inhibition of Morphine Tolerance. *Biomolecules* **2023**, *13*, 233. https://doi.org/10.3390/biom13020233

Academic Editor: Salvatore Cuzzocrea

Received: 2 December 2022
Revised: 10 January 2023
Accepted: 13 January 2023
Published: 25 January 2023

1. Introduction

Pain management is still one of the most significant problems that affect patient quality of life [1]. Pain can result from injury [2], surgery [3] or disease, such as type 1 and 2 diabetes, osteoarthritis as well as cancer [4–6] and it can also arise as a side effect of anti-cancer chemotherapies [7]. Therapies for pain relief depend on the intensity or persistence of pain; for mild pain acetaminophen, non-steroidal anti-inflammatory drugs (NSAIDs) such as aspirin, ibuprofen and naproxen are recommended, while in the case of severe pain, opioids such as morphine are the first choice [8].

The analgesic effect of morphine is mainly exerted by μ-receptor [9], although morphine also interacts, albeit weakly, with the κ and δ receptors [10]. Morphine receptors are distributed in the peripheral and central nervous system (PNS and CNS), especially in the periaqueductal gray matter and posterior horns of the spinal cord, but they are also present in different peripheral tissues and organs, being responsible for several side effects, including vomiting, nausea, constipation, pruritus, and respiratory depression. These effects typically arise within the first weeks of treatment and are likely to become more severe with time, when cognitive failure, tolerance and addiction may develop [11].

In particular, tolerance is defined as the necessity to increase the dose of morphine in order to obtain a good analgesic effect after repeated treatment and can be managed with opioid rotation, as long as the side effects do not lead to treatment withdrawal [12]. In recent decades, morphine tolerance has been attributed to the desensitization/internalization of μ-receptors and the involvement of glial cell overactivation has been recently suggested [13,14].

Glial cells (i.e., astrocytes and microglia together with oligodendrocytes and Schwann cells), not only exert supportive functions, but are increasingly considered to play a pivotal role in the transmission of pain and regulation of numerous pathophysiological processes [15–17]. In rats daily administered with morphine 10 mg/kg i.p. a significant increase in the number of glial cells was observed in the dorsal horn of the spinal cord, once tolerance has developed (i.e., after 6 days) [18]. Accordingly, it has been recently demonstrated that glial selective inhibitors as fluorocitrate, minocycline, propentofylline, and pentoxifylline delay morphine tolerance [19–22].

N-palmitoylethanolamine (PEA) is a glia modulator and a pro-homeostatic lipid mediator, belonging to the family of fatty acid ethanolamines, which are produced on demand from the lipid bilayer [23]. Recent studies have demonstrated the efficacy of PEA against persistent pain [24] and neurological disorders [25] with an optimal safety profile in humans [26]. Increasing information is currently available on the pleiotropic mechanism of action of PEA, which behaves as a cannabinomimetic compound, stimulating the effects of endocannabinoids through the so-called entourage effect [27]. Moreover, PEA is an agonist of the peroxisome proliferator-activated receptor type α (PPAR-α), a nuclear transcription factor able to inhibit both pro-inflammatory cytokine release [28] and the activity of enzymes such as cyclooxygenase [29]. In a model of chronic constriction injury, we have previously demonstrated that pain-relieving effects of PEA depend, at least in part, on a PPAR-α-mediated mechanism [30].

Notably, treatment with a bioavailable PEA formulation (i.e., ultramicronized PEA) was shown to delay the onset of morphine tolerance and prevent the increase of glial cells number in the dorsal horn of the spinal cord [18]. The mechanisms sustaining this effect of PEA are still unclear. It is known that morphine activates mast cells (MCs) in vivo and enhances histamine release [31–33], while PEA down-modulates MC degranulation [34] and astrocyte activity [35].

Based on this background, the aim of the present study was to investigate if a crosstalk exists between astrocytes and MCs during morphine treatment and if PEA can regulate it. Despite the well-known limitations of 2D cell cultures, the present study might provide novel insights in the mechanism(s) underlying PEA-induced delay of morphine tolerance and non-neuronal cell(s) mainly involved in. The results could thus improve our understanding on the management of opioid tolerance and related neuroinflammatory disorders.

2. Materials and Methods

2.1. Materials

Ultramicronized PEA (99.9% powder particle size <6 μm) was kindly provided by Epitech Group (Saccolongo, Italy) and for simplicity will be referred to as PEA hereafter. PEA was dissolved in dimethyl sulfoxide (DMSO, Merck, Milan, Italy) at 10 mM, and 100 μM was used with a final concentration of 1% DMSO. Morphine hydrochloride was purchased from S.A.L.A.R.S. (Como, Italy) and dissolved in sterile water.

2.2. Animals

The Sprague-Dawley rats were purchased from the Envigo company (Varese, Italy) around the seventeenth day of pregnancy. They were housed in a 26 cm × 41 cm cage at the Laboratory Animal Housing Center (Ce.SAL) of the University of Florence, fed with a standard diet and water ad libitum and kept at 23 ± 1 °C with a light cycle/12 h dark, with light at 7 am. The rats were kept alive until delivery and the pups were sacrificed between day 1 and day 3 of life to isolate a primary culture of rat cortical astrocytes. All handling of animals were carried out in accordance with the guidelines of the European Community for the care of animals (DL 116/92, application of the directives of the Council of the European Community of 24 November 1986; 86/609/EEC). The ethics policy of the University of Florence complies with the decree Guide for the Care and Use of Laboratory Animals of the US National Institutes of Health (NIH Publication number 85-23, revised 1996; University of Florence assurance number: A5278-01). Formal approval to conduct the experiments described was obtained from the Italian Ministry of Health (No. 333/2020-PR) and from the Animal Subjects Review Board of the University of Florence. All efforts were made to minimize animal suffering and reduce the number of animals used.

2.3. Cell Culture

The rat basophilic leukemia cells (RBL)-2H3, purchased from American Type Culture Collection (ATCC, Manassas, VA, USA), were cultured in Eagle's Minimum Essential Medium (EMEM)-High Glucose growth medium (ATCC, Manassas, VA, USA), containing L-glutamine (2 mM), supplemented with 15% Heat Inactivated Fetal Bovine Serum (FBS, Euroclone, Milan, Italy), penicillin (100 U/mL) and streptomycin (100 μg/mL; Merck, Milan, Italy). The RBL-2H3 cell line was used as a commonly accepted model of MCs [36].

Primary cultures of cortical astrocytes were obtained according to the method described by McCarthy and de Vellis [37]. Briefly, the cerebral cortex of new-born Sprague-Dawley rats (Envigo, Varese, Italy) post-natal day 1–3 was dissociated in Hanks balanced salt solution containing 0.5% trypsin, 0.2% EDTA and 1% DNase (Merck, Milan, Italy) for 30 min at 37 °C. The suspension was mechanically homogenized and filtered with 70 μm filters. Cells were plated in Dulbecco's Modified Eagle's Medium (DMEM)-High Glucose (Merck, Milan, Italy) supplemented with 20% FBS (Euroclone, Milan, Italy), L-glutamine (2 mM), penicillin (100 U/mL) and streptomycin (100 μg/mL; Merck, Milan, Italy). Confluent primary glial cultures were used to isolate astrocytes, removing microglia and oligodendrocytes by shaking. After 21 days of culture, astrocytes were plated according to experimental requirements. Cells were maintained in incubator, at 37 °C, and in a humidified atmosphere of 5% CO_2.

2.4. Cell Viability Assay

The RBL-2H3 cells and astrocytes were plated at the density of 4×10^3 and 5×10^3 cells/well respectively in a 96-well plate. After the appropriate treatments at 24 and 48 h, the medium was removed and the cells incubated with a 1 mg/mL solution of Thiazolyl Blue Tetrazolium Bromide (MTT; Merck, Milan, Italy), at 37 °C, for 1 h and a half. The dissolution of the crystals was carried out with 200 μL of DMSO per well, and the absorbance was read at 590 nm using the EnSight Multimode Plate Reader (Perkin Elmer) [38].

2.5. Cell Count

The RBL-2H3 cells were plated in 6-well plate, at a density of 4×10^5 cells/well, sensitized with anti-DNP IgE (50 ng/mL) and concurrently treated with PEA 100 μM. After 24 and 48 h, cells were detached and counted using a Burker's chamber. A mean of two counts for condition with 4 replicates were analyzed [39].

2.6. β-hexosaminidase Assay

The RBL-2H3 cells were plated in a 24-well plate, at a density of 5×10^4 cells/well, and the next day cells were sensitized with anti-dinitrophenyl (DNP)-IgE (50 ng/mL) at 37 °C overnight. After 18 h, two washes with modified Tyrode's (MT) buffer (20 mM HEPES, 135 mM NaCl, 5 mM KCl, 1.8 mM $CaCl_2$, 1 mM $MgCl_2$, 5.6 mM glucose, and 0.05% BSA, pH 7.4) were performed and then cells were treated in 250 μL of MT buffer containing morphine 30 μM, PEA 10–100 μM or morphine 30 μM + PEA 100 μM for 30 min at 37 °C. Cells were then stimulated with DNP-BSA (625 ng/mL) for 5, 10, 15 and 30 min at 37 °C.

After stimulation, the supernatant was collected and the cells lysed with 250 μL of MT buffer containing 0.1% Triton X-100, and then sonicated for 30 s on ice. Both supernatants and cell lysates were transferred to a new 96-well plate, 50 μL per sample, and incubated for 1 h at 37 °C with 50 μL/well of 1 mM p-nitrophenyl-N-acetyl-d-glucosaminide (NAG; Merck, Milan, Italy) dissolved in 0.1 M citrate buffer (pH 4.5). The enzymatic reaction was then stopped by adding 100 μL/well of 0.1 M carbonate buffer (pH 10.0). The absorbance (A) was read at 405 nm using EnSight Multimode Plate Reader (Perkin Elmer). The percentage of β-hexosaminidase released was calculated as follows: 100 × [(A supern − A white)/(A supern − A white) + (A pellet − A white pellet)] [40].

2.7. Histamine Assay

The amount of histamine in RBL-2H3 media was measured by isotopic dilution (ID) bidimentional high performance liquid chromatography (2D-HPLC) with tandem mass spectrometry (MS/MS) method. The instrument employed was a Varian (Palo Alto, CA, USA) triple quadrupole 1200 L system (able to perform MS/MS experiments) coupled with three Prostar 210 pumps (each one can manage one solvent line), a Prostar 410 autosampler and an electro-spray ionization (ESI) source. All the raw data collected were processed by Varian Workstation (version 6.8) software. The system was equipped with two 6-port valves. The following columns were used: (1) SeQuant® ZIC-HILIC 20 × 2.1 mm as loading column and (2) SeQuant® ZIC-HILIC 50 × 2.1 mm as analytic column, both with a particle size of 3.5 μm. The elution was performed with the following solvents: (1) solvent A, mQ water:acetonitrile 9:1 solution added with 5 mM formic acid and 15 mM ammonium formate; (2) solvent B, mQ water:acetonitrile 1:9 solution added with 15 mM formic acid and 5 mM ammonium formate; and (3) solvent C (used for sample loading), mQ water:acetonitrile 1:9 solution added with 17.5 mM formic acid 2.5 mM ammonium formate. The loading time was carried out in isocratic elution with the solvent C for 2 min at flow of 0.5 mL min^{-1}. It was demonstrated that in 2 min the analyte is completely retained by the loading column. After 2 min the valve changed position and the gradient of A and solvents started in counter-flow from the loading to the analytic columns. The gradient started from 90% to 20% of solvent B in 7 min and held for 3 min, then it returned at 90% of solvent B in 0.1 min and restored the initial condition for 13 min, for a total run time of 25 min [41]. More details on the method are reported in Supplementary Materials (Tables S1–S5; Figure S1).

2.8. Collection of Mast Cell-Conditioned Medium

RBL-2H3 media were collected after appropriate treatment and centrifuged two times for 5 min at 200× *g* to eliminate floating cells and debris before treating astrocytes.

2.9. Timeline of Astrocytes Treatment with Mast Cell-Conditioned Medium

Primary astrocytes were plated in 6-well plate at a density of 2.5 × 10^4 cells/well and subsequently treated for 48 h with 30 μM morphine in DMEM with 1% FBS. In parallel, RBL-2H3 were plated in 6-well plate at a density of 4 × 10^5 cells/well and pre-sensitized the next day with anti DNP-IgE 50 ng/mL ± PEA 100 μM.

The next day, after two washes in phosphate buffer solution (PBS) to eliminate DNP-IgE and PEA, the RBL-2H3 were incubated in DMEM with 1% heat inactivate FBS and 0.05% BSA and then stimulated for 24 h with DNP-BSA 625 ng/mL. Conditioned media of RBL-2H3 were collected as previously described. At the end of 48 h of morphine treatment, astrocytes were incubated for 6 h with media collected from control and PEA pre-treated RBL-2H3 (see Scheme 1).

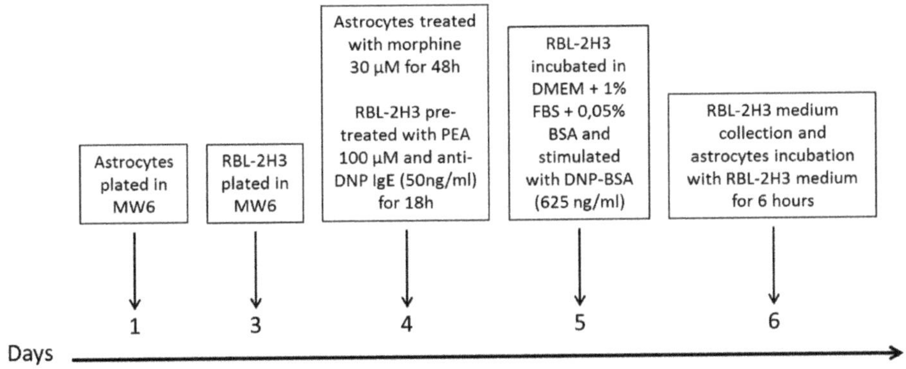

Scheme 1. Experimental scheme.

2.10. Protein Extraction, Electrophoresis, and Western Blot

Cells were lysed with 50 mM Tris-HCl pH 8.0, 150 mM NaCl, 1 mM EDTA, 0.5% Triton X-100 and complete protease inhibitors (Roche, Milan, Italy).

The lysates were collected, sonicated for 3 min in ice, vortexed for 1 min and finally centrifuged at 12,000× *g* for 10 min, at 4 °C, and the supernatant collected. Total protein concentrations were quantified by bicinchoninic acid test (Merck, Milan, Italy). An amount of 40 μg of proteins was resolved with precast polyacrylamide gel (BOLT 4–12% Bis-Tris Plus gel; Thermo Fisher Scientific, Monza, Italy) before electrophoretic transfer to nitrocellulose membranes (Bio-Rad, Milan, Italy). The membranes were blocked with 5% Blotto, non-fat dry milk (Santa Cruz Biotechnology, Dallas, TX, USA) dissolved in PBS-0.1% Tween (PBST; Merck, Milan, Italy) and then probed overnight at 4 °C with primary antibodies specific for mouse MOR-1 (1:1000; cod. sc-515933, Santa Cruz Biotechnology, Dallas, TX, USA) and mouse GAPDH (1:5000; cod. sc-32233, Santa Cruz Biotechnology, Dallas, TX, USA). The membranes were then incubated for 1 h in PBST containing the anti-mouse secondary antibody HRP conjugated (1:5000; Santa Cruz Biotechnology, Dallas, TX, USA). ECL (Enhanced Chemiluminescence Pierce, Rockford, IL, USA) was used to visualize peroxidase-coated bands. Densitometric analysis was performed using the ImageJ analysis software (ImageJ; NIH, Bethesda, MD, USA). Normalization with housekeeping gene GAPDH content was performed [42].

2.11. Immunofluorescence

The RBL-2H3 cells and astrocytes were plated at the density of 4×10^5 and 5×10^4 cells/well respectively on 22 × 22 mm cover slides placed in a 6-well plate. After the appropriate treatments, the cells were fixed with 4% paraformaldehyde for 15 min at room temperature. Permeabilization was performed by incubation for 1 h with PBS 1X-0.3% Triton and blocking phase with PBS 1X-0.3% Triton+0.5% BSA followed by overnight incubation at 4 °C with rabbit anti-mast cell tryptase (1:200, GeneTex, TX, USA) or rabbit anti-glial fibrillary acidic protein (GFAP; 1:200, cod. Z0334, Dako Agilent, Santa Clara, CA, USA) diluted in the blocking solution. The next day, after appropriate washing in PBS, the slides were incubated for 2 h with the Alexa Fluor ® 568 goat anti-rabbit conjugated secondary antibody (1:500, Life Milan, Italy) at room temperature, diluted in the blocking solution. Finally, nuclei were labeled with 4′,6-diamidine-2-phenylindole (DAPI; Life Technologies, Milan, Italy) for 5 min, diluted in PBS to a final concentration of 500 ng/mL. After washing in double distilled water, the slides were placed on a specimen slide using the Fluoromount™ aqueous mounting medium (Life Technologies, Milan, Italy). The images were acquired using a Leica DM6000 B motorized fluorescence microscope equipped with a DFC350FX camera (Leica, Mannheim, Germany). The intensity was calculated using imageJ and analyzing 10 fields per slide acquired randomly with a 20× objective [42].

2.12. Real Time Polymerase Chain Reaction (RT-PCR)

Total RNA was isolated from cells using NucleoSpin® RNA (Macherey-Nagel, Düren, Germany) following manufacturer's instruction. An amount of 500 nanograms of RNA was retrotranscribed using iScript cDNA Synthesis kit (Bio-Rad, Milan, Italy). RT-PCR was performed using SsoAdvanced Universal SYBR® Green Supermix (Bio-Rad, Milan, Italy) following the thermal profile suggested by the kit.

The following primers were used for detection:

rSerpina3n: forward 5′-CTTTCTGCAGTATGTGGGAATCACTTGG-3′ and reverse 5′-GGCTGCATTGCTCTAAGTAGGAGTGC-3′ (Invitrogen);

rCCL2: forward 5′-TCTTCCTCCACCACTATGCAGGTCTC-3′ and reverse 5′-TCTTT-GGGACACCTGCTGCTGGTG-3′ (Invitrogen);

rGFAP forward 5′-CTGACACACGTTGTGTTCAAGCAGCC-3′ and reverse 5′-CTG-AAGGTTAGCAGAGGTGACAAGGG-3′ (Invitrogen).

Validated primers to detect rIL-6 (qRnoCID0053166), rIL-1β (qRnoCID0004680), rEAAT2 (qRnoCED0005967), rTNFα (qRnoCED0009117), and rβ2microglobulin as housekeeping (qRnoCED0056999), were purchased from Bio-Rad (Bio-Rad, Milan, Italy). The differential expression of the transcripts was normalized on the housekeeping gene.

2.13. Measurement of Intracellular Ca^{2+}

Dynamic intracellular cytosolic Ca^{2+} ($[Ca^{2+}]i$) was evaluated in astrocytes plated on round glass coverslips (diameter 25 mm) and after the appropriate treatments they were labeled with 4 μM Fluo-4AM (Life technologies, Milan, Italy) for 45 min, at 37 °C. Cells were washed and kept in Hepes buffer (NaCl 140 mM, $NaHCO_3$ 12 mM, Hepes 10 mM, KCl 2.9 mM, NaH_2PO_4 0.5 mM, $CaCl_2$ 1.5 mM, glucose 10 mM, $MgCl_2$ 1.2 mM, pH = 7.4). The coverslips were mounted in a perfusion chamber and placed on the stage of an inverted reflected light fluorescence microscope (Zeiss Axio Vert. A1 FL-LED) equipped with fluorescence excitation (475 nm) with LED. Cells were incubated for at least 5 min in the control solution and then stimulated with 100 μM ATP. The dynamics of calcium was recorded for at least 3 min after administration of the agonist. The fluorescence of Fluo-4AM was recorded with a Tucsen Dhyana 400D CMOS camera (Tucsen Photonics Co., Ltd., Fuzhou, China) with a resolution of 1024 × 1020 $pixels^2$. Ca^{2+} dynamics was measured by single cell imaging analysis at 35 °C. The images were recorded using the Dhyana SamplePro software, recording 13 frames per second and dynamically analysed with the open-source community software for bioimaging Icy (Pasteur Institute, Paris, France). A signal-to-noise ratio of at least 5 arbitrary units of fluorescence (A.U) was considered as an agonist-induced increase in Ca^{2+}. The following parameters were evaluated: the ratio between the agonist-induced maximum change in fluorescence and the basal fluorescence ($\Delta F/F$, measured as A.U.), the time to reach the maximum calcium value and the area under the fluorescence calibrated curve baseline (AUC/F, evaluated as A.U.). The experiments were repeated in 3 different cell preparations, analysing at least 6 cells per optical field (using a 40× magnification objective) [43].

2.14. Statistical Analysis

Results were expressed as mean ± S.E.M. and analysis of variance (ANOVA) was performed.

A Bonferroni significant difference procedure was used as post hoc comparison. Data were analysed using the "Origin 8.1" software (OriginLab, Northampton, MA, USA). Differences among groups were considered significant at values of $p < 0.05$. * p-value < 0.05; ** p-value < 0.01; *** p-value < 0.001.

3. Results

3.1. Morphine and PEA, Used Solely or in Combination, Do Not Affect Mast Cell Viability

We first verified by MTT test that none of the investigated concentrations of morphine (0.3–300 µM) was toxic on RBL-2H3 cells at either 24 or 48 h treatment (Figure 1a). The 30 µM concentration was therefore chosen since it was observed to induce a consistent degranulation without any toxic effects. Then, increasing concentrations of PEA (10 nM–100 µM) together with 30 µM morphine were tested on RBL-2H3 cells by MMT. Figure 1b,c show that PEA at the concentration range 10 nM–10 µM, alone or in combination with 30 µM morphine, does not affect cell viability in a significant manner, either at 24 or 48 h. Although a decrease in RBL-2H3 cell viability was observed with PEA 50 µM and 100 µM, with the effect being worsened by morphine, PEA 100 µM (24 h treatment) did not affect the cell count, suggesting that it was not toxic at the higher concentration (Figure 1d). A decrease in the number of cells was observed at 48 h, probably due to an impaired proliferation more than a toxic effect of PEA (Figure 1d). For this reason, 24 h treatment was selected for further experiments. The total protein content at 24 h in control and PEA 100 µM-treated RBL-2H3 cells confirmed the findings from the cell count assay (Figure 1e).

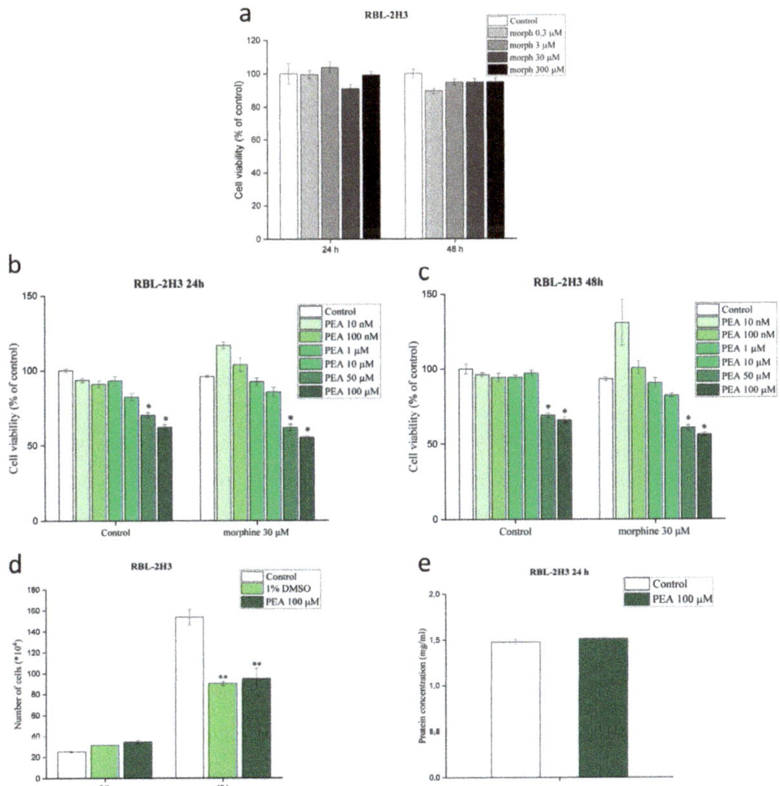

Figure 1. Effect of morphine and *N*-palmitoylethanolamine (PEA) on MC viability. (**a**) MTT of RBL-2H3 cells treated with increasing concentrations of morphine (0.3–300 µM) for 24 and 48 h; (**b**,**c**) MTT of RBL-2H3 treated with increasing concentrations of PEA (10 nM-100 µM) alone or in combination with 30 µM morphine for 24 and 48 h; control condition was arbitrarily set as 100%; (**d**) Cell count of RBL-2H3 cells treated with PEA 100 µM and 1% DMSO at 24 and 48 h; (**e**) Protein content in RBL-2H3 cells treated for 24 h with PEA 100 µM. Values are expressed as the mean ± S.E.M. of three experiments. * $p < 0.05$; ** $p < 0.01$ vs. control.

3.2. *RBL-2H3 Cells Express the μ-Receptor and Morphine Augments Mast Cell Degranulation*

Since opioid treatment is known to activate MCs [33], but no information is available on the molecular mechanism(s), we first evaluated the expression of morphine receptors on RBL-2H3 cells. Western blot analysis of MOR-1 demonstrated, for the first time, that RBL-2H3 cells express morphine μ-receptor (Figure 2a and Figure S2). Given the consistent basal degranulation, the RBL-2H3 cell activation protocol had first to be optimized (Figure S3) before investigating morphine-induced degranulation. Similar to what reported by other groups [44,45], RBL-2H3 cells were pre-sensitized with anti-DNP IgE (50 ng/mL) for 18 h. The next day, cells were incubated with morphine (0.3–300 μM) for 30 min in a buffer containing 0.05% BSA and then stimulated with DNP-BSA (625 ng/mL) for 10 min [46,47]. In these conditions, a controlled granule release was achieved. As reported in Figure 2b, morphine induced a concentration dependent decrease of tryptase intensity, suggesting an increased tryptase release and thus degranulation. To better investigate morphine-induced degranulation, the release of β-hexosaminidase was also studied and the percentage of β-hexosaminidase in the medium was compared to that inside the cells. A significant 15% increase in the release of β-hexosaminidase was observed in sensitized RBL-2H3 cells challenged with DNP-BSA for 5, 10, 15 and 30 min and treated with 30 μM morphine for 30 min (Figure 2c). Lengthening morphine incubation time up to 24 h did not generate any statistically significant increase (Figure 2d), probably due to the very short time needed by morphine to induce maximum RBL-2H3 cell degranulation.

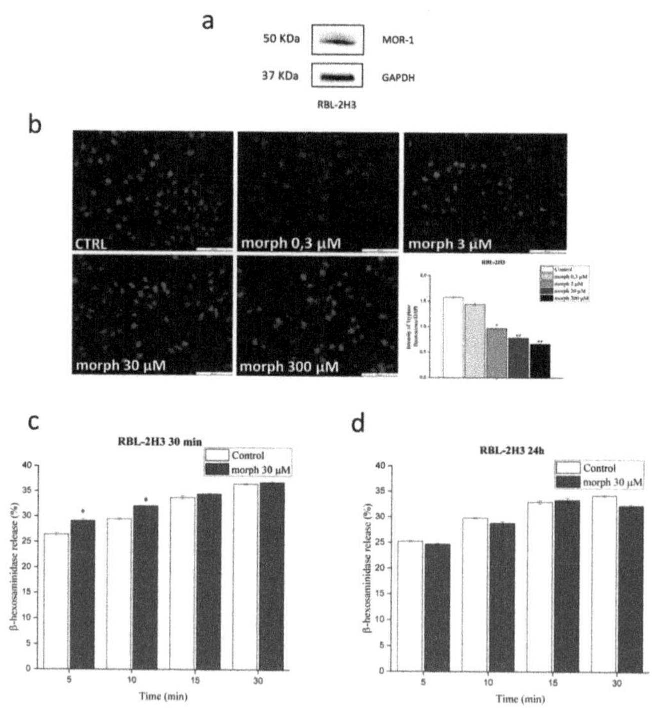

Figure 2. Effect of morphine on MC degranulation. (**a**) Western blot of MOR-1 (μ-receptor) on RBL-2H3 cells. The expression was compared to GAPDH levels. (**b**) Immunofluorescence for tryptase on RBL-2H3 cells pre-sensitized with anti-DNP IgE 50 ng/mL, treated the next day with increasing concentrations of morphine (0.3–300 μM) for 30 min and then stimulated with DNP-BSA for 10 min. Cell nuclei were labeled with DAPI. Scale bar = 50 μm. (**c**) β-hexosaminidase assay of RBL-2H3 cells pre-sensitized with anti DNP-IgE 50 ng/mL, treated for 30 min with morphine 30 μM and stimulated with DNP-BSA for 5, 10, 15, 30 min; (**d**) β-hexosaminidase assay of RBL-2H3 cells pre-sensitized with

anti DNP-IgE 50 ng/mL, treated for 24 h with morphine 30 μM and stimulated with DNP-BSA for 5, 10, 15, 30 min. DNP-BSA was used at a final concentration of 625 ng/mL in all experiments. Values are expressed as the mean ± S.E.M. of three experiments. * $p < 0.05$; ** $p < 0.01$ vs. control.

3.3. PEA Counteracts Morphine-Augmented Mast Cell Degranulation

Since a previous in vivo study in rats demonstrated that one-week pretreatment with PEA enhanced the analgesic effect of morphine, thus exerting an opioid sparing effect [48], we choose to pre-treat RBL-2H3 cells with PEA during anti DNP-IgE sensitization (18 h) and then incubate cells with PEA for 30 min the following day. The incubation time and doses were based on preliminary experiments (Supplementary Material). Briefly, the β-hexosaminidase assay revealed that PEA 10 μM was effective in reducing RBL-2H3 cell degranulation up to 40%, while PEA at 100 μM achieved 80% reduction (Figure S4). The ability of 100 μM PEA to counteract morphine-induced RBL-2H3 cell degranulation was analysed by β-hexosaminidase assay. RBL-2H3 cells were challenged with DNP-BSA for 5, 10, 15 and 30 min. Interestingly, the effect of PEA was found to be statistically significant not only compared to morphine, but also with regard to the DNP-BSA-challenged RBL-2H3 cells (Figure 3a). Moreover, the second pre-treatment period with PEA (100 μM for 30 min, after the first 18 h pre-treatment) appeared not to be necessary, since the inhibitory activity was maintained after a single pre-treatment period both in the absence and presence of morphine 30 μM (Figure 3a). To confirm these data, the level of histamine released in the medium was also quantified by HPLC with the same experimental setting. The assay allowed to identify a slight, albeit non-significant, increase in histamine concentrations in 30 μM morphine-treated RBL-2H3 cells compared to controls. Conversely, pre-treatment with PEA 100 μM significantly decreased histamine release, both in the absence and presence of morphine 30 μM (Figure 3b).

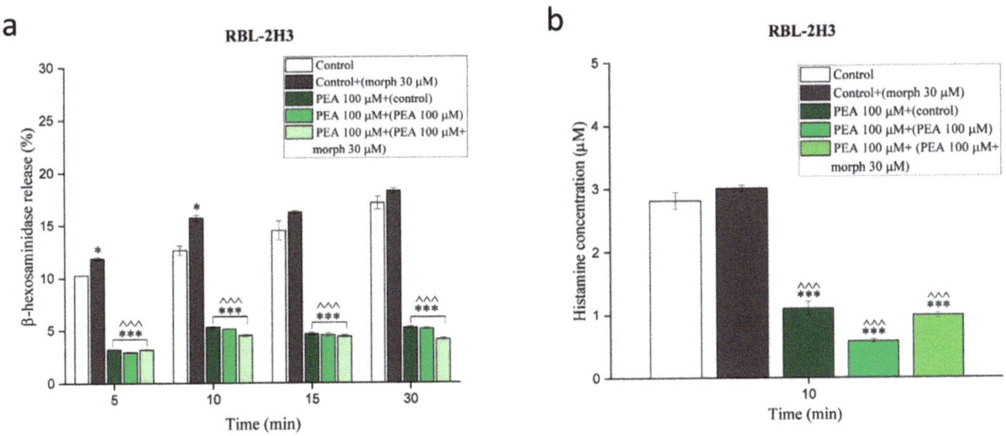

Figure 3. Effect of PEA on morphine-stimulated MC degranulation. RBL-2H3 sensitized with anti-DNP IgE 50 ng/mL and pre-treated with PEA 100 μM for 18 h, incubated the next day with 30 μM morphine, 100 μM PEA, the combination of PEA and morphine and medium only, for 30 min, then stimulated with DNP-BSA for 5, 10, 15 and 30 min for β-hexosaminidase assay (**a**) and 10 min for histamine dosage in the medium by HPLC (**b**). DNP-BSA was used at a final concentration of 625 ng/mL in all experiments. Values are expressed as the mean ± S.E.M. of three experiments. * $p < 0.05$; *** $p < 0.001$ vs. control. ^^^ $p < 0.001$ vs. control + (morphine 30 μM).

3.4. PEA Down-Modulation of Mast Cell Degranulation Is Long-Lasting

To investigate whether down-modulation of RBL-2H3 cells by PEA (100 μM) was long-lasting, we challenged cells with DNP-BSA for 30 min, 2, 6 and 24 h the day after anti DNP-IgE sensitization and measured the content of histamine in the medium at every timepoint. After each challenge, PEA-treated RBL-2H3 cells released a significantly lower amount of histamine compared to controls. As shown in Figure 4a, histamine concentration (μM) in treated and control cells was respectively 2.55 ± 0.1 vs. 10.2 ± 0.13 at 30 min, 6.86 ± 0.71 vs. 15.07 ± 0.5 at 2 h, 24.57 ± 0.1 vs. 39.03 ± 0.7 at 6 h and finally 89.98 ± 0.9 vs. 126.92 ± 0.9 at 24 h. PEA 100 μM significantly increased also tryptase immunofluorescence, thus indicating a decreased degranulation (Figure 4b). Since MC activation is known to induce not only the release of pre-formed mediators (e.g., histamine) but also the release of de-novo synthesized cytokines and chemokines, the effect of PEA on newly synthesized mediators was also investigated. Using RT-PCR, we found that 18 h pre-treatment with PEA 100 μM before 6 h DNP-BSA challenge, efficiently counteracted the increase in the expression of the CCL2 (0.1 ± 0.002 vs. 1 ± 0.02) and TNF-α (0.63 ± 0.015 vs. 1 ± 0.02) genes, compared to controls (Figure 4c).

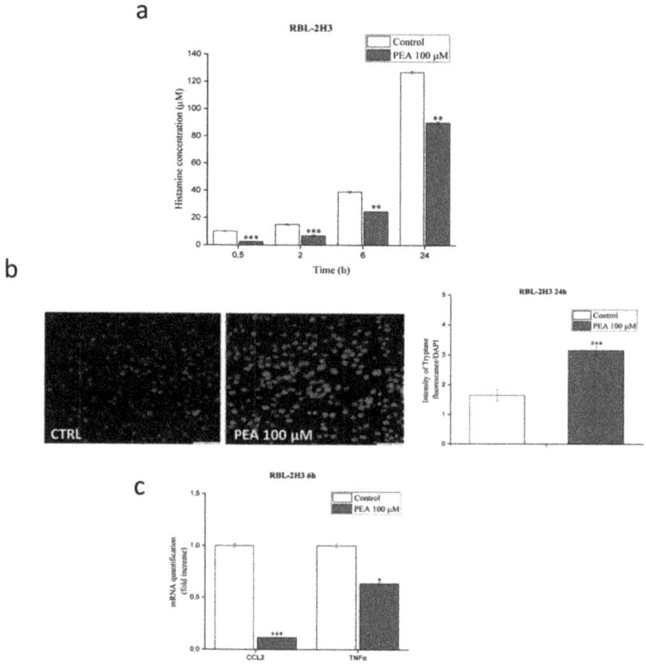

Figure 4. Long-lasting effect of PEA on MC degranulation. (**a**) Histamine concentration in the medium of RBL-2H3 cells pre-sensitized and pre-treated with PEA 100 μM for 18 h and stimulated with DNP-BSA the next day for 30 min, 2, 6 and 24 h. (**b**) Immunofluorescence of tryptase on RBL-2H3 cells pre-treated with PEA 100 μM for 18 h together with the pre-sensitization, and the next day stimulated with DNP-BSA for 24 h. Cell nuclei were labeled with DAPI. Scale bar = 50 μm. (**c**) RT-PCR of RBL-2H3 cells pre-sensitized and pre-treated with PEA 100 μM for 18 h and the next day stimulated with DNP-BSA for 6 h. DNP-BSA was used at a final concentration of 625 ng/mL in all experiments. Control condition was arbitrarily set at 1, values are expressed as the mean ± S.E.M. of three experiments. * $p < 0.05$; ** $p < 0.01$; *** $p < 0.001$ vs. control.

3.5. PEA Down-Modulates Mast Cell-Induced Astrocyte Expression of Genes Involved in Inflammation and Pain

First, the effect of morphine and PEA on primary cortical astrocyte viability was evaluated as better detailed in the Supplementary section. Briefly, cell viability was not affected by increasing concentrations of morphine (1–30 μM) while PEA (10–30 μM) affected astrocyte viability, either used alone or in combination with morphine (Figure S5).

Second, the presence and functional expression of μ-receptors was studied, using western blot and calcium influx respectively, confirming that primary cortical astrocytes express a functional μ-receptor (Figure S6). The effect of RBL-2H3 cell mediators (with or without PEA treatment) on astrocyte viability and activity (with or without morphine treatment) was then studied by incubating the primary cortical astrocytes with RBL-2H3 cell medium. Briefly, we showed that RBL-2H3 cell-conditioned medium does not affect astrocytes viability (Figure 5a). Furthermore, by using RT-PCR we showed that morphine treatment does not alter the astrocyte expression of any of the investigated genes, i.e., GFAP, EAAT2, Serpina3n, IL-1β, IL-6, CCL2. On the contrary, control astrocytes incubated with control RBL-2H3 cell medium showed a significant increase in the expression of all the genes but EAAT2 (Figure 5b,c). Cell medium from PEA pre-treated RBL-2H3 cells significantly reversed the increased expression for all the investigated genes, except GFAP that was furtherly increased (Figure 5b,c). The morphine-treated astrocytes incubated with control RBL-2H3 cell media showed a significant increase in the expression of all the genes examined, including EAAT2, with values exceeding those in control astrocytes treated with the same medium. The incubation with cell medium from PEA pre-treated RBL-2H3 cells significantly reversed the increased expression of all genes, except for GFAP (Figure 5b,c). In order to better understand the apparently contradictory findings on GFAP gene expression, immunofluorescence analysis of GFAP was performed. In contrast to RT-PCR findings, increased GFAP staining was observed in 30 μM morphine-treated astrocytes. Similar findings were also seen in control (and even more so in morphine-treated astrocytes) incubated with control RBL-2H3 cell medium. In line with RT-PCR findings, the cell medium from PEA pre-treated RBL-2H3 cells further enhanced GFAP staining compared to control RBL-2H3 cell medium, both in control and morphine-treated astrocytes (Figure 6). Summarizing, morphine alone did not change the gene expression profile of astrocytes, while increasing GFAP immunoreactivity. On the contrary, a significant increase of the investigated genes was evident if mediators from RBL-2H3 cells were added. Finally, pre-treatment of RBL-2H3 cells with PEA significantly down-modulated the expression of astrocyte genes involved in neuropathic pain and inflammation.

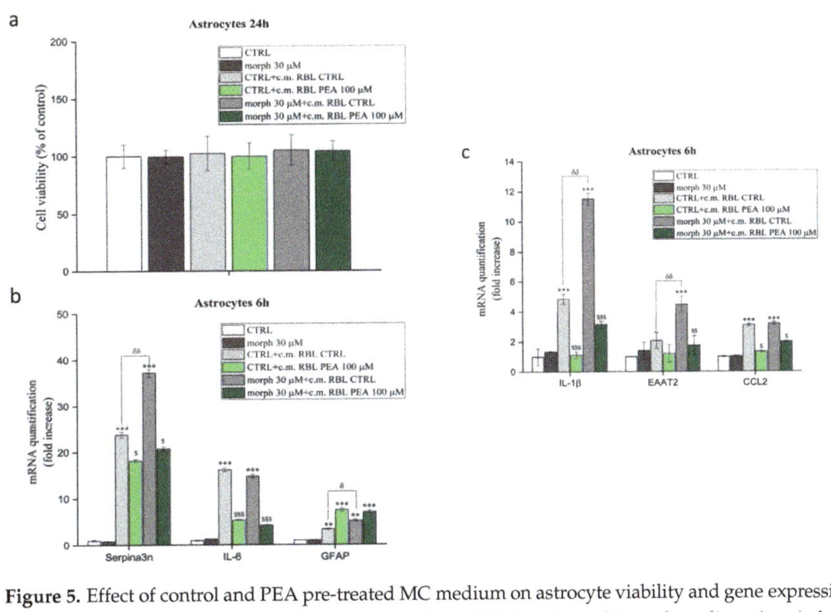

Figure 5. Effect of control and PEA pre-treated MC medium on astrocyte viability and gene expression. (**a**) MTT of astrocytes treated with 30 μM morphine (48 h) and conditioned medium (c.m.) from RBL-2H3 (24 h) stimulated with DNP-BSA (24 h) and pre-treated or not (control) with 100 μM PEA.

Control condition was arbitrarily set at 100, values are expressed as the mean ± S.E.M. of two experiments. (**b**,**c**) RT-PCR of astrocytes treated with 30 µM morphine (48 h) and conditioned medium (c.m.) from RBL-2H3 (6 h) stimulated with DNP-BSA (24 h) and pre-treated or not (control) with 100 µM PEA. Control condition was arbitrarily set at 1, values are expressed as the mean ± S.E.M. of three experiments. DNP-BSA was used at a final concentration of 625 ng/mL in all experiments. ** $p < 0.01$; *** $p < 0.001$ vs. control astrocytes. $^{\$}$ $p < 0.05$; $^{\$\$}$ $p < 0.01$; $^{\$\$\$}$ $p < 0.001$ vs astrocytes + c.m. RBL-2H3 CTRL. $^{\delta}$ $p < 0.05$; $^{\delta\delta}$ $p < 0.01$ morphine astrocytes + c.m. RBL-2H3 CTRL vs. control astrocytes + c.m. RBL-2H3 CTRL.

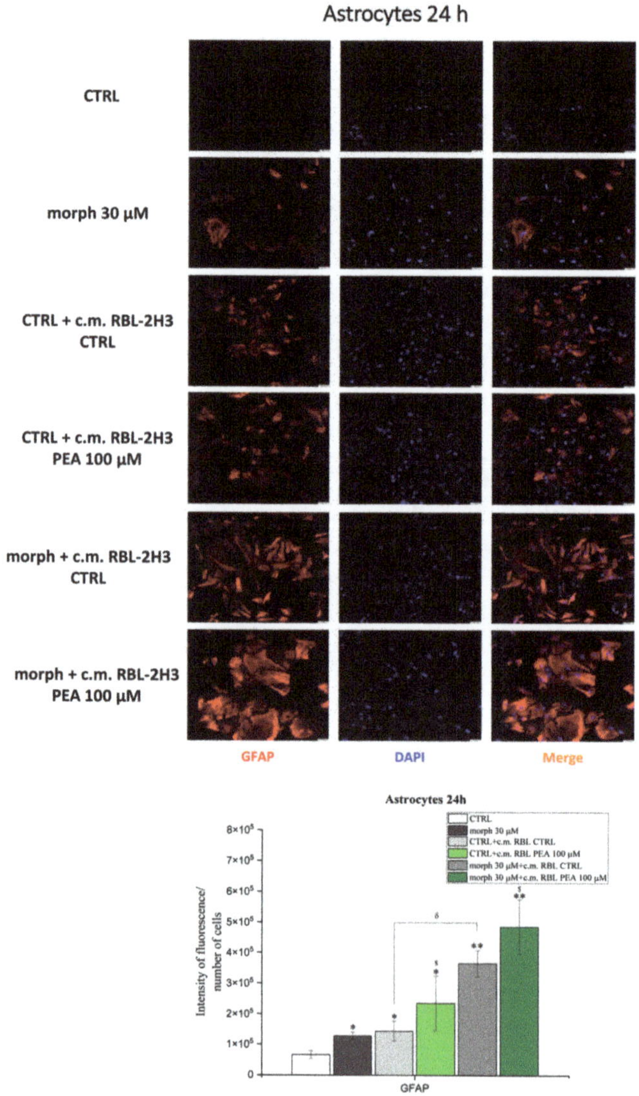

Figure 6. Effect of control and PEA pre-treated RBL-2H3 medium on astrocyte GFAP expression. Immunofluorescence of GFAP on astrocytes treated with 30 µM morphine (48 h) and c.m. from RBL-2H3 (24 h) stimulated with DNP-BSA (625 ng/mL, 24 h) and pre-treated or not (control) with

100 μM PEA. Cell nuclei were labeled with DAPI. Values are expressed as the mean ± S.E.M. of three experiments. Scale bar = 50 μm. * $p < 0.05$; ** $p < 0.01$ vs. control astrocytes. $^{\$}$ $p< 0.05$ vs. astrocytes + c.m. RBL-2H3 CTRL. $^{\delta}$ $p < 0.05$; morphine astrocytes + c.m. RBL-2H3 CTRL vs. control astrocytes + c.m. RBL-2H3 CTRL.

4. Discussion

The management of persistent pain is a major challenge for clinicians as well as patients. Opioids are the treatment of choice, but their use is limited by the occurrence of side effects such as tolerance. We have recently shown that ultramicronized PEA effectively delays tolerance to morphine, tramadol and oxycodone [18,41]. Moreover, ultramicronized PEA was shown to enhance the analgesic effect of opioids, thus limiting their dosage and side effects, both in naïve and neuropathic animals [48,49]. However, the mechanisms underlying the ability of PEA to reduce opioid tolerance are still unclear.

In light of the evidence that (i) PEA down-modulates MC degranulation [50] and (ii) glial cells are involved in the onset of tolerance [19], we hypothesized and proved that RBL-2H3 mediators activate astrocytes, while PEA counteracts this activation by down-modulating RBL-2H3 degranulation.

Mast cells are immune cells increasingly recognized to play pivotal roles not only in allergy, but also in host defense, innate and acquired immunity, homeostatic responses, [51] and inflammation [52]. Mast cells are located in almost all tissues, particularly around blood vessels and in proximity of nerve endings [52], with the same distribution being observed at meningeal level, both in the spinal cord and brain [53,54]. The critical role of MCs in the generation and maintenance of neuropathic pain has recently been investigated [55]. In particular, MC degranulated mediators, including nerve growth factor [56], have been shown to activate and sensitize nociceptors [57,58] as well as stimulate particular pain pathways [59].

Here we used a validated model of rat MCs, i.e., the RBL-2H3 cell line, and confirmed that sensitization with anti-dinitrophenyl (DNP)-IgE is needed to induce a tunable degranulation, as previously reported by other groups [47,60–62]. Although opioids, such as morphine, are known to activate MCs in vivo [61–64] and codeine was shown to induce human MC degranulation in vitro [65], the effect of morphine on the RBL-2H3 cell line was not investigated yet. Here we showed that morphine was able to significantly and dose-dependently increase degranulation of DNP-BSA-primed RBL-2H3 cells, without exerting any cytotoxic effect at the 0.3–300 μM dose range. Interestingly, morphine-induced degranulation was not only monitored by the release of a well-known MC mediator, i.e., β-hexosaminidase [66], but also by immunofluorescence for tryptase, a serine protease stored in MC granules [67] and released upon MC activation [52,68]. Although the presence of tryptase in RBL-2H3 cells has been questioned [69], our finding is in line with previous studies confirming tryptase release in RBL-2H3 after appropriate stimulation [47,70].

Moreover, here we showed for the first time that RBL-2H3 cells express the μ-receptor. This is consistent with what has been previously shown in human MCs (both primary and cell line LAD2) [65]. Although the exact involvement of μ-receptors in morphine-induced MC degranulation was out of the scope of the present study and was not further investigated, it could be assumed that morphine effect depended on μ-receptors, even if alternative pathways were identified (e.g., Mas-related G protein-coupled receptor) [71].

In the present study, the ability of ultramicronized PEA to down-modulate MC degranulation, in response to DNP-BSA challenge, either alone or associated with morphine, was confirmed. In particular, the effect of ultramicronized PEA was monitored by β-hexosaminidase assay, tryptase immunofluorescence, histamine quantification and RT-PCR for newly synthesized mediators (i.e., CCL2 and TNF-α).

Within pre-formed granules, histamine is one of the main molecules produced by MCs [72], acting as a pro-inflammatory factor [73]. Our results demonstrated a time-dependent histamine release in RBL-2H3 cells, counteracted by PEA treatment up to 24 h of DNP-BSA stimulation. The result parallels with that obtained by another group in which histamine was measured by ELISA assay [50]. Moreover, the ability of ultramicronized PEA to decrease histamine levels was also observed by our group in a rat model of chronic neuropathic pain [49].

Notably, TNF-α may activate MCs as well as microglia and astrocytes, being involved in neuroinflammation and chronic pain [74], while CCL2 regulates recruitment of inflammatory cells [75] and acts as a chemoattractant to MCs [76]. The present results concerning the effect of PEA on RBL-2H3 are consistent with previous findings showing that this lipid amide negatively controls releasability of activated MCs, both in vitro [50,77,78], ex vivo [79] and in vivo [80,81], in accordance with the so-called ALIA mechanism originally postulated by the Nobel laureate Rita Levi Montalcini [82].

MCs can interact with virtually all cell populations in the CNS, particularly glial cells [53]. Interestingly, several lines of evidence indicate that chronic morphine treatment may directly or indirectly activate glial cells [13,19,83], resulting in the release of pro-inflammatory cytokines (such as TNF-α, IL-1β, IL-6 and chemokines), which in turn further sensitize the NMDA receptors, promoting the onset of morphine tolerance [84]. In particular, astrocytes are emerging as relevant players in the development of opioid tolerance [19,85,86] and morphine repeated treatment was found to significantly activate astrocytes in the spinal horn [87]. Consistent with previous findings [88,89], we here showed that primary rat cortical astrocytes express the μ-receptor and release calcium in response to morphine, thus confirming previous findings on the astrocyte-activating effect of morphine [90]. Notably, the lack of morphine toxicity towards astrocytes was also shown.

The existence of a crosstalk between MCs and astrocyte in opioid tolerance development is an intriguing hypothesis that has not yet been evaluated. In the CNS, MCs are located at the meningeal [54] and blood brain barrier levels, in apposition to astrocytes and neurons and may enter the parenchyma when barrier permeability is increased [53,91]. Moreover, the powerful mediators released by MCs in the periphery can signal to the CNS compartment including astrocytes [92]. To study this relationship, morphine-activated cortical astrocytes were incubated with the conditioned medium of control or PEA pre-treated MCs. First, it was confirmed that MC medium did not alter astrocyte viability. Untreated MC medium led to a significant increase in gene expression of IL-1β, IL-6, CCL2, Serpina3n, EAAT2 and GFAP, in control and even more so in morphine-treated astrocytes, thus confirming the involvement of these non-neuronal cells in neuroinflammation [84]. Interestingly, the increase in the expression of astrocyte neuroinflammatory (but not proliferative, i.e., GFAP) genes was effectively counteracted by pre-treatment of MCs with ultramicronized PEA.

IL-1β is one of the first cytokines that have been involved in the onset of morphine tolerance: selective blockade of the signaling associated with IL-1β prevented the development of tolerance following chronic administration of morphine [93]. Moreover, once tolerance was established, intrathecal administration of an IL-1 receptor antagonist reversed hyperalgesia and prevented the additional development of tolerance and allodynia [53]. In addition, IL-1β knockout mice showed an increased morphine analgesia and prevention of morphine tolerance [93]. Accordingly, the decrease of IL-1β gene expression in response to incubation of morphine-treated astrocytes with PEA-treated MC medium might be considered as a protective effect of PEA against opioid-associated neuroinflammation.

IL-6, whose production is induced by IL-1β and TNF-α, is also involved in morphine tolerance, as well as TNF-α itself, whose inhibition suppress the development of morphine tolerance in rats [94]. High levels of spinal IL-6 have been shown in tolerant rats [95] and repeated intrathecal administration of melanocortin 4 receptor antagonist reduced IL-6 expression and decreased astrocyte activation in morphine tolerant rats, counteracting the loss of morphine analgesic effect [96]. The decreased astrocytic gene expression of IL-6 in response to PEA-treated MC medium might again confirm the protective effect of PEA against neuroinflammation. As far as the chemokine CCL2 concerns, similar considerations could be drawn. Indeed, CCL2 is up-regulated during neuropathic pain conditions, while its neutralization (through intrathecal antibodies) improved morphine-induced analgesia [97].

Beside inflammatory cytokines and chemokines, other markers of astrocyte activation were analyzed here, to evaluate astrocyte phenotype and functions. SERPINA3 is an acute phase protein [98] belonging to the serpin superfamily of serine protease inhibitors [99]. SERPINA3 is synthesized mainly in the liver, lungs, and brain [100]. Here, activated astrocytes represent the main producers of SERPINA3 [101], whose expression is increased in response to IL-1 and TNF-α [102]. Previous experiments have shown that mice lacking serpina3n gene develop a more severe mechanical allodynia compared to wild-type mice [103]. Interestingly, Serpina3n was found to be upregulated in postmortem mid brain tissues from chronic cocaine abusers [104], and the astrogliosis induced by ischemic stroke-associated neuroinflammation was characterized by increased Serpina3n mRNA expression [105]. Furthermore, Serpina3n is involved in neuropathic pain [103] and was actually found to be upregulated in dorsal root ganglia after peripheral nerve injury [106]. With this in mind, the down-regulation of Serpina3n expression in morphine-treated astrocytes following incubation with the medium of PEA-treated MCs can be considered a further proof of the ability of ultramicronized PEA to rebalance the astrocyte-MC crosstalk. A possible explanation for this finding could be that ultramicronized PEA downmodulated TNF-α release from MCs, with this cytokine being responsible for the onset of morphine tolerance in vivo [94]. The hypothesis is supported by the in vivo evidence that TNF-α levels are lower in rats co-treated with morphine and PEA compared to those treated with morphine only [18].

Excitatory amino acid transporter 2 (EAAT2) is primarily localized on astrocytes and responsible for up to 90% of glutamate uptake in the brain [107]. Some evidence correlates a down-regulation of EAAT2 to chronic morphine treatment and pain-related conditions, although a defined role of this transporter is still unclear. In our experiments, EAAT2 over-expression on astrocytes may be viewed as an attempt to enhance glutamate up-take during neuroinflammation.

Finally, the up-regulation of GFAP, i.e., the main intermediate filament protein of the astrocyte cytoskeletal compartment [108], is a key feature of astrocyte activation [108] responsible for both pathogenetic and protective roles [109,110]. While in vivo studies have mainly shown GFAP up-regulation in response to morphine chronic treatment [18,19,111], in vitro findings have been more inconsistent. In the present study, an up-regulation of GFAP in morphine-treated astrocytes was observed, with MC conditioned medium further increasing its expression. Surprisingly, the incubation of astrocytes with the conditioned medium from PEA-treated MCs resulted in an opposite effect on the expression of GFAP (i.e., increase) compared to all the other investigated genes, which indeed were decreased. In the light of the observed down-regulation of inflammatory genes, it might be inferred that in the evaluated in vitro system, GFAP increase in astrocytes is part of the PEA-mediated protective response to morphine potential damage.

5. Conclusions

In conclusion, the present study showed that morphine induced MC degranulation leads to astrocyte activation, with MC mediators inducing astrocyte over-expression of genes involved in pain and inflammation. Pre-treatment with ultramicronized PEA decreased MC degranulation and reduced the expression of these astrocyte genes accordingly. Given the known involvement of astrocytes in the development and maintenance of morphine tolerance, our findings highlight one of the pathways through which PEA may control this phenomenon, balancing the crosstalk between MCs and astrocytes. This research lays the foundations for a more in-depth study on the molecular mechanisms sustaining PEA-induced MC down-regulation and modulation of glial cell behavior.

Supplementary Materials: The following supporting information can be downloaded at: https://www.mdpi.com/article/10.3390/biom13020233/s1, Tables S1. Chromatographic gradient with valve position switch time. Tables S2. MRM parameters. Tables S3. Linear regression coefficients and LOD. Tables S4. Accuracy & Precision (RSD) values for histamine. Tables S5. ME & RE values for histamine. Figure S1. The injection valve position in 2D-HPLC system. Figure S2. Entire Western blot of MOR-1 in RBL-2H3. Figure S3. Optimization of RBL-2H3 degranulation protocol. Figure S4. Effect of PEA 10 and 100 µM on RBL-2H3 degranulation. Figure S5. Effect of morphine and PEA on astrocyte viability. Figure S6. Effect of morphine on astrocyte calcium release [112–114].

Author Contributions: Conceptualization, C.G. (Carla Ghelardini) and L.D.C.M.; methodology, A.T., V.F., G.B. and M.P.; software, F.M. and P.F.; validation, C.C. and L.M.; formal analysis, C.C.; investigation, A.T. and E.L.; resources, L.M. and C.G. (Chiara Gomiero); data curation, A.T. and E.L.; writing—original draft preparation, A.T. and C.G. (Chiara Gomiero); writing—review and editing, A.T. and L.D.C.M.; visualization, P.F.; supervision, C.G. (Carla Ghelardini).; project administration, C.G. (Carla Ghelardini); funding acquisition, L.D.C.M. and C.G. (Carla Ghelardini). All authors have read and agreed to the published version of the manuscript.

Funding: This research was supported by the Italian Ministry of University and Research and by the University of Florence.

Institutional Review Board Statement: The study was conducted according to ARRIVE guidelines and approved by the Animal Subjects Review Board of the University of Florence and by the Italian Ministry of Health (No. 498/2017-PR).

Informed Consent Statement: Not applicable.

Data Availability Statement: The data presented in this study are available on request from the corresponding author.

Conflicts of Interest: C.G. (Chiara Gomiero) discloses a contract with Epitech Group SpA, which had no role in study design and execution, data analysis and interpretation, in the writing of the report, and in the decision to submit the paper for publication. The remaining authors declare no conflict of interest.

References

1. Gatchel, R.J.; McGeary, D.D.; McGeary, C.A.; Lippe, B. Interdisciplinary Chronic Pain Management: Past, Present, and Future. *Am. Psychol.* **2014**, *69*, 119–130. [CrossRef] [PubMed]
2. Yaksh, T.L. Pain after Injury: Some Basic Mechanisms. *J. Fla. Med. Assoc.* **1997**, *84*, 16–19.
3. White, P.F.; Kehlet, H. Improving Postoperative Pain Management: What Are the Unresolved Issues? *Anesthesiology* **2010**, *112*, 220–225. [CrossRef]
4. Sierra-Silvestre, E.; Somerville, M.; Bisset, L.; Coppieters, M.W. Altered Pain Processing in Patients with Type 1 and 2 Diabetes: Systematic Review and Meta-Analysis of Pain Detection Thresholds and Pain Modulation Mechanisms. *BMJ Open Diabetes Res. Care* **2020**, *8*, e001566. [CrossRef] [PubMed]
5. Dieppe, P.A.; Lohmander, L.S. Pathogenesis and Management of Pain in Osteoarthritis. *Lancet* **2005**, *365*, 965–973. [CrossRef]
6. Portenoy, R.K.; Ahmed, E. Cancer Pain Syndromes. *Hematol. Oncol. Clin. N. Am.* **2018**, *32*, 371–386. [CrossRef]
7. Farquhar-Smith, P. Chemotherapy-Induced Neuropathic Pain. *Curr. Opin. Support. Palliat. Care* **2011**, *5*, 1–7. [CrossRef]
8. Shaheed, C.A.; Machado, G.C.; Underwood, M. Drugs for Chronic Pain. *Br. J. Gen. Pract.* **2020**, *70*, 576–577. [CrossRef]
9. Haghjooy-Javanmard, S.; Ghasemi, A.; Laher, I.; Zarrin, B.; Dana, N.; Vaseghi, G. Influence of Morphine on TLR4/NF-KB Signaling Pathway of MCF-7 Cells. *Bratisl. Lek. Listy* **2018**, *119*, 229–233. [CrossRef]
10. Ghelardini, C.; Di Cesare Mannelli, L.; Bianchi, E. The Pharmacological Basis of Opioids. *Clin. Cases Miner. Bone Metab.* **2015**, *12*, 219–221. [CrossRef]
11. Vanegas, G.; Ripamonti, C.; Sbanotto, A.; de Conno, F. Side Effects of Morphine Administration in Cancer Patients. *Cancer Nurs.* **1998**, *21*, 289–297. [CrossRef] [PubMed]
12. Dumas, E.O.; Pollack, G.M. Opioid Tolerance Development: A Pharmacokinetic/Pharmacodynamic Perspective. *AAPS J.* **2008**, *10*, 537–551. [CrossRef] [PubMed]
13. Mika, J.; Wawrzczak-Bargiela, A.; Osikowicz, M.; Makuch, W.; Przewlocka, B. Attenuation of Morphine Tolerance by Minocycline and Pentoxifylline in Naive and Neuropathic Mice. *Brain Behav. Immun.* **2009**, *23*, 75–84. [CrossRef]
14. Davis, M. Modulation of Microglia Can Attenuate Neuropathic Pain Symptoms and Enhance Morphine Effectiveness. *J. Pain Palliat. Care Pharmacother.* **2009**, *23*.
15. Mahmoud, S.; Gharagozloo, M.; Simard, C.; Gris, D. Astrocytes Maintain Glutamate Homeostasis in the CNS by Controlling the Balance between Glutamate Uptake and Release. *Cells* **2019**, *8*, 184. [CrossRef]

16. Blanco-Suárez, E.; Caldwell, A.L.M.; Allen, N.J. Role of Astrocyte-Synapse Interactions in CNS Disorders. *J. Physiol.* **2017**, *595*, 1903–1916. [CrossRef]
17. Duran, J.; Gruart, A.; López-Ramos, J.C.; Delgado-García, J.M.; Guinovart, J.J. Glycogen in Astrocytes and Neurons: Physiological and Pathological Aspects. *Adv. Neurobiol.* **2019**, *23*, 311–329. [CrossRef] [PubMed]
18. Di Cesare Mannelli, L.; Corti, F.; Micheli, L.; Zanardelli, M.; Ghelardini, C. Delay of Morphine Tolerance by Palmitoylethanolamide. *Biomed. Res. Int.* **2015**, *2015*, 894732. [CrossRef]
19. Song, P.; Zhao, Z.Q. The Involvement of Glial Cells in the Development of Morphine Tolerance. *Neurosci. Res.* **2001**, *39*, 281–286. [CrossRef]
20. Mika, J.; Osikowicz, M.; Makuch, W.; Przewlocka, B. Minocycline and Pentoxifylline Attenuate Allodynia and Hyperalgesia and Potentiate the Effects of Morphine in Rat and Mouse Models of Neuropathic Pain. *Eur. J. Pharmacol.* **2007**, *560*, 142–149. [CrossRef]
21. Cui, Y.; Liao, X.X.; Liu, W.; Guo, R.X.; Wu, Z.Z.; Zhao, C.M.; Chen, P.X.; Feng, J.Q. A Novel Role of Minocycline: Attenuating Morphine Antinociceptive Tolerance by Inhibition of P38 MAPK in the Activated Spinal Microglia. *Brain Behav. Immun.* **2008**, *22*, 114–123. [CrossRef]
22. Raghavendra, V.; Tanga, F.Y.; DeLeo, J.A. Attenuation of Morphine Tolerance, Withdrawal-Induced Hyperalgesia, and Associated Spinal Inflammatory Immune Responses by Propentofylline in Rats. *Neuropsychopharmacology* **2003**, *29*, 327–334. [CrossRef]
23. D'aloia, A.; Molteni, L.; Gullo, F.; Bresciani, E.; Artusa, V.; Rizzi, L.; Ceriani, M.; Meanti, R.; Lecchi, M.; Coco, S.; et al. Palmitoylethanolamide Modulation of Microglia Activation: Characterization of Mechanisms of Action and Implication for Its Neuroprotective Effects. *Int. J. Mol. Sci.* **2021**, *22*, 3054. [CrossRef] [PubMed]
24. Re, G.; Barbero, R.; Miolo, A.; di Marzo, V. Palmitoylethanolamide, Endocannabinoids and Related Cannabimimetic Compounds in Protection against Tissue Inflammation and Pain: Potential Use in Companion Animals. *Vet. J.* **2007**, *173*, 21–30. [CrossRef] [PubMed]
25. Beggiato, S.; Tomasini, M.C.; Ferraro, L. Palmitoylethanolamide (PEA) as a Potential Therapeutic Agent in Alzheimer's Disease. *Front. Pharmacol.* **2019**, *10*, 821. [CrossRef] [PubMed]
26. Gabrielsson, L.; Mattsson, S.; Fowler, C.J. Palmitoylethanolamide for the Treatment of Pain: Pharmacokinetics, Safety and Efficacy. *Br. J. Clin. Pharmacol.* **2016**, *82*, 932–942. [CrossRef]
27. Artukoglu, B.B.; Beyer, C.; Zuloff-Shani, A.; Brener, E.; Bloch, M.H. Efficacy of Palmitoylethanolamide for Pain: A Meta-Analysis. *Pain Physician* **2017**, *20*, 353.
28. Rankin, L.; Fowler, C.J. The Basal Pharmacology of Palmitoylethanolamide. *Int. J. Mol. Sci.* **2020**, *21*, 7942. [CrossRef]
29. Gabrielsson, L.; Gouveia-Figueira, S.; Häggström, J.; Alhouayek, M.; Fowler, C.J. The Anti-Inflammatory Compound Palmitoylethanolamide Inhibits Prostaglandin and Hydroxyeicosatetraenoic Acid Production by a Macrophage Cell Line. *Pharmacol. Res. Perspect.* **2017**, *5*, e00300. [CrossRef]
30. Di Cesare Mannelli, L.; D'Agostino, G.; Pacini, A.; Russo, R.; Zanardelli, M.; Ghelardini, C.; Calignano, A. Palmitoylethanolamide Is a Disease-Modifying Agent in Peripheral Neuropathy: Pain Relief and Neuroprotection Share a PPAR-Alpha-Mediated Mechanism. *Mediat. Inflamm.* **2013**, *2013*, 328797. [CrossRef]
31. Nguyen, J.; Luk, K.; Vang, D.; Soto, W.; Vincent, L.; Robiner, S.; Saavedra, R.; Li, Y.; Gupta, P.; Gupta, K. Morphine Stimulates Cancer Progression and Mast Cell Activation and Impairs Survival in Transgenic Mice with Breast Cancer. *Br. J. Anaesth.* **2014**, *113* (Suppl. 1), i4–i13. [CrossRef]
32. Grosman, N. Histamine Release from Isolated Rat Mast Cells: Effect of Morphine and Related Drugs and Their Interaction with Compound 48/80. *Agents Actions* **1981**, *11*, 196–203. [CrossRef]
33. Vincent, L.; Vang, D.; Nguyen, J.; Gupta, M.; Luk, K.; Ericson, M.E.; Simone, D.A.; Gupta, K. Mast Cell Activation Contributes to Sickle Cell Pathobiology and Pain in Mice. *Blood* **2013**, *122*, 1853–1862. [CrossRef]
34. Mazzari, S.; Canella, R.; Petrelli, L.; Marcolongo, G.; Leon, A. N-(2-Hydroxyethyl) Hexadecanamide Is Orally Active in Reducing Edema Formation and Inflammatory Hyperalgesia by down-Modulating Mast Cell Activation. *Eur. J. Pharmacol.* **1996**, *300*, 227–236. [CrossRef]
35. Beggiato, S.; Borelli, A.C.; Ferraro, L.; Tanganelli, S.; Antonelli, T.; Tomasini, M.C. Palmitoylethanolamide Blunts Amyloid-B42-Induced Astrocyte Activation and Improves Neuronal Survival in Primary Mouse Cortical Astrocyte-Neuron Co-Cultures. *J. Alzheimer's Dis.* **2018**, *61*, 389–399. [CrossRef] [PubMed]
36. Blank, U.; Varin-Blank, N. La Lignée Mastocytaire RBL: Modèle Expérimental in Vitro et Application En Pharmacologie Clinique. *Rev. Fr. Allergol. Immunol. Clin.* **2004**, *44*, 51–56. [CrossRef]
37. McCarthy, K.D.; de Vellis, J. Preparation of Separate Astroglial and Oligodendroglial Cell Cultures from Rat Cerebral Tissue. *J. Cell Biol.* **1980**, *85*, 890–902. [CrossRef] [PubMed]
38. Berridge, M.V.; Tan, A.S. Characterization of the Cellular Reduction of 3-(4,5-Dimethylthiazol-2-Yl)-2,5-Diphenyltetrazolium Bromide (MTT): Subcellular Localization, Substrate Dependence, and Involvement of Mitochondrial Electron Transport in MTT Reduction. *Arch. Biochem. Biophys.* **1993**, *303*, 474–482. [CrossRef]
39. Jönsson, D.; Nilsson, B.O. The Antimicrobial Peptide LL-37 Is Anti-Inflammatory and Proapoptotic in Human Periodontal Ligament Cells. *J. Periodontal Res.* **2012**, *47*, 330–335. [CrossRef]
40. Kawahara, T. Establishment and Characterization of Mouse Bone Marrow-Derived Mast Cell Hybridomas. *Exp. Cell Res.* **2012**, *318*, 2385–2396. [CrossRef]

41. Micheli, L.; Lucarini, E.; Toti, A.; Ferrara, V.; Ciampi, C.; Parisio, C.; Bartolucci, G.; Mannelli, L.D.C.; Ghelardini, C. Effects of Ultramicronized N-Palmitoylethanolamine Supplementation on Tramadol and Oxycodone Analgesia and Tolerance Prevention. *Pharmaceutics* **2022**, *14*, 403. [CrossRef] [PubMed]
42. Micheli, L.; Parisio, C.; Lucarini, E.; Vona, A.; Toti, A.; Pacini, A.; Mello, T.; Boccella, S.; Ricciardi, F.; Maione, S.; et al. VEGF-A/VEGFR-1 Signalling and Chemotherapy-Induced Neuropathic Pain: Therapeutic Potential of a Novel Anti-VEGFR-1 Monoclonal Antibody. *J. Exp. Clin. Cancer Res.* **2021**, *40*, 320. [CrossRef] [PubMed]
43. Nistri, S.; Mannelli, L.D.C.; Ghelardini, C.; Zanardelli, M.; Bani, D.; Failli, P. Pretreatment with Relaxin Does Not Restore NO-Mediated Modulation of Calcium Signal in Coronary Endothelial Cells Isolated from Spontaneously Hypertensive Rats. *Molecules* **2015**, *20*, 9524–9535. [CrossRef]
44. Passante, E.; Ehrhardt, C.; Sheridan, H.; Frankish, N. RBL-2H3 Cells Are an Imprecise Model for Mast Cell Mediator Release. *Inflamm. Res.* **2009**, *58*, 611–618. [CrossRef] [PubMed]
45. Passante, E.; Frankish, N. The RBL-2H3 Cell Line: Its Provenance and Suitability as a Model for the Mast Cell. *Inflamm. Res.* **2009**, *58*, 737–745. [CrossRef] [PubMed]
46. Qian, F.; Guo, G.; Li, Y.; Kulka, M. A Novel Eremophilane Lactone Inhibits FcεRI-Dependent Release of pro-Inflammatory Mediators: Structure-Dependent Bioactivity. *Inflamm. Res.* **2016**, *65*, 303–311. [CrossRef]
47. Rujitharanawong, C.; Yoodee, S.; Sueksakit, K.; Peerapen, P.; Tuchinda, P.; Kulthanan, K.; Thongboonkerd, V. Systematic Comparisons of Various Markers for Mast Cell Activation in RBL-2H3 Cells. *Cell Tissue Res.* **2022**, *390*, 413–428. [CrossRef]
48. Di Cesare Mannelli, L.; Micheli, L.; Lucarini, E.; Ghelardini, C. N-Palmitoylethanolamine Supplementation for Long-Lasting, Low-Dosed Morphine Antinociception. *Front. Pharmacol.* **2018**, *9*, 473. [CrossRef]
49. Micheli, L.; Lucarini, E.; Nobili, S.; Bartolucci, G.; Pallecchi, A.; Toti, A.; Ferrara, V.; Ciampi, C.; Ghelardini, C.; Di Cesare Mannelli, L. Ultramicronized N-Palmitoylethanolamine Contributes to Morphine Efficacy against Neuropathic Pain: Implication of Mast Cells and Glia. *Curr. Neuropharmacol.* **2022**, *21*. [CrossRef]
50. Petrosino, S.; Schiano Moriello, A.; Verde, R.; Allarà, M.; Imperatore, R.; Ligresti, A.; Mahmoud, A.M.; Peritore, A.F.; Iannotti, F.A.; di Marzo, V. Palmitoylethanolamide Counteracts Substance P-Induced Mast Cell Activation in Vitro by Stimulating Diacylglycerol Lipase Activity. *J. Neuroinflamm.* **2019**, *16*, 274. [CrossRef]
51. Moon, T.C.; Dean Befus, A.; Kulka, M. Mast Cell Mediators: Their Differential Release and the Secretory Pathways Involved. *Front. Immunol.* **2014**, *5*, 569. [CrossRef]
52. Payne, V.; Kam, P.C.A. Mast Cell Tryptase: A Review of Its Physiology and Clinical Significance. *Anaesthesia* **2004**, *59*, 695–703. [CrossRef]
53. Mittal, A.; Sagi, V.; Gupta, M.; Gupta, K. Mast Cell Neural Interactions in Health and Disease. *Front. Cell. Neurosci.* **2019**, *13*, 110. [CrossRef]
54. Sayed, B.A.; Christy, A.L.; Walker, M.E.; Brown, M.A. Meningeal Mast Cells Affect Early T Cell Central Nervous System Infiltration and Blood-Brain Barrier Integrity through TNF: A Role for Neutrophil Recruitment? *J. Immunol.* **2010**, *184*, 6891–6900. [CrossRef] [PubMed]
55. Kaur, G.; Singh, N.; Jaggi, A.S. Mast Cells in Neuropathic Pain: An Increasing Spectrum of Their Involvement in Pathophysiology. *Rev. Neurosci.* **2017**, *28*, 759–766. [CrossRef] [PubMed]
56. Koda, H.; Mizumura, K. Sensitization to Mechanical Stimulation by Inflammatory Mediators and by Mild Burn in Canine Visceral Nociceptors in Vitro. *J. Neurophysiol.* **2002**, *87*, 2043–2051. [CrossRef] [PubMed]
57. Skaper, S.D.; Facci, L.; Giusti, P. Mast Cells, Glia and Neuroinflammation: Partners in Crime? *Immunology* **2014**, *141*, 314–327. [CrossRef] [PubMed]
58. Xanthos, D.N.; Gaderer, S.; Drdla, R.; Nuro, E.; Abramova, A.; Ellmeier, W.; Sandkühler, J. Central Nervous System Mast Cells in Peripheral Inflammatory Nociception. *Mol. Pain* **2011**, *7*, 1744–8069. [CrossRef] [PubMed]
59. Levy, D.; Kainz, V.; Burstein, R.; Strassman, A.M. Mast Cell Degranulation Distinctly Activates Trigemino-Cervical and Lumbosacral Pain Pathways and Elicits Widespread Tactile Pain Hypersensitivity. *Brain Behav. Immun.* **2012**, *26*, 311–317. [CrossRef]
60. Lian, Q.; Cheng, Y.; Zhong, C.; Wang, F. Inhibition of the IgE-Mediated Activation of RBL-2H3 Cells by TIPP, a Novel Thymic Immunosuppressive Pentapeptide. *Int. J. Mol. Sci.* **2015**, *16*, 2252. [CrossRef]
61. Fu, H.; Cheng, H.; Cao, G.; Zhang, X.; Tu, J.; Sun, M.; Mou, X.; Shou, Q.; Ke, Y. The Inhibition of Mast Cell Activation of Radix Paeoniae Alba Extraction Identified by TCRP Based and Conventional Cell Function Assay Systems. *PLoS ONE* **2016**, *11*, e0155930. [CrossRef] [PubMed]
62. Okahashi, N.; Nakata, M.; Hirose, Y.; Morisaki, H.; Kataoka, H.; Kuwata, H.; Kawabata, S. Streptococcal H2O2 Inhibits IgE-Triggered Degranulation of RBL-2H3 Mast Cell/Basophil Cell Line by Inducing Cell Death. *PLoS ONE* **2020**, *15*, e0231101. [CrossRef]
63. Yaksh, T.L.; Allen, J.W.; Veesart, S.L.; Horais, K.A.; Malkmus, S.A.; Scadeng, M.; Steinauer, J.J.; Rossi, S.S. Role of Meningeal Mast Cells in Intrathecal Morphine-Evoked Granuloma Formation. *Anesthesiology* **2013**, *118*, 664–678. [CrossRef] [PubMed]
64. Dybendal, T.; Guttormsen, A.B.; Elsayed, S.; Askeland, B.; Harboe, T.; Florvaag, E. Screening for Mast Cell Tryptase and Serum IgE Antibodies in 18 Patients with Anaphylactic Shock during General Anaesthesia. *Acta Anaesthesiol. Scand.* **2003**, *47*, 1211–1218. [CrossRef]
65. Sheen, C.H.; Schleimer, R.P.; Kulka, M. Codeine Induces Human Mast Cell Chemokine and Cytokine Production: Involvement of G-Protein Activation. *Allergy* **2007**, *62*, 532–538. [CrossRef]

66. Fukuishi, N.; Murakami, S.; Ohno, A.; Yamanaka, N.; Matsui, N.; Fukutsuji, K.; Yamada, S.; Itoh, K.; Akagi, M. Does β-Hexosaminidase Function Only as a Degranulation Indicator in Mast Cells? The Primary Role of β-Hexosaminidase in Mast Cell Granules. *J. Immunol.* **2014**, *193*, 1886–1894. [CrossRef] [PubMed]
67. Alanazi, S.; Grujic, M.; Lampinen, M.; Rollman, O.; Sommerhoff, C.P.; Pejler, G.; Melo, F.R. Mast Cell β-Tryptase Is Enzymatically Stabilized by DNA. *Int. J. Mol. Sci.* **2020**, *21*, 5065. [CrossRef] [PubMed]
68. Atiakshin, D.; Buchwalow, I.; Samoilova, V.; Tiemann, M. Tryptase as a Polyfunctional Component of Mast Cells. *Histochem. Cell Biol.* **2018**, *149*, 461–477. [CrossRef]
69. Shim, J.K.; Kennedy, R.H.; Weatherly, L.M.; Abovian, A.V.; Hashmi, H.N.; Rajaei, A.; Gosse, J.A. Searching for Tryptase in the RBL-2H3 Mast Cell Model: Preparation for Comparative Mast Cell Toxicology Studies with Zebrafish. *J. Appl. Toxicol.* **2019**, *39*, 473–484. [CrossRef]
70. Li, Y.; Sun, X.; Juan, Z.; Guan, X.; Wang, M.; Meng, Y.; Ma, R. Propofol Pretreatment Alleviates Mast Cell Degranulation by Inhibiting SOC to Protect the Myocardium from Ischemia–Reperfusion Injury. *Biomed. Pharmacother.* **2022**, *150*, 113014. [CrossRef]
71. Yaksh, T.L.; Eddinger, K.A.; Kokubu, S.; Wang, Z.; Dinardo, A.; Ramachandran, R.; Zhu, Y.; He, Y.; Weren, F.; Quang, D.; et al. Mast Cell Degranulation and Fibroblast Activation in the Morphine-Induced Spinal Mass: Role of Mas-Related G Protein-Coupled Receptor Signaling. *Anesthesiology* **2019**, *131*, 132–147. [CrossRef] [PubMed]
72. Thangam, E.B.; Jemima, E.A.; Singh, H.; Baig, M.S.; Khan, M.; Mathias, C.B.; Church, M.K.; Saluja, R. The Role of Histamine and Histamine Receptors in Mast Cell-Mediated Allergy and Inflammation: The Hunt for New Therapeutic Targets. *Front. Immunol.* **2018**, *9*, 1873. [CrossRef] [PubMed]
73. Carlos, D.; Sá-Nunes, A.; de Paula, L.; Matias-Peres, C.; Jamur, M.C.; Oliver, C.; Serra, M.F.; Martins, M.A.; Faccioli, L.H. Histamine Modulates Mast Cell Degranulation through an Indirect Mechanism in a Model IgE-Mediated Reaction. *Eur. J. Immunol.* **2006**, *36*, 1494–1503. [CrossRef]
74. Gu, Y.; Yang, D.K.; Spinas, E.; Kritas, S.K.; Saggini, A.; Caraffa, A.; Antinolfi, P.; Saggini, R.; Conti, P. Role of TNF in Mast Cell Neuroinflammation and Pain. *J. Biol. Regul. Homeost. Agents* **2015**, *29*, 787–791.
75. Castellani, M.L.; Bhattacharya, K.; Tagen, M.; Kempuraj, D.; Perrella, A.; de Lutiis, M.; Boucher, W.; Conti, P.; Theoharides, T.C.; Cerulli, G.; et al. Anti-Chemokine Therapy for Inflammatory Diseases. *Int. J. Immunopathol. Pharmacol.* **2016**, *20*, 447–453. [CrossRef]
76. Bischoff, S.C.; Krieger, M.; Brunner, T.; Rot, A.; Tscharner, V.V.; Baggiolini, M.; Dahinden, C.A. RANTES and Related Chemokines Activate Human Basophil Granulocytes through Different G Protein-coupled Receptors. *Eur. J. Immunol.* **1993**, *23*, 761–767. [CrossRef] [PubMed]
77. Facci, L.; Dal Toso, R.; Romanello, S.; Buriani, A.; Skaper, S.D.; Leon, A. Mast Cells Express a Peripheral Cannabinoid Receptor with Differential Sensitivity to Anandamide and Palmitoylethanolamide. *Proc. Natl. Acad. Sci. USA* **1995**, *92*, 3376–3380. [CrossRef]
78. Cerrato, S.; Brazis, P.; della Valle, M.F.; Miolo, A.; Puigdemont, A. Effects of Palmitoylethanolamide on Immunologically Induced Histamine, PGD2 and TNFalpha Release from Canine Skin Mast Cells. *Vet. Immunol. Immunopathol.* **2010**, *133*, 9–15. [CrossRef] [PubMed]
79. Abramo, F.; Lazzarini, G.; Pirone, A.; Lenzi, C.; Albertini, S.; della Valle, M.F.; Schievano, C.; Vannozzi, I.; Miragliotta, V. Ultramicronized Palmitoylethanolamide Counteracts the Effects of Compound 48/80 in a Canine Skin Organ Culture Model. *Vet. Dermatol.* **2017**, *28*, 104–456. [CrossRef]
80. Bettoni, I.; Comelli, F.; Colombo, A.; Bonfanti, P.; Costa, B. Non-Neuronal Cell Modulation Relieves Neuropathic Pain: Efficacy of the Endogenous Lipid Palmitoylethanolamide. *CNS Neurol. Disord. Drug Targets* **2013**, *12*, 34–44. [CrossRef]
81. Scarampella, F.; Abramo, F.; Noli, C. Clinical and Histological Evaluation of an Analogue of Palmitoylethanolamide, PLR 120 (Comicronized Palmidrol INN) in Cats with Eosinophilic Granuloma and Eosinophilic Plaque: A Pilot Study. *Vet. Dermatol.* **2001**, *12*, 29–39. [CrossRef]
82. Aloe, L.; Leon, A.; Levi-Montalcini, R. A Proposed Autacoid Mechanism Controlling Mastocyte Behaviour. *Agents Actions* **1993**, *39*, C145–C147. [CrossRef] [PubMed]
83. Watkins, L.R.; Hutchinson, M.R.; Johnston, I.N.; Maier, S.F. Glia: Novel Counter-Regulators of Opioid Analgesia. *Trends Neurosci.* **2005**, *28*, 661–669. [CrossRef]
84. Liu, D.-Q.; Zhou, Y.-Q.; Gao, F. Targeting Cytokines for Morphine Tolerance: A Narrative Review. *Curr. Neuropharmacol.* **2019**, *17*, 366–376. [CrossRef]
85. Dozio, V.; Daali, Y.; Desmeules, J.; Sanchez, J.C. Deep Proteomics and Phosphoproteomics Reveal Novel Biological Pathways Perturbed by Morphine, Morphine-3-Glucuronide and Morphine-6-Glucuronide in Human Astrocytes. *J. Neurosci. Res.* **2022**, *100*, 220–236. [CrossRef]
86. Ma, R.; Kutchy, N.A.; Hu, G. Astrocyte-Derived Extracellular Vesicle-Mediated Activation of Primary Ciliary Signaling Contributes to the Development of Morphine Tolerance. *Biol. Psychiatry* **2021**, *90*, 575–585. [CrossRef]
87. Congiu, M.; Micheli, L.; Santoni, M.; Sagheddu, C.; Muntoni, A.L.; Makriyannis, A.; Malamas, M.S.; Ghelardini, C.; Di Cesare Mannelli, L.; Pistis, M. N-Acylethanolamine Acid Amidase Inhibition Potentiates Morphine Analgesia and Delays the Development of Tolerance. *Neurotherapeutics* **2021**, *18*, 2722–2736. [CrossRef] [PubMed]
88. Ruzicka, B.B.; Fox, C.A.; Thompson, R.C.; Meng, F.; Watson, S.J.; Akil, H. Primary Astroglial Cultures Derived from Several Rat Brain Regions Differentially Express μ, δ and κ Opioid Receptor MRNA. *Mol. Brain Res.* **1995**, *34*, 209–220. [CrossRef] [PubMed]

89. Nam, M.H.; Han, K.S.; Lee, J.; Bae, J.Y.; An, H.; Park, S.; Oh, S.J.; Kim, E.; Hwang, E.; Bae, Y.C.; et al. Expression of M-Opioid Receptor in CA1 Hippocampal Astrocytes. *Exp. Neurobiol.* **2018**, *27*, 120. [CrossRef]
90. Sil, S.; Periyasamy, P.; Guo, M.L.; Callen, S.; Buch, S. Morphine-Mediated Brain Region-Specific Astrocytosis Involves the ER Stress-Autophagy Axis. *Mol. Neurobiol.* **2018**, *55*, 6713. [CrossRef]
91. Esposito, E.; Paterniti, I.; Mazzon, E.; Genovese, T.; di Paola, R.; Galuppo, M.; Cuzzocrea, S. Effects of Palmitoylethanolamide on Release of Mast Cell Peptidases and Neurotrophic Factors after Spinal Cord Injury. *Brain Behav. Immun.* **2011**, *25*, 1099–1112. [CrossRef]
92. Kempuraj, D.; Thangavel, R.; Selvakumar, G.P.; Zaheer, S.; Ahmed, M.E.; Raikwar, S.P.; Zahoor, H.; Saeed, D.; Natteru, P.A.; Iyer, S.; et al. Brain and Peripheral Atypical Inflammatory Mediators Potentiate Neuroinflammation and Neurodegeneration. *Front. Cell. Neurosci.* **2017**, *11*, 216. [CrossRef] [PubMed]
93. Shavit, Y.; Wolf, G.; Goshen, I.; Livshits, D.; Yirmiya, R. Interleukin-1 Antagonizes Morphine Analgesia and Underlies Morphine Tolerance. *Pain* **2005**, *115*, 50–59. [CrossRef] [PubMed]
94. Shen, C.H.; Tsai, R.Y.; Shih, M.S.; Lin, S.L.; Tai, Y.H.; Chien, C.C.; Wong, C.S. Etanercept Restores the Antinociceptive Effect of Morphine and Suppresses Spinal Neuroinflammation in Morphine-Tolerant Rats. *Anesth. Analg.* **2011**, *112*, 454–459. [CrossRef]
95. Raghavendra, V.; Rutkowski, M.D.; Deleo, J.A. The Role of Spinal Neuroimmune Activation in Morphine Tolerance/Hyperalgesia in Neuropathic and Sham-Operated Rats. *J. Neurosci.* **2002**, *22*, 9980–9989. [CrossRef]
96. Niu, Z.J.; Ma, J.G.; Chu, H.C.; Zhao, Y.; Feng, W.; Cheng, Y.W. Melanocortin 4 Receptor Antagonists Attenuates Morphine Antinociceptive Tolerance, Astroglial Activation and Cytokines Expression in the Spinal Cord of Rat. *Neurosci. Lett.* **2012**, *529*, 112–117. [CrossRef] [PubMed]
97. Kwiatkowski, K.; Popiolek-Barczyk, K.; Piotrowska, A.; Rojewska, E.; Ciapała, K.; Makuch, W.; Mika, J. Chemokines CCL2 and CCL7, but Not CCL12, Play a Significant Role in the Development of Pain-Related Behavior and Opioid-Induced Analgesia. *Cytokine* **2019**, *119*, 202–213. [CrossRef]
98. Zhang, Y.; He, J.; Zhao, J.; Xu, M.; Lou, D.; Tso, P.; Li, Z.; Li, X. Effect of ApoA4 on SERPINA3 Mediated by Nuclear Receptors NR4A1 and NR1D1 in Hepatocytes. *Biochem. Biophys. Res. Commun.* **2017**, *487*, 327–332. [CrossRef]
99. Grek, C.L.; Townsend, D.M.; Uys, J.D.; Manevich, Y.; Coker, W.J.; Pazoles, C.J.; Tew, K.D. S-Glutathionylated Serine Proteinase Inhibitors as Plasma Biomarkers in Assessing Response to Redox-Modulating Drugs. *Cancer Res.* **2012**, *72*, 2383–2393. [CrossRef]
100. Park, J.; Masaki, T.; Mezaki, Y.; Yokoyama, H.; Nakamura, M.; Maehashi, H.; Fujimi, T.J.; Gouraud, S.S.; Nagatsuma, K.; Nakagomi, M.; et al. Alpha-1 Antichymotrypsin Is Involved in Astrocyte Injury in Concert with Arginine-Vasopressin during the Development of Acute Hepatic Encephalopathy. *PLoS ONE* **2017**, *12*, e0189346. [CrossRef] [PubMed]
101. Kanemaru, K.; Meckelein, B.; Marshall, D.C.L.; Sipe, J.D.; Abraham, C.R. Synthesis and Secretion of Active Alpha 1-Antichymotrypsin by Murine Primary Astrocytes. *Neurobiol. Aging* **1996**, *17*, 767–771. [CrossRef] [PubMed]
102. Kiss, D.L.; Xu, W.; Gopalan, S.; Buzanowska, K.; Wilczynska, K.M.; Rydel, R.E.; Kordula, T. Duration of A1-Antichymotrypsin Gene Activation by IL-1 Is Determined by Efficiency of IkBα Resynthesis in Primary Human Astrocytes. *J. Neurochem.* **2005**, *92*, 730. [CrossRef]
103. Vicuña, L.; Strochlic, D.E.; Latremoliere, A.; Bali, K.K.; Simonetti, M.; Husainie, D.; Prokosch, S.; Riva, P.; Griffin, R.S.; Njoo, C.; et al. The Serine Protease Inhibitor SerpinA3N Attenuates Neuropathic Pain by Inhibiting T Cell–Derived Leukocyte Elastase. *Nat. Med.* **2015**, *21*, 518–523. [CrossRef] [PubMed]
104. Bannon, M.J.; Johnson, M.M.; Michelhaugh, S.K.; Hartley, Z.J.; Halter, S.D.; David, J.A.; Kapatos, G.; Schmidt, C.J. A Molecular Profile of Cocaine Abuse Includes the Differential Expression of Genes That Regulate Transcription, Chromatin, and Dopamine Cell Phenotype. *Neuropsychopharmacology* **2014**, *39*, 2191–2199. [CrossRef] [PubMed]
105. Zamanian, J.L.; Xu, L.; Foo, L.C.; Nouri, N.; Zhou, L.; Giffard, R.G.; Barres, B.A. Genomic Analysis of Reactive Astrogliosis. *J. Neurosci.* **2012**, *32*, 6391–6410. [CrossRef] [PubMed]
106. Sun, W.; Kou, D.; Yu, Z.; Yang, S.; Jiang, C.; Xiong, D.; Xiao, L.; Deng, Q.; Xie, H.; Hao, Y. A Transcriptomic Analysis of Neuropathic Pain in Rat Dorsal Root Ganglia Following Peripheral Nerve Injury. *Neuromol. Med.* **2020**, *22*, 250–263. [CrossRef]
107. Maragakis, N.J.; Dykes-Hoberg, M.; Rothstein, J.D. Altered Expression of the Glutamate Transporter EAAT2b in Neurological Disease. *Ann. Neurol.* **2004**, *55*, 469–477. [CrossRef]
108. Middeldorp, J.; Hol, E.M. GFAP in Health and Disease. *Prog. Neurobiol.* **2011**, *93*, 421–443. [CrossRef]
109. Pekny, M.; Wilhelmsson, U.; Pekna, M. The Dual Role of Astrocyte Activation and Reactive Gliosis. *Neurosci. Lett.* **2014**, *565*, 30–38. [CrossRef]
110. Linnerbauer, M.; Rothhammer, V. Protective Functions of Reactive Astrocytes Following Central Nervous System Insult. *Front. Immunol.* **2020**, *11*, 573256. [CrossRef]
111. Beitner-Johnson, D.; Guitart, X.; Nestler, E.J. Glial Fibrillary Acidic Protein and the Mesolimbic Dopamine System: Regulation by Chronic Morphine and Lewis-Fischer Strain Differences in the Rat Ventral Tegmental Area. *J. Neurochem.* **1993**, *61*, 1766–1773. [CrossRef] [PubMed]
112. European Medicines Agency. Guideline Bioanalytical Method Validation. 2015. Available online: https://www.ema.europa.eu/en/documents/scientific-guideline/guideline-bioanalytical-method-validation_en.pdf (accessed on 12 January 2023).

113. ICH Q2(R2) Validation of Analytical Procedures. Available online: https://www.ema.europa.eu/en/ich-q2r2-validation-analytical-procedures (accessed on 12 January 2022).
114. Matuszewski, B.K.; Constanzer, M.L.; Chavez-Eng, C.M. Strategies for the Assessment of Matrix Effect in Quantitative Bioanalytical Methods Based on HPLC−MS/MS. *Anal. Chem.* **2003**, *75*, 3019–3030. [CrossRef] [PubMed]

biomolecules

MDPI

Article

First Evidence of the Protective Effects of 2-Pentadecyl-2-Oxazoline (PEA-OXA) in In Vitro Models of Acute Lung Injury

Aniello Schiano Moriello [1,2], Fiorentina Roviezzo [3], Fabio Arturo Iannotti [1], Giuseppina Rea [4], Marco Allarà [1,2], Rosa Camerlingo [5], Roberta Verde [1], Vincenzo Di Marzo [1,6,*] and Stefania Petrosino [1,2,*]

1 Endocannabinoid Research Group, Institute of Biomolecular Chemistry, National Research Council, 80078 Pozzuoli, Italy
2 Epitech Group SpA, Saccolongo, 35100 Padova, Italy
3 Department of Pharmacy, University of Naples Federico II, 80138 Naples, Italy
4 Microenvironment Molecular Targets, National Cancer Institute G. Pascale Foundation, IRCCS, 80131 Naples, Italy
5 Cellular Biology and Biotherapy-Research Department, National Cancer Institute G. Pascale Foundation, IRCCS, 80131 Naples, Italy
6 Canada Excellence Research Chair on the Microbiome-Endocannabinoidome Axis in Metabolic Health, CRIUCPQ and INAF, Faculties of Medicine and Agriculture and Food Sciences, Université Laval, Quebec City, QC G1V 4G5, Canada
* Correspondence: vdimarzo@icb.cnr.it (V.D.); spetrosino@icb.cnr.it (S.P.)

Abstract: Acute respiratory distress syndrome (ARDS) is a serious inflammatory lung disorder and a complication of SARS-CoV-2 infection. In patients with severe SARS-CoV-2 infection, the transition to ARDS is principally due to the occurrence of a cytokine storm and an exacerbated inflammatory response. The effectiveness of ultra-micronized palmitoylethanolamide (PEA-um) during the earliest stage of COVID-19 has already been suggested. In this study, we evaluated its protective effects as well as the effectiveness of its congener, 2-pentadecyl-2-oxazoline (PEA-OXA), using in vitro models of acute lung injury. In detail, human lung epithelial cells (A549) activated by polyinosinic–polycytidylic acid (poly-(I:C)) or Transforming Growth Factor-beta (TGF-β) were treated with PEA-OXA or PEA. The release of IL-6 and the appearance of Epithelial–Mesenchymal Transition (EMT) were measured by ELISA and immunofluorescence assays, respectively. A possible mechanism of action for PEA-OXA and PEA was also investigated. Our results showed that both PEA-OXA and PEA were able to counteract poly-(I:C)-induced IL-6 release, as well as to revert TGF-β-induced EMT. In addition, PEA was able to produce an "entourage" effect on the levels of the two endocannabinoids AEA and 2-AG, while PEA-OXA only increased PEA endogenous levels, in poly-(I:C)-stimulated A549 cells. These results evidence for the first time the superiority of PEA-OXA over PEA in exerting protective effects and point to PEA-OXA as a new promising candidate in the management of acute lung injury.

Keywords: acute respiratory distress syndrome; anti-inflammatory; endocannabinoids; fibrosis; lung epithelial cells; palmitoylethanolamide; 2-pentadecyl-2-oxazoline

Citation: Schiano Moriello, A.; Roviezzo, F.; Iannotti, F.A.; Rea, G.; Allarà, M.; Camerlingo, R.; Verde, R.; Di Marzo, V.; Petrosino, S. First Evidence of the Protective Effects of 2-Pentadecyl-2-Oxazoline (PEA-OXA) in In Vitro Models of Acute Lung Injury. *Biomolecules* **2023**, *13*, 33. https://doi.org/10.3390/biom13010033

Academic Editor: Myron R. Szewczuk

Received: 1 December 2022
Revised: 20 December 2022
Accepted: 21 December 2022
Published: 24 December 2022

1. Introduction

Acute respiratory distress syndrome (ARDS) is the most devastating condition of acute lung injury, characterized by pulmonary edema, severe hypoxemia and impaired ability to eliminate CO_2, and represents one among the most challenging clinical disorders of critical care medicine with high mortality [1–3]. A wide spectrum of risk factors associated with ARDS, such as pneumonia, bacterial or viral infection, transfusion of blood components, trauma, acute pancreatitis and drug reaction, can cause direct or indirect acute lung injury [2]. Among viral infections the current SARS-CoV-2 pandemic has arisen as a new risk factor in as much as it can affect the lower respiratory tract by causing ARDS [4]. In particular, in the early stage of COVID-19, the first cells infected by SARS-CoV-2 are the nasal ciliated cells [5]. Successively, if innate or adaptive responses are not able to clear the virus, the latter can spread from the nasal cavity to the lung via inhalation and thus infect alveolar cells [6,7], causing a diffuse alveolar damage that might progress to ARDS. Patients with severe COVID-19 exhibit systemic hyper-inflammation characterized by a cytokine storm, including an excessive release of pro-inflammatory mediators, such as interleukin (IL)-1, IL-6, IL-8 and tumor necrosis factor-alpha (TNF-α) [8], which in turn is responsible for ARDS. In addition, in the lungs of patients with severe COVID-19 the expansion of fibroblasts, which determines a degree of fibrosis that increases over the course of the disease [8], was found, causing or worsening lung injury and failure.

Palmitoylethanolamide (PEA), originally identified in egg yolk and subsequently in a wide variety of food sources, is highly recognized as an endogenous bioactive lipid, given its presence in most cells and tissues, from both animals and humans [9]. PEA is synthesized "on demand" in conditions of current or potential damage and is endowed with anti-inflammatory, analgesic and neuroprotective properties [10], which are mediated by several molecular and cellular mechanisms. One of these is the Autacoid Local Inflammation Antagonism (ALIA), through which PEA downregulates the degranulation of mast cells [11]. PEA is known to directly activate the peroxisome proliferator-activated receptor- (PPAR-α) [12] and the orphan G-protein-coupled receptor 55 (GPR55) [13] or to indirectly activate the cannabinoid receptors CB1 and CB2 [14,15] and the transient receptor potential vanilloid 1 (TRPV1) [16–18]. The indirect interaction of PEA with cannabinoid and vanilloid receptors is known as the "entourage effect", since it is due to PEA increasing the levels of the endocannabinoids and endovanilloids, i.e., anandamide (AEA) and 2-arachidonoyl-glycerol (2-AG). The effect depends upon the inhibition or down-regulation of the AEA-hydrolyzing enzyme fatty acid amide hydrolase (FAAH) [14] or the stimulation of the activity of the biosynthesizing enzyme diacylglycerol lipase (DAGL) [15]. PEA tissue concentrations are altered during different neuro-inflammatory disorders, suggesting that (i) increased levels might represent a compensatory mechanism to restore homeostasis, while (ii) decreased levels might contribute to the etiology of the disease [9,10,19]. For these reasons, the exogenous application of PEA could be required to potentiate the endogenous protective mechanisms, when the endogenous production of PEA is insufficient o lacking [20,21].

Recently, a natural congener of PEA, 2-pentadecyl-2-oxazoline (PEA-OXA), identified in both green and roasted coffee beans [22], has been reported to have anti-inflammatory and anti-nociceptive properties in an experimental model of acute inflammatory pain [23], to reduce neuroinflammation in an experimental model of Parkinson's disease [24] and to exert neuroprotective effects in different neuroinflammatory conditions associated with spinal and brain trauma in mice [25]. It has been hypothesized that PEA-OXA could exert its protective role by inhibiting the enzyme responsible for PEA catabolism, *N*-acyl-ethanolamine-hydrolyzing acid amidase (NAAA), resulting in an increase in the endogenous levels of PEA [23].

Therefore, given the ability of PEA and its congener PEA-OXA to exert important anti-inflammatory and protective effects, the present work aimed to investigate their potential effectiveness in in vitro models of acute lung injury, reproducing the clinical conditions of SARS-CoV-2 infection. In particular, this study was based on two currently well-established facts, i.e., i) IL-6 is a key contributor of the cytokine storm observed in SARS-CoV-2 infection-associated hyperinflammation and multiorgan failure [8,26], suggesting that this cytokine is a promising marker and an efficacious therapeutic target for the treatment of COVID-19 [26,27]; and ii) ultramicronized PEA (PEA-um) is a promising adjuvant to be assumed in the earliest stage of COVID-19, since a reduction in the inflammatory state has been demonstrated both in cultured murine alveolar macrophages activated by the SARS-CoV-2 Spike Protein [28] and in a randomized clinical trial [29] carried out with this formulation of PEA.

To achieve our objectives, here we used the human lung epithelial cell line A549 to reproduce: (i) a viral infection resulting in IL-6 cytokine release and (ii) Epithelial–Mesenchymal Transition (EMT) mechanisms resulting in lung fibrosis.

2. Materials and Methods

All reagents were purchased from Sigma-Aldrich (Milano, Italy) unless otherwise stated. Pentadecyl-2-oxazoline (PEA-OXA) and palmitoylethanolamide in an ultra-micronized formulation (referred to as PEA hereafter) were obtained from Epitech Group SpA (Saccolongo, Padova, Italy). Polyinosinic–polycytidylic acid (poly-(I:C)) was purchased from InvivoGen (Aurogene, Roma, Italy). 5′-Iodoresiniferatoxin (IRTX) and GW6471 were purchased from Tocris Bioscience (Space Import-Export, Milano, Italy). The deuterated standards—$[^2H]_8$-AEA, $[^2H]_5$-2-AG and $[^2H]_4$-PEA—were purchased from Cayman Chemical (Cabru, Arcore, Italy). The human lung epithelial cell line (A549) was purchased from LGC Standards (Milano, Italy). The human IL-6 ELISA Kit and transforming growth factor beta (TGF-β) were purchased from Abcam (Prodotti Gianni, Milano, Italy). Total mRNA was isolated from A549 cells using Trizol Reagent (Thermo Fisher, Milano, Italy) following the manufacturer's instructions. cDNA preparation from RNA was performed using iScript Reverse Transcription enzyme (Biorad, Milano, Italy). Specific primer sequences were designed using Primer3 Software (https://primer3.ut.ee/, accessed on 28 November 2022) and synthetized by Eurofin (Milano, Italy).

2.1. Cell Culture

A549 cells were grown in Dulbecco's Modified Eagle Medium (DMEM) complemented with penicillin (400 U mL^{-1}), streptomycin (50 mg mL^{-1}) and 10% Fetal Bovine Serum (FBS), in the presence of a 5% CO_2 atmosphere at 37 °C, plated on 100 mm diameter Petri dishes.

2.2. Poly-(I:C)-Induced Inflammatory Response in A549 Cells

A549 cells were plated into 24-well culture dishes at a cell density of 2×10^5 cells per well for 1 day at 37 °C in a 5% CO_2 atmosphere. After 1 day, A549 cells were stimulated with poly-(I:C) (100 μg mL^{-1}) or vehicle (water) and incubated for 6 h at 37 °C in a 5% CO_2 atmosphere. Poly-(I:C)-stimulated A549 cells were treated with PEA-OXA (0.1, 1 and 10 μM), PEA (0.1, 1 and 10 μM) or vehicle (dimethyl sulfoxide or methanol, respectively) and incubated for the indicated time. Poly-(I:C)-stimulated A549 cells were also treated with a TRPV1 antagonist, IRTX (0.1 μM), or PPAR-α antagonist, GW6471 (1 μM), in the presence or absence of PEA-OXA (10 μM) or PEA (10 μM) and incubated for the indicated time. After 6 h, the supernatants were collected, and the amounts of produced IL-6 were measured by using a human IL-6 ELISA kit according to the manufacturer's instructions (Abcam) and by using a reader Glomax® Explorer (Promega, Milano, Italy). Data are expressed as picograms per milliliter of IL-6.

2.3. RNA Extraction and Quantitative PCR (qPCR)

Total RNA was isolated from A549 cells using TRIzol Reagent (cat# 15596018, Life Technologies, Milano, Italy) and reacted with DNase-I (cat# AMPD1, Merk, Milano, Italy) for 15 min at room temperature, followed by spectrophotometric quantification. Subsequently, the RNA integrity number (RIN) for each RNA sample was analyzed on an Agilent 2100 bioanalyzer (Roma, Italy). Purified RNA was reverse-transcribed by the use of the iScript cDNA Synthesis Kit (cat# 1708841, Bio-Rad, Milano, Italy). Quantitative PCR (qPCR) was carried out in a real-time PCR system CFX384 (Bio-Rad, Milano, Italy) using the SYBR Green PCR Kit (Cat# 1725274, Bio-Rad for mRNAs) detection technique and specific primer sequences (Table 1).

Quantitative PCR was performed on independent biological samples (n = 3). Each sample was amplified simultaneously in quadruplicate in a one-assay run with a non-template control blank for each primer pair to control for contamination or primer–dimer formation, and the cycle threshold (Ct) value for each experimental group was determined.

The housekeeping gene ribosomal protein S16 was used to normalize the Ct values, using the $2^{-\Delta Ct}$ formula.

Table 1. List of primer sequences used in the qPCR analysis.

gene	Forward (5′–3′)	Reverse (5′–3′)
CNR2	TAGTGCTGAGAGGACCCA	CGCTATCCACCTTCCTACAA
TRPV1	CTGCCCGACCATCACAGTC	CTGCGATCATAGAGCCTGAGG
PPARα	TTCGCAATCCATCGGCGAG	CCACAGGATAAGTCACCGAGG
NAAA	TGACAGTGGATGTGCAATTCTT	GCCTTTATCTCGTTCATCACCAG
S16	TCGGACGCAAGAAGACAGCGA	AGCGTGCGCGGCTCAATCAT

2.4. Quantification by Liquid Chromatography–Atmospheric Pressure Chemical Ionization–Mass Spectrometry (LC-APCI-MS) of the Endogenous AEA, 2-AG and PEA Levels in A549 Cells

A549 cells, plated in 6-well culture dishes at a cell density of 9×10^5 cells per well, were stimulated with poly-(I:C) (100 μg mL^{-1}) or vehicle (water) and treated in the presence or absence of PEA-OXA (10 μM), PEA (10 μM) or vehicle (dimethyl sulfoxide and methanol, respectively) and incubated for 6 h at 37 °C in a 5% CO_2 atmosphere. After 6 h, cells and supernatants were collected and homogenized in a solution of chloroform/methanol/Tris-HCl 50 mM, pH 7.4 (2:1:1 by vol.) containing 10 pmol of $[^2H]_8$-AEA and 50 pmol of $[^2H]_5$-2-AG and $[^2H]_4$-PEA as internal standards. The lipid-containing organic phase was dried, weighed and pre-purified by open-bed chromatography on silica gel. Fractions derived by eluting the column with a solution of chloroform/methanol (90:10 by vol) were analyzed by LC-APCI-MS by using a Shimadzu (Shimadzu, Kyoto, Japan) High Performance Liquid Chromatography (HPLC) apparatus (LC-10ADVP) coupled with a Shimadzu (LCMS-2020) quadrupole MS via a Shimadzu APCI interface. LC-APCI-MS analyses of AEA, 2-AG and PEA were performed in the selected ion-monitoring (SIM) mode [30,31], using m/z values of 356 and 348 (molecular ion +1 for deuterated and undeuterated AEA), 384.35 and 379.35 (molecular ion + 1 for deuterated and undeuterated 2-AG), and 304 and 300 (molecular ion + 1 for deuterated and undeuterated PEA). The AEA, 2-AG and PEA levels were determined on the basis of their area ratio with the internal standard signal areas to provide the amounts in pmol mg^{-1} of the lipid extract.

2.5. TGF-β-Induced Epithelial–Mesenchymal Transition in A549 Cells

A549 cells, harvested at 80% confluence and plated into 24-well culture dishes, were stimulated with TGF-β1 (2 ng mL^{-1}) or vehicle (PBS) and incubated for 72 h at 37 °C in a 5% CO_2 atmosphere. TGF-β-stimulated A549 cells were treated with PEA-OXA (10 μM), PEA (10 μM) or vehicle (dimethyl sulfoxide and methanol, respectively) and incubated for the indicated time. After 72 h, A549 cells were fixed with 70% ethanol/0.1% Triton for 30 min at 4 °C, treated with 5% BSA for 60 min at room temperature and then stained with primary antibodies, rabbit anti-human cytokeratin (clone ab9377, Abcam) and mouse anti-human vimentin V9 (clone ab8069, Abcam), overnight at 4 °C. The secondary antibodies, goat anti-rabbit Alexa fluor488 (Cell Signaling Technology Danvers, MA, USA) and goat anti-mouse Alexa fluor594 (Cell Signaling Technology Danvers, MA, USA), were incubated for 60 min at 4 °C, and DAPI (Sigma, Milano, Italy), used to stain the nucleus, was incubated for 7 min at room temperature. Appropriate isotype controls were used. The images were acquired with a fluorescence microscope (Zeiss, Milano, Italy) and AxionCam MRc5 (Zeiss, Milano, Italy).

2.6. Data Analysis

Each experiment was performed in at least 3–4 independent biological samples for each group. Data were expressed as means ± standard error of the mean (SEM). Statistical analyses were performed using GraphPad Prism software version 9.0 (GraphPad Software Inc., San Diego, CA, USA). One-way analysis of variance (ANOVA) followed by Tukey's multiple comparison test was used for the analysis. p values < 0.05 were considered statistically significant. Figures were generated in GraphPad Prism software version 9.0.

3. Results

3.1. PEA-OXA and PEA Reduce Poly-(I:C)-Induced Release of Il-6 in Lung Epithelial Cells

A549 cells stimulated with poly-(I:C) (100 μg mL^{-1} for 6 h) and treated with the vehicle of PEA-OXA or PEA significantly released IL-6, as compared to vehicle-stimulated A549 cells (Figure 1). PEA-OXA and PEA, at the highest concentration tested (10 μM), reduced the release of IL-6 from poly-(I:C)-stimulated A549 cells, as compared to poly-(I:C)-stimulated A549 cells treated with the vehicle of PEA-OXA or PEA (Figure 1). PEA-OXA (10 μM) was more effective (1.6-fold) than PEA (10 μM) in reducing the release of IL-6 from poly-(I:C)-stimulated A549 cells (53% and 33% of inhibition, respectively) (Figure 1). No effect on IL-6 release was observed when A549 cells were treated with PEA-OXA or PEA alone (0.1, 1 and 10 μM), i.e., in the absence of poly-(I:C), as compared to vehicle-treated A549 cells (Figure 1).

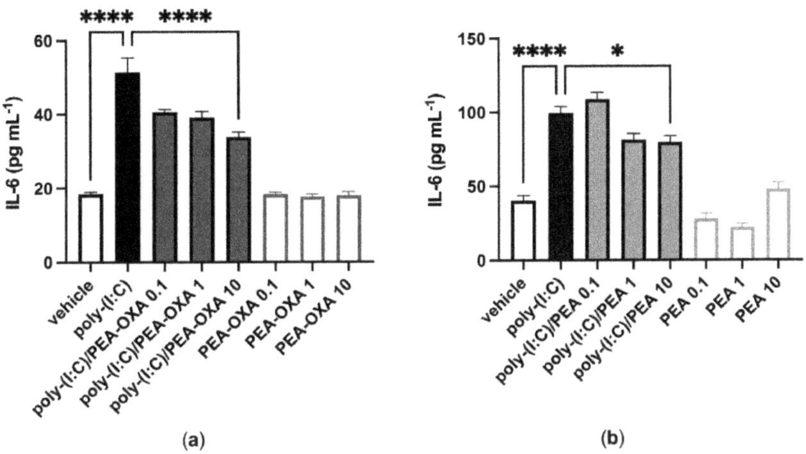

Figure 1. PEA-OXA and PEA reduce IL-6 release from poly-(I:C)-stimulated A549 cells. (**a**) IL-6 release was measured after stimulation of A549 cells with poly-(I:C) (100 μg mL^{-1}) in the presence or absence of PEA-OXA (0.1, 1 and 10 μM) for 6 h at 37 °C in a 5% CO_2 atmosphere; (**b**) IL-6 release was measured after stimulation of A549 cells with poly-(I:C) (100 μg mL^{-1}) in the presence or absence of PEA (0.1, 1 and 10 μM) for 6 h at 37 °C in a 5% CO_2 atmosphere. Each bar shows the mean ± SEM of independent experiments (n = 4). p-values were determined by ANOVA followed by Tukey's multiple comparisons test. **** $p < 0.0001$ and * $p < 0.05$.

3.2. The Anti-Inflammatory Effect of PEA-OXA and PEA Is Not Reverted by Antagonism at the TRPV1 or PPAR-α Receptors in Lung Epithelial Cells

In untreated A549 cells, we a found robust mRNA expression of PPARα and a low relative mRNA expression of TRPV1 (Figure 2). Instead, no mRNA expression of CB2 was found in untreated A549 cells (Figure 2). Importantly, the highest mRNA expression levels in this cell line were found for NAAA (Figure 2).

When A549 cells were stimulated with poly-(I:C) (100 μg mL^{-1} for 6 h) and treated with a selective TRPV1 (IRTX, 0.1 μM) or PPAR-α (GW6471, 1 μM) receptor antagonist, IL-6 release was comparable to that observed in poly-(I:C)-stimulated A549 cells treated with the vehicle (Figure 3a). When poly-(I:C)-stimulated A549 cells were co-treated with PEA-OXA (10 μM) and IRTX (0.1 μM) or GW6271 (1 μM), IL-6 release was comparable to that observed in poly-(I:C)-stimulated A549 cells treated with PEA-OXA (10 μM) (Figure 3a). Likewise, when poly-(I:C)-stimulated A549 cells were co-treated with PEA (10 μM) and IRTX (0.1 μM) or GW6271 (1 μM), IL-6 release was comparable to that observed in poly-(I:C)-stimulated A549 cells treated PEA (10 μM) (Figure 3b). No effect was observed on IL-6 release when A549 cells were treated with the antagonists alone, i.e., in the absence of poly-(I:C), as compared to vehicle-treated A549 cells (Figure 3a).

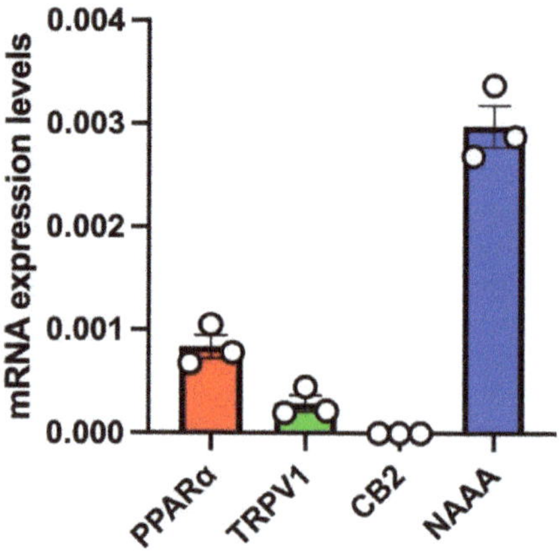

Figure 2. mRNA expression levels of PEA targets (PPARα, TRPV1, CB2) and PEA-catabolizing enzyme (NAAA) in A549 cells. Bar chart with individual points showing the mRNA expression levels of the indicated proteins (PPARα, TRPV1, CB2 and NAAA) measured in A549 cells. Each bar shows the mean ± SEM of 3 independent biological samples. Data are expressed using the $2^{-\Delta ct}$ formula.

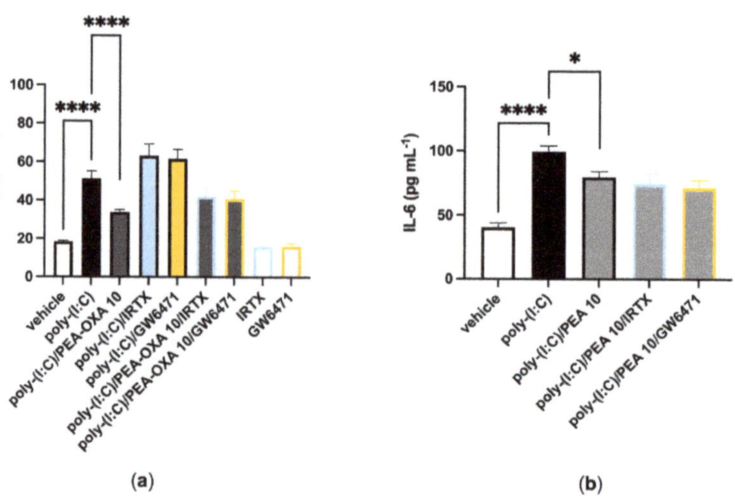

Figure 3. TRPV1 and PPAR-α antagonists do not revert the anti-inflammatory effect of PEA-OXA and PEA in poly-(I:C)-stimulated A549 cells. (**a**) IL-6 release was measured after that A549 cells were stimulated with poly-(I:C) (100 μg mL^{-1}) and treated with IRTX (0.1 μM) or GW6471 (1 μM) in the presence or absence of PEA-OXA (10 μM), for 6 h at 37 °C in a 5% CO_2 atmosphere; (**b**) IL-6 release was measured after A549 cells were stimulated with poly-(I:C) (100 μg mL^{-1}) and treated with IRTX (0.1 μM) or GW6471 (1 μM) in the presence or absence of PEA (10 μM), for 6 h at 37 °C in a 5% CO_2 atmosphere. Each bar shows the mean ± SEM of independent experiments (n = 4). The p-values were determined by ANOVA followed by Tukey's multiple comparisons test. **** $p < 0.0001$ and * $p < 0.05$.

3.3. Effect of PEA and PEA-OXA Treatment on AEA, 2-AG and PEA Endogenous Levels, in Poly-(I:C)-Stimulated A549 Cells

When A549 cells were stimulated with poly-(I:C) under the same conditions shown above (100 μg mL^{-1} for 6 h), the endogenous levels of AEA, 2-AG and PEA did not change, as compared to those in A549 cells stimulated with vehicle (Figure 4). By contrast, when poly-(I:C)-stimulated A549 cells were treated with PEA (10 μM), the endogenous levels of AEA and 2-AG were significantly increased by 4-fold and 1.5-fold, respectively, compared to those in poly-(I:C)-stimulated A549 cells treated with the PEA vehicle (Figure 4b,c). In addition, the levels of PEA were significantly increased by 96-fold when poly-(I:C)-stimulated A549 cells were treated with PEA (10 μM), as compared to those in poly-(I:C)-stimulated A549 cells treated with the vehicle of PEA (Figure 4a). The endogenous levels of PEA were significantly increased by 18-fold when poly-(I:C)-stimulated A549 cells were treated with PEA-OXA (10 μM), as compared to those in poly-(I:C)-stimulated A549 cells treated with the vehicle of PEA-OXA (Figure 4a). No statistically significant variation in the endogenous levels of AEA and 2-AG was observed when poly-(I:C)-stimulated A549 cells were treated with PEA-OXA (10 μM) (Figure 4a), compared to poly-(I:C)-stimulated A549 cells treated with the PEA-OXA vehicle, although a trend towards the enhancement of the AEA levels was observed (Figure 4a).

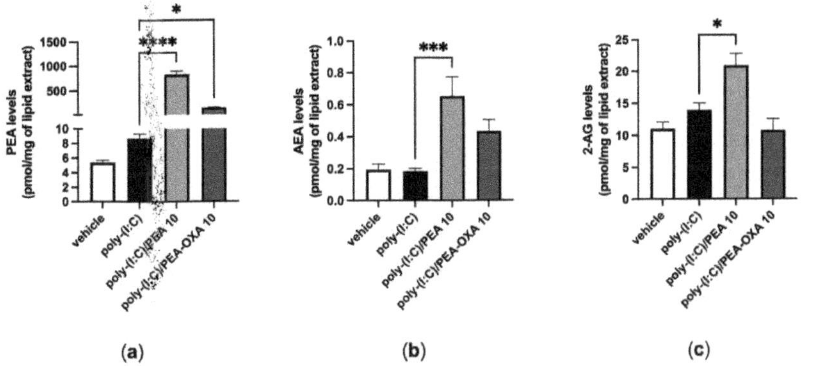

Figure 4. Variation of the levels of PEA, AEA and 2-AG in poly-(I:C)-stimulated A549 cells treated with PEA and PEA-OXA. PEA (**a**), AEA (**b**) and 2-AG (**c**) levels were quantified by LC–MS; after that, A549 cells were stimulated with poly-(I:C) (100 μg mL^{-1}) in the presence or absence of PEA (10 μM) and PEA-OXA (10 μM) for 6 h at 37 °C in a 5% CO_2 atmosphere. Each bar shows the mean ± SEM of independent experiments (n = 3). The p-values were determined by ANOVA followed by Tukey's multiple comparisons test. **** $p < 0.0001$, *** $p < 0.001$ and * $p < 0.05$.

3.4. PEA-OXA and PEA Block TGF-β-Induced Epithelial–Mesenchymal Transition in Lung Epithelial Cells

We previously reported that A549 cells cultured in the absence of TGF-β1 maintain a classic epithelial morphology appearing short, spindle-shaped and triangle-shaped, whereas following incubation with TGF-β1 (2 ng mL^{-1}) for 72 h, these cells assume an elongated shape, with many cells losing contact with their neighboring cells and displaying a long spindle shape typical of a fibroblast-like morphology, an effect known as Epithelial–Mesenchymal Transition (EMT) [32]. Consistent with these morphological observations, alterations in the expression and distribution of cytokeratin and vimentin were evidenced here by immunofluorescence assays. Indeed, these observations showed that, under the same experimental conditions, TGF-β1 induced the down-regulation of cytokeratin (Figure 5a, panels B and T; Figure 5b) and the up-regulation of vimentin (Figure 5a, panels H and T; Figure 5c), when compared to vehicle incubated A549 cells (Figure 5a, panels A, G and S; Figure 5b,c). Treatment of TGF-β1-incubated A549 cells with PEA-OXA (10 μM) or PEA (10 μM) reverted the down-regulation of cytokeratin (Figure 5a, panels C, U, D and V; Figure 5b) and the up-regulation of vimentin (Figure 5a, panels I, U, J and V; Figure 5c). No cell alteration in the expression and distribution of cytokeratin and vimentin was evidenced by immunofluorescence assay when A549 cells were treated with PEA-OXA (10 μM) alone (Figure 5a, panels E, K and W; Figure 5b,c), i.e., in the absence of TGF-β1, as compared to vehicle-treated A549 cells (Figure 5a, panels A, G and S; Figure 5b,c). Likewise, no cell alteration in the expression and distribution of vimentin was evidenced by immunofluorescence assay when A549 cells were treated with PEA (10 μM) alone (Figure 5a, panels L and X; Figure 5c), as compared to vehicle-treated A549 cells (Figure 5a, panels G and S; Figure 5c). Instead, PEA (10 μM) alone showed a per se effect on the expression and distribution of cytokeratin (Figure 5a, panels F and X; Figure 5b), as compared to vehicle-treated A549 cells (Figure 5a, panels A and S; Figure 5b).

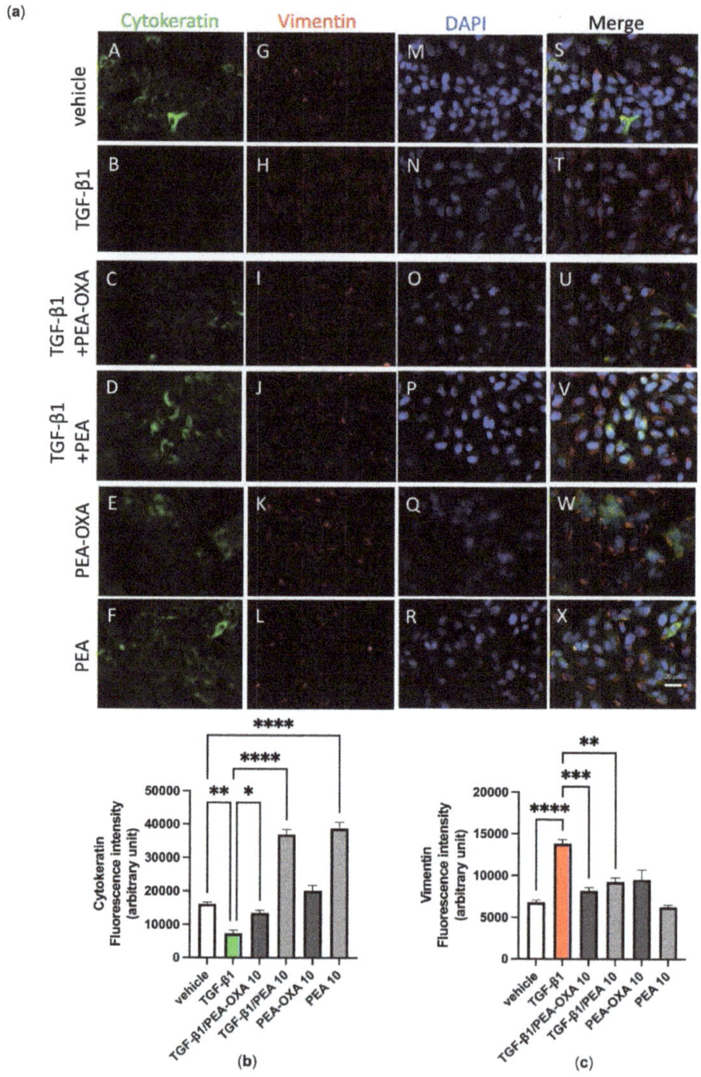

Figure 5. TGF-β1-induced epithelial–mesenchymal transition in A549 cells as evidenced by immunofluorescence approaches is avoided by the treatment with PEA-OXA or PEA. (**a**) Immunofluorescence staining was examined for the following markers: cytokeratin, an epithelial marker (green fluorescence, panels A–F), and vimentin, a mesenchymal marker (red fluorescence, panels G–L). DAPI staining was included to visualize the cell nucleus (blue fluorescence, panels M–R). In the merged images the co-expression and co-distribution of the markers are visualized (panels S–X). Quantifications of cytokeratin (**b**) and vimentin (**c**). The cells were captured with a 40 × microscope objective (Bar = 20 μm). Each bar shows the mean ± SEM of independent experiments (n = 3). The p-values were determined by ANOVA followed by Tukey's multiple comparisons test. **** $p < 0.0001$, *** $p < 0.001$, ** $p < 0.01$ and * $p < 0.05$.

4. Discussion

Since the effectiveness of ultra-micronized PEA i) in attenuating acute lung inflammation, reducing immune cell infiltration and cytokine release in an acute lung injury model induced by lipopolysaccharide (LPS) [33] and ii) in inhibiting the pro-inflammatory response activated by SARS-CoV-2 spike protein in cultured murine alveolar macrophages has been already demonstrated [28], in this study we investigated for the first time if PEA-OXA, a congener of PEA endowed with anti-inflammatory activity [23], could be as effective as PEA in counteracting the IL-6 production and lung cell failure typical of COVID-19. For this purpose, we used two different in vitro models of acute lung injury. In the first in vitro model, we used a human lung epithelial cell line (A549) activated by a stable synthetic double-stranded RNA (poly-(I:C)) that can bind to toll-like receptor 3 (TLR3) with high affinity [34] to stimulate the pathophysiological viral disease state and reproduce the cell signaling pathways typical of the cytokine storm in terms of IL-6 overproduction. Our results demonstrated that both PEA-OXA and PEA were able to counteract IL-6 release induced by poly-(I:C) in A549 cells. Interestingly, PEA-OXA resulted to be more efficacious than PEA (at the same concentration tested and in terms of percent of the effect of poly-(I:C) alone) at counteracting poly-(I:C)-induced IL-6 cytokine release. It is possible that PEA-OXA exerts a stronger protective effect against IL-6 than PEA because it might act through a dual mechanism of action, both PEA-mediated and PEA-independent. In fact, we also investigated the mechanism of action through which PEA-OXA and PEA could exert their anti-inflammatory effects in A549 cells. Our results indicated that the inhibitory effect of PEA-OXA and PEA on poly-(I:C)-induced IL-6 cytokine release was not reverted by an antagonism at TRPV1 and PPAR-α receptors, although these receptors, and particularly the latter one, were strongly expressed in A549 cells. This suggests a non-TRPV1- and non-PPAR-α-mediated mechanism for the actions of the two molecules. Therefore, we investigated the ability of PEA-OXA and PEA to modulate the endogenous levels of endocannabinoids (AEA and 2-AG), as well as, in the case of PEA-OXA, the endogenous levels of PEA, in poly-(I:C)-stimulated A549 cells. Our results confirmed the existence of an "entourage" effect of PEA on the endogenous levels of 2-AG [15] and demonstrated for the first time the ability of PEA to also increase the endogenous levels of AEA in an inflammatory condition. In addition, we confirmed the ability of PEA-OXA to increase the endogenous levels of PEA in inflammatory conditions, an effect that might be exerted by inhibiting the enzyme responsible for PEA degradation (i.e. NAAA) [23]. In line with this hypothesis, we found a strong mRNA expression of NAAA in untreated A549 cells. This finding was not surprising, since NAAA is known to be abundantly expressed in lung cells and tissues [35]. Intriguingly, despite the fact that PEA-OXA significantly (18-fold) elevated the endogenous levels of PEA, this was not sufficient for this compound to also trigger the elevation of the endocannabinoid levels, suggesting that the elevation of PEA cellular levels beyond a certain threshold is necessary to induce an "entourage" effect. Indeed, and not surprisingly, exogenous PEA administration to the cells caused a much higher elevation of the cellular PEA levels (96-fold) than PEA-OXA at the same concentration. These results, taken together, suggest that the anti-inflammatory effects of PEA-OXA and PEA may be partially mediated by bioactive endogenous lipids (i.e., endocannabinoids in the case of PEA, and PEA in the case of PEA-OXA). Further studies will be needed to further clarify the mechanism(s) of the anti inflammatory actions described for PEA and PEA-OXA in lung cells.

In the second in vitro model, we used A549 cells activated by TGF-β in order to reproduce the lung fibrosis that is a critical feature of chronic lung diseases and a serious complication of SARS-CoV-2 infection. In particular, TGF-β modulates lung tissue morphogenesis and differentiation by inducing the development of EMT, an important cellular process in chronic respiratory diseases. Our results suggest that following exposure to TGF-β, A549 cells acquire a fibroblast-like morphology characterized by a decrease of the epithelial marker expression and an increase of the mesenchymal marker expression. The immunofluorescence analysis showed the presence of EMT, characterized by a reduction in cytokeratin-positive staining and an increase in vimentin-positive staining, which was inhibited by the treatment with both PEA-OXA and PEA. These results are in agreement with previous data, in which ultra-micronized PEA inhibited the inflammation response and lung fibrosis in mice subjected to idiopathic pulmonary fibrosis [36].

5. Conclusions

In summary, in this study we reported for the first time the protective effects of PEA-OXA and PEA in counteracting the inflammatory response induced by poly-(I:C), as well as in reverting the fibrosis induced by TGF-β. Our results also evidence a greater effectiveness of PEA-OXA over PEA and point to PEA-OXA as a new and promising candidate in the management of acute lung injury caused by conditions induced by a cytokine storm.

Author Contributions: Conceptualization, A.S.M., V.D. and S.P.; methodology, G.R., M.A., R.C. and R.V.; formal analysis, F.A.I., A.S.M. and S.P.; investigation, A.S.M., F.R., F.A.I. and S.P.; data curation, S.P.; writing—original draft preparation, A.S.M., F.R. and S.P.; writing—review and editing, V.D. and S.P.; supervision, V.D. and S.P.; project administration, S.P. All authors have read and agreed to the published version of the manuscript.

Funding: This research received no external funding.

Institutional Review Board Statement: Not applicable for studies not involving humans or animals.

Data Availability Statement: Not applicable here.

Acknowledgments: The authors thank Francesco Della Valle for his foresight, excellent and invaluable contribution to scientific research, opening new perspectives from pharmacological research to the pharmaceutical industry. S.P. also dedicates this work to the memory of her beloved mother who died of COVID-19.

Conflicts of Interest: A.S.M., M.A. and S.P. are employees of the Epitech Group SpA. S.P. and V.D. are co-inventors on patents on Adelmidrol and/or PEA, respectively, which are unrelated to the present study. The other authors declare no other conflict of interest.

References

1. Su, C.-F.; Kao, S.J.; Chen, H.I. Acute Respiratory Distress Syndrome and Lung Injury: Pathogenetic Mechanism and Therapeutic Implication. *World J. Crit. Care Med.* **2012**, *1*, 50–60. [CrossRef] [PubMed]
2. Rezoagli, E.; Fumagalli, R.; Bellani, G. Definition and Epidemiology of Acute Respiratory Distress Syndrome. *Ann. Transl. Med.* **2017**, *5*, 282. [CrossRef] [PubMed]
3. Gragossian, A.; Siuba, M.T. Acute Respiratory Distress Syndrome. *Emerg. Med. Clin. North Am.* **2022**, *40*, 459–472. [CrossRef] [PubMed]
4. Khodadadi, H.; Salles, É.L.; Jarrahi, A.; Chibane, F.; Costigliola, V.; Yu, J.C.; Vaibhav, K.; Hess, D.C.; Dhandapani, K.M.; Baban, B. Cannabidiol Modulates Cytokine Storm in Acute Respiratory Distress Syndrome Induced by Simulated Viral Infection Using Synthetic RNA. *Cannabi. annabinoid. Res.* **2020**, *5*, 197–201. [CrossRef] [PubMed]
5. Ahn, J.H.; Kim, J.; Hong, S.P.; Choi, S.Y.; Yang, M.J.; Ju, Y.S.; Kim, Y.T.; Kim, H.M.; Rahman, M.D.T.; Chung, M.K.; et al. Nasal Ciliated Cells Are Primary Targets for SARS-CoV-2 Replication in the Early Stage of COVID-19. *J. Clin. Invest.* **2021**, *131*, 148517. [CrossRef] [PubMed]
6. Hou, Y.J.; Okuda, K.; Edwards, C.E.; Martinez, D.R.; Asakura, T.; Dinnon, K.H.; Kato, T.; Lee, R.E.; Yount, B.L.; Mascenik, T.M.; et al. SARS-CoV-2 Reverse Genetics Reveals a Variable Infection Gradient in the Respiratory Tract. *Cell* **2020**, *182*, 429–446.e14. [CrossRef]
7. Justet, A.; Zhao, A.Y.; Kaminski, N. From COVID to Fibrosis: Lessons from Single-Cell Analyses of the Human Lung. *Hum. Genomics* **2022**, *16*, 20. [CrossRef]
8. Lamers, M.M.; Haagmans, B.L. SARS-CoV-2 Pathogenesis. *Nat. Rev. Microbiol.* **2022**, *20*, 270–284. [CrossRef]
9. Petrosino, S.; Schiano Moriello, A. Palmitoylethanolamide: A Nutritional Approach to Keep Neuroinflammation within Physiological Boundaries-A Systematic Review. *Int. J. Mol. Sci.* **2020**, *21*, 9526. [CrossRef]
10. Petrosino, S.; Di Marzo, V. The Pharmacology of Palmitoylethanolamide and First Data on the Therapeutic Efficacy of Some of Its New Formulations. *Br. J. Pharmacol.* **2017**, *174*, 1349–1365. [CrossRef]
11. Aloe, L.; Leon, A.; Levi-Montalcini, R. A Proposed Autacoid Mechanism Controlling Mastocyte Behaviour. *Agents Actions* **1993**, *39 Spec. No.*, C145–C147. [CrossRef]
12. Lo Verme, J.; Fu, J.; Astarita, G.; La Rana, G.; Russo, R.; Calignano, A.; Piomelli, D. The Nuclear Receptor Peroxisome Proliferator-Activated Receptor-Alpha Mediates the Anti-Inflammatory Actions of Palmitoylethanolamide. *Mol. Pharmacol.* **2005**, *67*, 15–19. [CrossRef] [PubMed]

13. Ryberg, E.; Larsson, N.; Sjögren, S.; Hjorth, S.; Hermansson, N.-O.; Leonova, J.; Elebring, T.; Nilsson, K.; Drmota, T.; Greasley, P.J. The Orphan Receptor GPR55 Is a Novel Cannabinoid Receptor. *Br. J. Pharmacol.* **2007**, *152*, 1092–1101. [CrossRef] [PubMed]
14. Di Marzo, V.; Melck, D.; Orlando, P.; Bisogno, T.; Zagoory, O.; Bifulco, M.; Vogel, Z.; De Petrocellis, L. Palmitoylethanolamide Inhibits the Expression of Fatty Acid Amide Hydrolase and Enhances the Anti-Proliferative Effect of Anandamide in Human Breast Cancer Cells. *Biochem. J.* **2001**, *358*, 249–255. [CrossRef] [PubMed]
15. Petrosino, S.; Schiano Moriello, A.; Verde, R.; Allarà, M.; Imperatore, R.; Ligresti, A.; Mahmoud, A.M.; Peritore, A.F.; Iannotti, F.A.; Di Marzo, V. Palmitoylethanolamide Counteracts Substance P-Induced Mast Cell Activation in Vitro by Stimulating Diacylglycerol Lipase Activity. *J. Neuroinflamm.* **2019**, *16*, 274. [CrossRef] [PubMed]
16. Ho, W.-S.V.; Barrett, D.A.; Randall, M.D. "Entourage" Effects of N-Palmitoylethanolamide and N-Oleoylethanolamide on Vasorelaxation to Anandamide Occur through TRPV1 Receptors. *Br. J. Pharmacol.* **2008**, *155*, 837–846. [CrossRef] [PubMed]
17. De Petrocellis, L.; Davis, J.B.; Di Marzo, V. Palmitoylethanolamide Enhances Anandamide Stimulation of Human Vanilloid VR1 Receptors. *FEBS Lett.* **2001**, *506*, 253–256. [CrossRef] [PubMed]
18. Petrosino, S.; Schiano Moriello, A.; Cerrato, S.; Fusco, M.; Puigdemont, A.; De Petrocellis, L.; Di Marzo, V. The Anti-Inflammatory Mediator Palmitoylethanolamide Enhances the Levels of 2-Arachidonoyl-Glycerol and Potentiates Its Actions at TRPV1 Cation Channels. *Br. J. Pharmacol.* **2016**, *173*, 1154–1162. [CrossRef]
19. Iannotti, F.A.; Di Marzo, V.; Petrosino, S. Endocannabinoids and Endocannabinoid-Related Mediators: Targets, Metabolism and Role in Neurological Disorders. *Prog. Lipid Res.* **2016**, *62*, 107–128. [CrossRef]
20. Skaper, S.D.; Di Marzo, V. Endocannabinoids in Nervous System Health and Disease: The Big Picture in a Nutshell. *Philos. Trans. R. Soc. Lond. B. Biol. Sci.* **2012**, *367*, 3193–3200. [CrossRef]
21. Skaper, S.D.; Facci, L. Mast Cell-Glia Axis in Neuroinflammation and Therapeutic Potential of the Anandamide Congener Palmitoylethanolamide. *Philos. Trans. R. Soc. Lond. B Biol. Sci.* **2012**, *367*, 3312–3325. [CrossRef] [PubMed]
22. Impellizzeri, D.; Cordaro, M.; Bruschetta, G.; Crupi, R.; Pascali, J.; Alfonsi, D.; Marcolongo, G.; Cuzzocrea, S. 2-Pentadecyl-2-Oxazoline: Identification in Coffee, Synthesis and Activity in a Rat Model of Carrageenan-Induced Hindpaw Inflammation. *Pharmacol. Res.* **2016**, *108*, 23–30. [CrossRef] [PubMed]
23. Petrosino, S.; Campolo, M.; Impellizzeri, D.; Paterniti, I.; Allarà, M.; Gugliandolo, E.; D'Amico, R.; Siracusa, R.; Cordaro, M.; Esposito, E.; et al. 2-Pentadecyl-2-Oxazoline, the Oxazoline of Pea, Modulates Carrageenan-Induced Acute Inflammation. *Front. Pharmacol.* **2017**, *8*, 308. [CrossRef] [PubMed]
24. Cordaro, M.; Siracusa, R.; Crupi, R.; Impellizzeri, D.; Peritore, A.F.; D'Amico, R.; Gugliandolo, E.; Di Paola, R.; Cuzzocrea, S. 2-Pentadecyl-2-Oxazoline Reduces Neuroinflammatory Environment in the MPTP Model of Parkinson Disease. *Mol. Neurobiol.* **2018**, *55*, 9251–9266. [CrossRef]
25. Impellizzeri, D.; Cordaro, M.; Bruschetta, G.; Siracusa, R.; Crupi, R.; Esposito, E.; Cuzzocrea, S. N-Palmitoylethanolamine-Oxazoline as a New Therapeutic Strategy to Control Neuroinflammation: Neuroprotective Effects in Experimental Models of Spinal Cord and Brain Injury. *J. Neurotrauma* **2017**, *34*, 2609–2623. [CrossRef]
26. Aliyu, M.; Zohora, F.T.; Anka, A.U.; Ali, K.; Maleknia, S.; Saffarioun, M.; Azizi, G. Interleukin-6 Cytokine: An Overview of the Immune Regulation, Immune Dysregulation, and Therapeutic Approach. *Int. Immunopharmacol.* **2022**, *111*, 109130. [CrossRef]
27. Rezaei Tolzali, M.M.; Noori, M.; Shokri, P.; Rahmani, S.; Khanzadeh, S.; Nejadghaderi, S.A.; Fazlollahi, A.; Sullman, M.J.M.; Singh, K.; Kolahi, A.-A.; et al. Efficacy of Tocilizumab in the Treatment of COVID-19: An Umbrella Review. *Rev. Med. Virol.* **2022**, e2388. [CrossRef]
28. Del Re, A.; Corpetti, C.; Pesce, M.; Seguella, L.; Steardo, L.; Palenca, I.; Rurgo, S.; De Conno, B.; Sarnelli, G.; Esposito, G. Ultramicronized Palmitoylethanolamide Inhibits NLRP3 Inflammasome Expression and Pro-Inflammatory Response Activated by SARS-CoV-2 Spike Protein in Cultured Murine Alveolar Macrophages. *Metabolites* **2021**, *11*, 592. [CrossRef]
29. Albanese, M.; Marrone, G.; Paolino, A.; Di Lauro, M.; Di Daniele, F.; Chiaramonte, C.; D'Agostini, C.; Romani, A.; Cavaliere, A.; Guerriero, C.; et al. Effects of Ultramicronized Palmitoylethanolamide (Um-PEA) in COVID-19 Early Stages: A Case-Control Study. *Pharmaceuticals* **2022**, *15*, 253. [CrossRef]
30. Di Marzo, V.; Goparaju, S.K.; Wang, L.; Liu, J.; Bátkai, S.; Járai, Z.; Fezza, F.; Miura, G.I.; Palmiter, R.D.; Sugiura, T.; et al. Leptin-Regulated Endocannabinoids Are Involved in Maintaining Food Intake. *Nature* **2001**, *410*, 822–825. [CrossRef]
31. Marsicano, G.; Wotjak, C.T.; Azad, S.C.; Bisogno, T.; Rammes, G.; Cascio, M.G.; Hermann, H.; Tang, J.; Hofmann, C.; Zieglgänsberger, W.; et al. The Endogenous Cannabinoid System Controls Extinction of Aversive Memories. *Nature* **2002**, *418*, 530–534. [CrossRef] [PubMed]
32. Riemma, M.A.; Cerqua, I.; Romano, B.; Irollo, E.; Bertolino, A.; Camerlingo, R.; Granato, E.; Rea, G.; Scala, S.; Terlizzi, M.; et al. Sphingosine-1-Phosphate/TGF-β Axis Drives Epithelial Mesenchymal Transition in Asthma-like Disease. *Br. J. Pharmacol.* **2022**, *179*, 1753–1768. [CrossRef] [PubMed]
33. Peritore, A.F.; D'Amico, R.; Siracusa, R.; Cordaro, M.; Fusco, R.; Gugliandolo, E.; Genovese, T.; Crupi, R.; Di Paola, R.; Cuzzocrea, S.; et al. Management of Acute Lung Injury: Palmitoylethanolamide as a New Approach. *Int. J. Mol. Sci.* **2021**, *22*, 5533. [CrossRef] [PubMed]
34. Stowell, N.C.; Seideman, J.; Raymond, H.A.; Smalley, K.A.; Lamb, R.J.; Egenolf, D.D.; Bugelski, P.J.; Murray, L.A.; Marsters, P.A.; Bunting, R.A.; et al. Long-Term Activation of TLR3 by Poly(I:C) Induces Inflammation and Impairs Lung Function in Mice. *Respir. Res.* **2009**, *10*, 43. [CrossRef] [PubMed]

35. Tsuboi, K.; Takezaki, N.; Ueda, N. The N-Acylethanolamine-Hydrolyzing Acid Amidase (NAAA). *Chem. Biodivers.* **2007**, *4*, 1914–1925. [CrossRef]
36. Di Paola, R.; Impellizzeri, D.; Fusco, R.; Cordaro, M.; Siracusa, R.; Crupi, R.; Esposito, E.; Cuzzocrea, S. Ultramicronized Palmitoylethanolamide (PEA-Um(®)) in the Treatment of Idiopathic Pulmonary Fibrosis. *Pharmacol. Res.* **2016**, *111*, 405–412. [CrossRef]

biomolecules

MDPI

Article

Palmitoylethanolamide Mitigates Paclitaxel Toxicity in Primary Dorsal Root Ganglion Neurons

Amira Elfarnawany [1,2] and Faramarz Dehghani [1,*]

[1] Department of Anatomy and Cell Biology, Medical Faculty, Martin Luther University Halle-Wittenberg, Grosse Steinstrasse 52, 06108 Halle (Saale), Germany
[2] Zoology Department, Faculty of Science, Tanta University, Tanta 31527, Egypt
* Correspondence: faramarz.dehghani@medizin.uni-halle.de

Abstract: Chemotherapy-induced peripheral neuropathy (CIPN) is a common side effect of several chemotherapeutic agents, such as Paclitaxel. The main symptoms of CIPN are pain and numbness in the hands and feet. Paclitaxel is believed to accumulate in the dorsal root ganglia and free nerve endings. Novel therapeutic agents might help to mitigate or prevent Paclitaxel toxicity on dorsal root ganglion (DRG) neurons. Thus, we used primary DRG neurons as a model to investigate the potential neuroprotective effects of the endocannabinoid-like substance, palmitoylethanolamide (PEA). DRG neurons were isolated from cervical to sacral segments of spinal nerves of Wister rats (6–8 weeks old). After isolation and purification of neuronal cell populations, different concentrations of Paclitaxel (0.01–10 μM) or PEA (0.1–10 μM) or their combination were tested on cell viability by MTT assay at 24 h, 48, and 72 h post-treatment. Furthermore, morphometric analyses of neurite length and soma size for DRG neurons were performed. Adverse Paclitaxel effects on cell viability were apparent at 72 h post-treatment whereas Paclitaxel significantly reduced the neurite length in a concentration-dependent manner nearly at all investigated time points. However, Paclitaxel significantly increased the size of neuronal cell bodies at all time windows. These phenotypic effects were significantly reduced in neurons additionally treated with PEA, indicating the neuroprotective effect of PEA. PEA alone led to a significant increase in neuron viability regardless of PEA concentrations, apparent improvements in neurite outgrowth as well as a significant decrease in soma size of neurons at different investigated time points. Taken together, PEA showed promising protective effects against Paclitaxel-related toxicity on DRG neurons.

Keywords: peripheral neuropathic pain; neurotoxicity; dorsal root ganglion neurons; palmitoylethanolamide; paclitaxel; neurite length; soma size; MTT assay

Citation: Elfarnawany, A.; Dehghani, F. Palmitoylethanolamide Mitigates Paclitaxel Toxicity in Primary Dorsal Root Ganglion Neurons. *Biomolecules* **2022**, *12*, 1873. https://doi.org/10.3390/biom12121873

Academic Editors: Rosalia Crupi and Salvatore Cuzzocrea

Received: 22 November 2022
Accepted: 10 December 2022
Published: 14 December 2022

1. Introduction

Chemotherapy-induced neuropathic pain (CINP) is a dose-limiting side effect of some anticancer drugs, such as bortezomib, cisplatin, oxaliplatin, paclitaxel, thalidomide, and vincristine [1]. The incidence of CINP in patients ranges from 12.1% to 96.2%, depending on the chemotherapeutic agent used and the type of cancer treated [2]. Taxanes are a class of chemotherapy drugs that promote tubulin polymerization into highly stable intracellular microtubules and cause cell death by intermixing with microtubules via normal cell division [3,4]. Paclitaxel is a Taxane derivative that has been used successfully as a first-line treatment for a variety of solid tumors, including ovarian cancer, breast cancer, cervical cancer, lung carcinomas, and other solid tumors [5–7].

Unfortunately, peripheral neuropathic pain (PNP) is a common side effect of Paclitaxel treatment affecting around 70.8% (95% CI 43.5–98.1) of patients [8]. The incidence ranges from 30 to 50% after a single dose and rises to more than 50% after a second dose [9]. Hyperalgesia, allodynia, and sporadic burning, shooting, numbness, spasm, and prickling sensations are some of CINP signs, and these can drastically lower the patient's quality of life [10,11]. Chemotherapy-induced peripheral neuropathy (CIPN) is predominately a sensory axonopathy and neuronopathy, and the sensory neurons residing in dorsal root ganglions (DRGs) are the primary targets. Therefore, DRG explants have been shown to represent a good, simple, and well-accepted model for studying peripheral neuropathy induced by antineoplastic agents [12]. The ability of DRG explants to outgrow neurites in vitro when exposed to nerve growth factor (NGF), as well as the interference with neurite elongation by toxic substances, is the basis for their use in drug neurotoxicity assessment [12–14].

The neurotoxic effect of Paclitaxel on neurite length of DRG was shown to be dose- and time-dependent [15,16], and DRG dissociated post-mitotic neurons were observed to die by necrosis [15]. Paclitaxel also caused the enlargement of neuronal cell bodies, and suppression of DRGs neuritis [17]. Paclitaxel has shown to demonstrate concentration- and time-dependent effects on vesicular trafficking and membrane localization of Nav1.7 in sensory axons of DRGs, providing a possible mechanistic explanation for increased excitability of primary afferents and pain [18]. Paclitaxel was reported to alter intracellular trafficking in both *Drosophila* and mouse models of CIPN by inducing recycling defects in mouse DRG neurons in vitro [19]. Currently, tricyclic antidepressants and analgesic drugs such as amitriptyline, morphine, gabapentin, and duloxetine display limited efficacy for preventing and alleviating paclitaxel-induced peripheral neuropathic pain and/or suffering of patients from serious side effects [20–23]. As a result, finding novel therapeutic agents that can mitigate or prevent Paclitaxel neurotoxicity on DRG neurons is very crucial.

The endocannabinoid system (ECS) is an important biological system that regulates and balances a wide range of physiological functions in the body, making it a target for many drugs and therapies [24]. Modulating the ECS activity showed promising therapeutic effects in a wide range of diseases and pathological conditions, including neurodegenerative, cardiovascular, and inflammatory disorders, obesity/metabolic syndrome, cachexia, chemotherapy-induced nausea and vomiting, tissue injury, and pain [25]. Palmitoylethanolamide (PEA), an endogenous fatty acid amide analogue of the endocannabinoid anandamide, has an important role in tissue protective mechanisms [26,27]. PEA was discovered nearly 5 decades ago in lipid extracts of various natural products, and its anti-inflammatory and antinociceptive properties were later described [28].

There is evidence for PEA synthesis during inflammation and tissue damage. PEA has a variety of beneficial effects, including the relief of inflammation and pruritus, and is effective in the control of neurogenic and neuropathic pain [29]. The hypothesized theories for PEA's mode of action include modulating endocannabinoid signaling and indirectly activating cannabinoid receptors via "entourage" effects [30–33].

PEA acts primarily through the direct activation of the nuclear receptor PPAR-α [34]. After the activation of PPAR-α receptor, a chain of events leads to suppression of pain and inflammatory signals, including the inhibition of the release of pro-inflammatory cytokines such as IL-1β and IL-6 [35]. Previous studies showed a PEA-mediated protection of dentate gyrus granule cells during secondary neuronal damage, which was mediated by PPAR-α activation and influenced by reduction in inflammatory processes [36]. In a chronic constriction injury model of neuropathic pain, repeated PEA treatment (30 mg/kg) not only decreased edema and macrophage infiltrates, but also declined the decrease in axon diameter and myelin thickness [37]. However, research on studying the protective role of PEA against the toxicity of Paclitaxel on DRG neurons is still lacking.

In the present study, the effects of different Paclitaxel and PEA concentrations were investigated, either individually or in combination, on the viability, morphology, and neurite length of primary DRG neurons at various time points. We hypothesized that PEA might reduce the neurotoxicity induced by Paclitaxel on DRG neurons in a concentration- or/and time-dependent manner.

2. Materials and Methods

2.1. Ethics Statement

All animal experiments were carried out in accordance with the policy on ethics and the policy on the use of animals in neuroscience research, as specified in directive 2010/63/EU of the European Parliament and of the Council of the European Union on the protection of animals used for scientific purposes and were approved by local authorities for laboratory animal care and use (State of Saxony-Anhalt, Germany, permission number: I11M27).

2.2. Materials

Experiments were conducted with Palmitoylethanolamide (PEA, Tocris Bioscience, cat No. 0879-10 mg, Bristol, UK), Paclitaxel (Taxol equivalent, Invitrogen, cat No. P3456-5 mg, Schwerte, Germany), Nerve Growth Factor-2.5S from the murine submaxillary gland (NGF, Sigma Aldrich, Merck, cat No. N6009-4X 25 μg, St. Louis, MO, USA) and glial cell-derived neurotrophic factor (GDNF, Sigma-Aldrich, cat No. SRP3309-10 μg, St. Louis, MO, USA), Uridine (Uridin, Sigma-Aldrich, U3003-5 g, Darmstadt, Germany), and 5-Fluoro-2-deoxyuridine (FudR, Sigma-Aldrich, cat No. F0503-100 mg, Darmstadt, Germany). PEA and Paclitaxelwere dissolved in DMSO to obtain stock solutions of 10 mM PEA and 1 mM Paclitaxel and stored at −20 °C, while NGF and GDNF dissolved in 0.1 % Bovine Serum Albumin (BSA, Sigma-Aldrich, cat No. A7906-10 g, St. Louis, MO, USA). A total of 20 mM uridine/5-fluorodeoxyuridine (UFdU) stock solution was prepared by mixing 48.8 mg uridine and 49.2 mg 5-fluorodeoxyuridine in 10 mL distilled water, and 100 μL aliquots were prepared and frozen at −20 °C. Notably, controls contained the similar highest concentration of DMSO (0.1%) to exclude any effects on investigated parameters.

2.3. Isolation and Preparation of DRG Neurons

DRG tissues isolated from 6–8 weeks of age Wister rats. In brief, rats were deeply anesthetized with isoflurane (Florene, 100% (*V*/*V*), 250 mL, Abcam, cat No. B506, Carros, France) by inhalation and sacrificed by decapitation with a commercial guillotine. Under aseptic conditions, the vertebral column was isolated and carefully cleared from all surrounding muscle, fat, and other soft tissue. The spinal cord was then exposed and scooped out. Following the dorsal roots. DRGs were localized, removed, collected from intervertebral foramina at both sides, and placed in a 3 mL sterile dish containing Hanks balanced salt solutions without Mg^{2+}/Ca^{2+} (HBSS, Invitrogen, REF. 24020-091, Schwerte, Germany). Dorsal root neuronal culture was prepared according to a previously published protocol [38] with some modifications. Briefly, isolated DRGs were enzymatically digested in the first enzymatic solution, which contained 60 U/mL papain solution (Sigma-Aldrich, cat No. P4762-100 mg, St. Louis, MO, USA), 3 μL of 80 mg/mL saturated sodium hydrogen carbonate solution (NaHCO3, Merck, cat No. k22399729, Darmstadt, Germany), and 0.6 mg/mL L-Cysteine (L-Cys, Sigma-Aldrich, Cat No. C7352-25 g, St. Louis, MO, USA) dissolved in 1.5 mL of HBSS without Mg^{2+}/ Ca^{2+}. Afterwards, DRGs were incubated in a papain solution for 15 min in a 37 °C water bath, then incubated in a second solution which consisted of 4 mg/mL collagenase type II solution (CLS2, Sigma-Aldrich, Cat No.C6885-1 gm, St. Louis, MO, USA) and 4.6 mg/mL dispase type II (Dispase II, Sigma-Aldrich, Cat No. D4693-1 gm, St. Louis, MO, USA) solution in 3 mL HBSS without Mg^{2+}/ Ca^{2+}. The DRGs were mixed gently in collagenase solution and incubated again for 15 min in a water bath at 37 °C.

The resulting cell suspension was centrifuged at 200× *g* for 1 min. The collagenase solution was carefully aspirated, and the DRGs were washed with 2ml of titration media consisting of high glucose Dulbecco's Modified Eagle Medium (DMEM, Invitrogen, Ref. 41965-039, Schwerte, Germany) containing 10% heat-inactivated Fetal Bovine Serum (FBS, Invitrogen, REF. 10270-106, Schwerte, Germany). The DRGs were triturated 10–15 times by using p1000 pipette tips until the cell suspension became cloudy. Bovine serum albumin (BSA) was used for purification (15% (*W*/*V*) BSA solution) to obtain nearly pure neurons without myelin debris. After trituration, single-cell suspensions from DRGs were centrifuged through 15% (*W*/*V*) BSA solution in DMEM, 3 mL of 15% BSA solution: 1 mL of cell suspension in a 15 mL conical tube at 300 g for 8 min at room temperature (RT) to separate sensory neurons in the pellet from non-neuronal cells and debris [39]. The BSA solution was removed and the pellet containing neurons was re-suspended in 1 mL of culture medium consisting of 445 mL of F12 medium (1X, Invitrogen, REF.11765-054, Schwerte, Germany), 50 mL of FBS and 5 mL of 0.1 mg/mL streptomycin/penicillin (Sigma Aldrich, cat No. P4333/100 mL, Darmstadt, Germany). The cell suspension was filtered by a 40 μm cell strainer (SARSTEDT, cat No. D-51588, Schwerte, Germany) to obtain single-cell suspensions and remove undigested tissue debris.

2.4. Seeding and Growth of DRG Neurons

Coverslips 12 mm round were pre-coated with 2 mg/mL Poly-D-lysine (PDL, Sigma Aldrich, cat No. P6407, St. Louis, MO, USA) and 0.2 mg/mL laminin (Sigma Aldrich, cat No. L2020-1 mg, St. Louis, MO, USA) for at least 1 h or overnight in 4 °C, then washed one time with distilled H_2O directly before seeding the cells in culture medium. DRG neuronal cells (5000 cells in 50 μL culture medium) were then pre-seeded onto the center of the coated coverslips for 2 h in an incubator with 37 °C and 5% CO_2. Then, 1 mL of warm culture medium adjusted at pH 7.4 containing 50 ng/mL NGF and 20 ng/mL GDNF (which is essential for growing neuritis of neurons) and 20 μM UFdU (for inhibiting the growth of any remains of supporting cells in culture) was gently added to the wells, and the cells were maintained again at 37 °C with 5% CO_2. The growth and morphology of neurons were monitored after 2, 24, 48, and 72 h to detect the suitable time of treatment (Figure 1a).

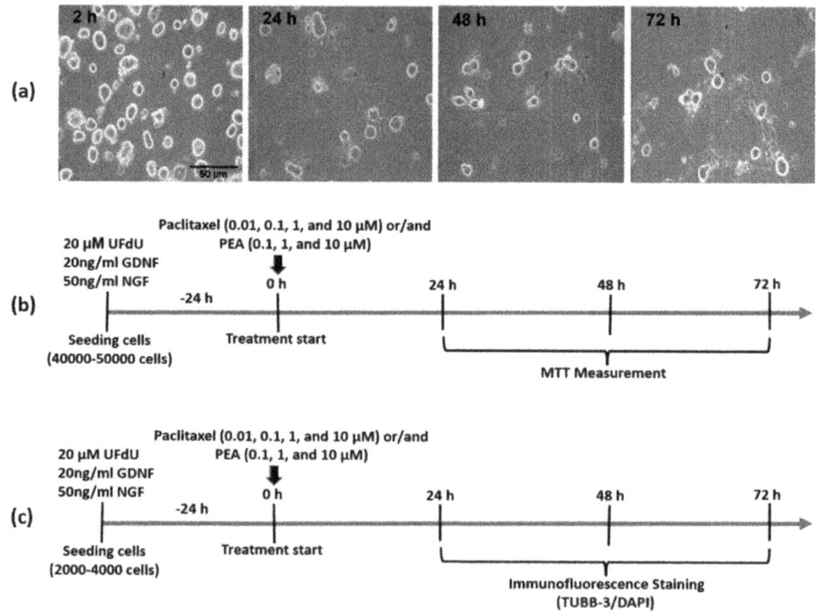

Figure 1. Morphological features and treatment protocols of DRG neurons. (**a**) Representative images show the morphology and growth of DRG neurons at different time points after BSA purification. Scale bars = 50 μm. (**b**,**c**) treatment protocols for studying the effects of Paclitaxel or /and palmitoylethanolamide (PEA) on DRG neurons viability and morphology (Neurite length measurement) respectively, at 24, 48, and 72-h post-treatment.

2.5. Cell Viability (MTT Assay)

DRG neurons were treated 24 h after seeding. Cells were treated with different concentrations of Paclitaxel (0.01, 0.1, 1, 10 μM) and PEA (0.1, 1, 10 μM), either individually or simultaneously combined to study the effects on cell viability. Paclitaxel concentrations were selected based on the literature [15–19,40,41] as well as PEA [42–44]. DRG neurons (4–5 × 10^4 cells/well) in 96 well plates were treated with different concentrations of Paclitaxel and PEA alone or in combination for 24, 48, and 72 h. (Figure 1b). Then, cell viability (%) was measured at the different time points using MTT assay. Four hours before termination of experiments at different time points, 3-(4,5-dimethylthiazol-2-yl)-2,5-diphenyltetrazolium bromide solution (MTT, Invitrogen, cat. No M6494, 5 mg/mL, Eugene, OR, USA) was added. Cells were further incubated for 4 h at 37 °C and 5% CO_2. After removing MTT solution, formazan crystals dissolved in 100 μL of dimethyl sulfoxide (DMSO, Sigma-Aldrich, cat No. D4540-500 mL, Lyon, France) were added and, after another 20 min absorbance values, were measured at wavelengths (540 nm and 720 nm) by a microplate reader (SynergyTMMx, BioTek Instruments, Winooski, VT, USA). DRG neurons cultured in normal media free of Paclitaxel or/and PEA were used as control groups. Controls contained the similar highest concentration of DMSO (0.1%) to exclude any solvent effects on cell viability. All experiments were performed three times independently with 2–3 technical replica for each treatment.

2.6. Immunofluorescence Staining and Microscopy

To investigate the effects of various treatments on the morphology of DRG neurons, cells (2–4 × 10^3 cells/well) were seeded on 12 mm sterile coverslips in a 24-well plate (Greiner Bio-One, Cat No. 662160, Frickenhausen, Germany), cultured for 24 h until most neurites outgrew, and then treated with different concentrations of Paclitaxel or PEA, either alone or in combination (Figure 1c). At the end of each time point, the cells were fixed with 4% paraformaldehyde (PFA, AppliChem, cat No.141451.1211, Darmstadt, Germany) for 15 min at RT and immediately subjected to immunofluorescence or stored in 1 × PBS at 4 °C until further use. For immunofluorescence staining, fixed cells were washed 3 times with 0.02 M PBS for 10 min before unspecific bindings were blocked by incubating cells in normal goat serum (NGS, Sigma Aldrich, cat No. G9023-10 mL, Taufkirchen, Germany, 1:20) in 0.02 M PBS/0.3% (*v*/*v*) Triton) for 30 min. Afterward, cells were incubated with neuronal marker mouse anti-β-III tubulin antibody (TUBB3, Biolegend, San Diego, cat No: 801201, CA, USA, 1:1000) overnight for labelling the cytoskeleton of neurons. Coverslips were thereafter washed thrice for 10 min in PBS, incubated with the secondary antibody goat anti-mouse Alexa Fluor® 488 conjugated (Life Technologies, cat No. 2066710, Darmstadt, Germany, 1:200) for 1 h, washed again 3 times with PBS, and stained with DAPI (4′,6-Diamin-2-phenylindol, Sigma-Aldrich, Munich, Germany, cat No. D9542) for visualization of nuclei. The stained cells were washed in distilled water and covered with DAKO fluorescence mounting medium (DAKO, Agilent Technologies, Inc., Santa Clara, CA 95051, USA). The DRG neurons photomicrographs were captured by using a Leica DMi8 (Wetzlar, Germany) microscope, and five images were randomly taken from each coverslip. The experiment was performed 3 times independently.

2.7. Image Analysis and Determination of Neurite Lengths and Soma Sizes

Measurement of neurite length as a marker for investigating the neurotoxicity of DRG neurons was assessed by using Neurite Tracer, a plugin for ImageJ software version v1.52 used for automated neurite tracing as previously described [15] with some adjustments (Figure S1). Briefly, a sample image pair from cultures of DRG neurons fluorescently labelled with TUBB3 as neuronal marker and DAPI as nuclear marker were opened in Image J (v1.46r (National Institutes of Health, Laboratory for Optical and Computational Instrumentation, University of Wisconsin, Madison, WI, USA) and converted to 8-bit grayscale, and then individually opened in neurite tracer plugin. Large bright objects (somats of neurons) were removed from all images by application of Fiji software version 2.9.0 (accessed 15 January 2022) (https://imagej.net/Fiji/Downloads). Thereafter, the resulting images were inserted to neurite tracer. Afterwards, the threshold was adjusted manually before starting the automated tracing of neuritis. Images with the traced neuritis were merged with RGB original images to ensure the reliability and accuracy of the tracing process. Afterwards, the number of neurons was determined by using a multi-point tool of ImageJ. Finally, traced neuritis lengths were normalized with the numbers of neurons to calculate the neurite length/cell. To determine the size of neuronal somata, soma areas of neurons were selected, and soma areas were measured. The results were normalized with those from the control group.

2.8. Statistical Analysis

Data analysis and visualization were carried out by using GraphPad Prism (GraphPad Software version 8.0.1 for Windows, La Jolla, CA, USA). The normal distribution of data was assessed by use of the Kolmogorov–Smirnov test. The effect of treatments on viability and neurite length of DRG neurons was assessed using one-way ANOVA (analysis of variance) followed by the Bonferroni multiple comparisons test ($p < 0.05$). An alpha level of 0.05 was used for all tests.

3. Results

3.1. Characterization of DRG Neuronal Cells

DRG neurons cultures were examined under a light microscope at various time points (2, 24, 48, and 72 h) to track their growth and morphology. After 2 hours, neuron somas appeared round, bright, and refractile, with a large nucleus (Figure 1a). Three distinct subpopulations (small, medium, and large neurons) based on soma diameter were observed, (Figure 1a). Most of the DRG neurons extended long thin neuritis after 24 h of cells seeding, while, after 48 and 72 h of culturing, all sensory neurons had long neuritis which connected and formed networks together (Figure 1a).

3.2. Effects of Paclitaxel or/and PEA on Cell Viability of DRG Neurons

DRG neurons were treated with different concentrations of Paclitaxel for 24, 48, and 72-h, and we found a significant reduction in the viability of cells at only 72 h post-treatment, regardless of Paclitaxel concentrations, compared to the untreated control group ($p < 0.001$) (Figure 2). Paclitaxel's effects on neuron viability were obviously time-dependent but not concentration-dependent. PEA, as expected, showed no statistically significant effect on the viability of cells in comparison to the untreated control group ($p > 0.05$) at 72 h post-treatment (Figure S2).

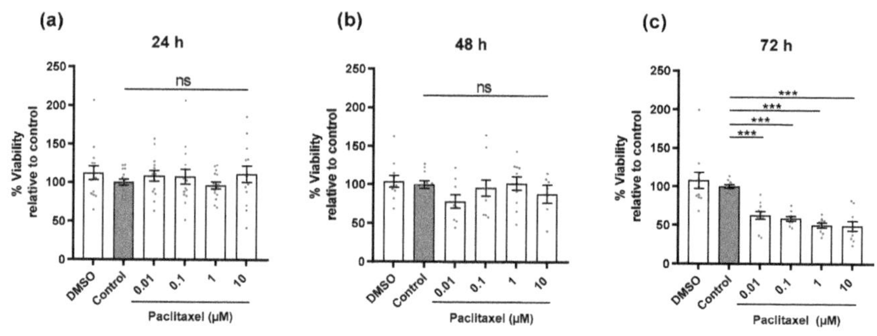

Figure 2. Effects of different concentrations of Paclitaxel on viability (%) of DRG neurons at different time points. Application of different concentrations of Paclitaxel showed no influence on the viability of neurons at (**a**) 24 h and (**b**) 48 h, whereas, at (**c**) 72 h post-treatment, Paclitaxel significantly reduced the viability of cells compared to control (*** $p < 0.001$). The asterisk denotes significant results regarding the respective measurement indicated with the bar. Values are served as mean ± SEM of three independent experiments performed in triplicate. SEM: Standard error mean.

The effects of combined treatments (Paclitaxel plus PEA) were compared to the effect of Paclitaxel alone on viability (%) at 72 h post-treatment. A significant increase was observed for almost all combinations of Paclitaxel (0.01–10 µM) plus PEA (0.1–10 µM) compared to cells treated with Paclitaxel alone. A significant effect was missed only for the combination (10 µM Paclitaxel + 1 µM PEA vs. 10 µM Paclitaxel) ($p < 0.05$) (Figure 3). Notably, the effect of PEA against Paclitaxel was clearly concentration independent.

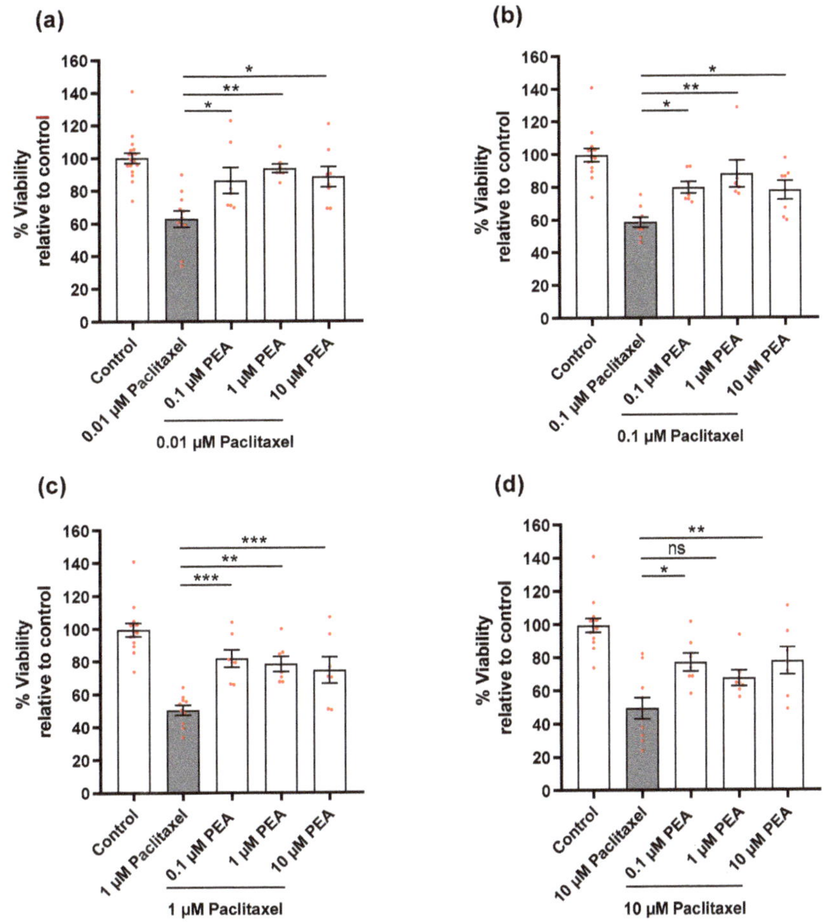

Figure 3. Effects of different concentrations of Paclitaxel (**a**) 0.01 µM, (**b**) 0.1 µM, (**c**) 1 µM, and (**d**) 10 µM either alone or in combination with different concentrations of PEA on viability (%) (mean ± SEM) of DRG neurons at 72 h post-treatment by using MTT assay. The asterisk indicates a significant increase in viability of DRG neurons treated with different Paclitaxel concentrations in combination with different concentrations of PEA at 72 h post-treatment compared to cells treated with Paclitaxel only (* $p < 0.05$, ** $p < 0.01$, and *** $p < 0.001$). Data are (mean ± SEM) of three independent experiments performed in duplicate.

3.3. Effects of Paclitaxel or/and PEA on Morphology of DRG Neurons

Toxic hallmarks of Paclitaxel were observed on the morphology of neurons such as suppression in neurite lengths of neurons, swelling of neuronal cell bodies, as well as retraction and blebbing formation at the distal endings of neurites (Figure 4a). To verify and quantify the Paclitaxel and PEA effects, two different endpoints were assessed, namely neurite length and soma size.

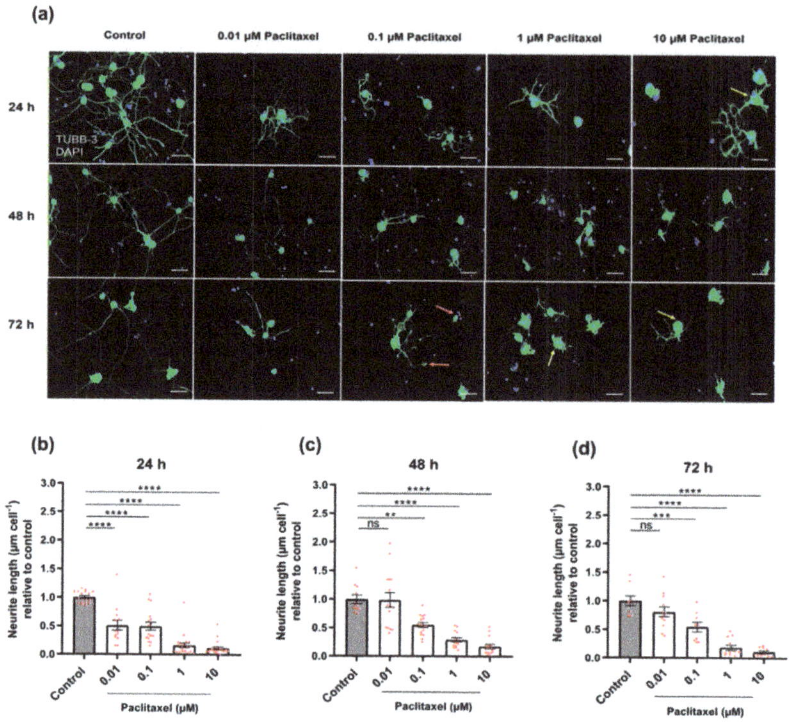

Figure 4. Effects of different Paclitaxel concentrations on morphology, neurite length, and soma size of DRG neurons. (**a**) Immunofluorescence staining of DRG neurons treated with different Paclitaxel concentrations (0.01, 0.1, 1, and 10 µM) labeled with anti-mouse beta III Tubulin antibody after 24, 48, and 72 h. Different concentrations of Paclitaxel had toxic effects leading to a reduction in neurite length and an increase in soma area (yellow arrows) at all time points. Additionally, other characteristics of Paclitaxel toxicity on neuronal morphology are visible, including swellings and blebbing at distal ends of neuritis (red arrows). Nuclei were counterstained with DAPI. Five to eight regions were recorded randomly per each coverslip. Scale bars = 75 µm. (**b**–**d**) significant suppression in neurite lengths of DRG neurons treated with different Paclitaxel concentrations in comparison with the control group (** $p < 0.01$, *** $p < 0.001$, **** $p < 0.0001$) at 24, 48, and 72-h post-treatment, respectively. Data are (mean ± SEM) of three independent experiments performed with (10–15) replicates.

3.3.1. Neurite Length

The treatment with the four different concentrations of Paclitaxel resulted in a significant reduction in neurite length 24 h after treatment when compared to the non-treated control group ($p < 0.05$) (Figure 4b). At 48 and 72 h post-treatment, all studied Paclitaxel groups had an apparent reduction in neurite length except for 0.01 µM Paclitaxel relative to control group (Figure 4c,d). Interestingly, Paclitaxel effects on neurite length were clearly time- and concentration-dependent. No alterations in morphology and neurite length were found in DRG neurons treated with PEA in comparison to the vehicle control group ($p > 0.05$) (Figure S3a–c).

The three different concentrations of PEA co-applied with 0.01 µM Paclitaxel had no significant protective effects on the neurite lengths of neurons at all investigated timelines compared to the 0.01 µM Paclitaxel group ($p > 0.05$; Figure S4). All combined groups of PEA with 0.1 µM Paclitaxel did not cause any significant increase in neurite length at any time point when compared to the 0.1 µM Paclitaxel group alone ($p > 0.05$); however, at 72 h after application, 1 μM PEA only plus 0.1 μM Paclitaxel resulted in an apparent increase in neurite length of DRG neurons compared to 0.1 μM Paclitaxel group ($p < 0.05$; Figure S5).

However, PEA concentrations (0.1, 1, and 10 μM) showed a significant protective effect on neurite outgrowth of DRG neurons when combined with 1 μM Paclitaxel and compared with 1 μM Paclitaxel alone at 24 and 72-h post-treatment. A total of 0.1 μM PEA combined with 1 μM Paclitaxel had a significant protective effect on neurite lengths of DRG neurons 48 h after treatment when compared to Paclitaxel alone ($p < 0.05$) (Figure 5).

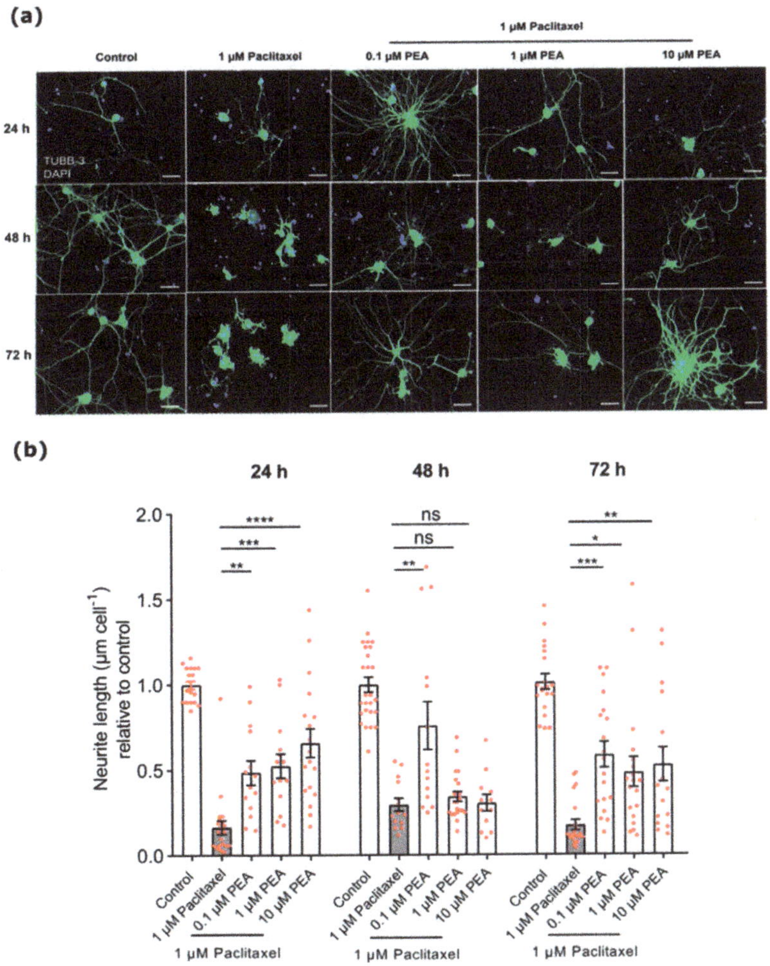

Figure 5. Showing protective effects of different PEA concentrations (0.1, 1, and 10 μM) co-applied with 1 μM Paclitaxel on neurite lengths of DRG neurons compared to 1 μM Paclitaxel alone at 24, 48, and 72 h post-treatment. (**a**) Representative microphotographs of DRG neurons stained with beta III Tubulin antibody for soma and neuritis (green) and DAPI for nuclei (blue). Scale bars = 75 μm. (**b**) Bar graphs indicated a significant increase in neurite lengths of neurons treated with different concentrations of PEA at 24 h and 72 h post-treatment compared to cells treated with Paclitaxel only, while at 48 h post-treatment only 0.1 μM PEA demonstrated a significant increase in neurite length against 1 μM Paclitaxel (* $p < 0.05$, ** $p < 0.01$, *** $p < 0.001$, **** $p < 0.0001$). Data are (mean ± SEM) of three independent experiments performed in 10–15 replicates. The asterisk denotes significant results regarding the respective measurement indicated with bar charts.

Regarding the protective effects of different PEA concentrations against 10 μM Paclitaxel, we found 0.1 or 1 μM PEA combined with 10 μM Paclitaxel showed a significant increase in neurite lengths of DRG neurons at 24, 48, and 72-h post-treatment compared to cells treated with 10 μM Paclitaxel only ($p < 0.05$) (Figure 6). Meanwhile, the 10 μM PEA plus 10 μM Paclitaxel group revealed a significant increase in neurite lengths only at 24 h post-treatment in comparison to the 10 μM Paclitaxel group ($p < 0.05$) (Figure 6).

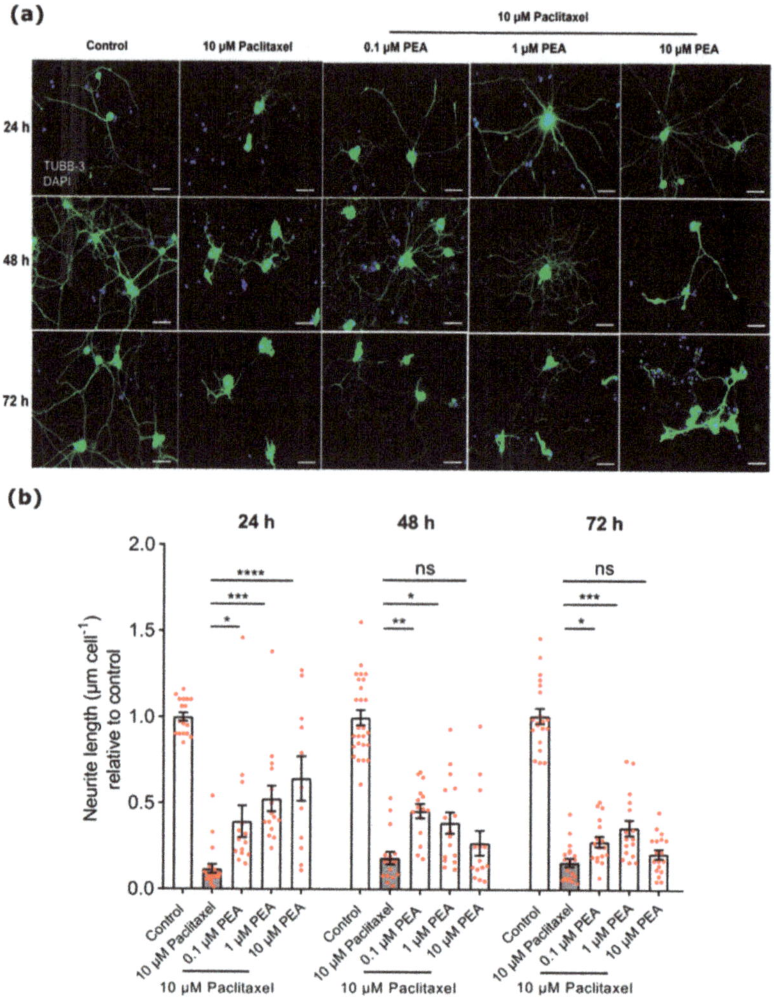

Figure 6. Effects of different PEA concentrations (0.1, 1, and 10 μM) combined with 10 μM Paclitaxel at 24, 48, and 72 h post-treatment. (**a**) Representative immunofluorescence images show DRG neurons labeled with beta III Tubulin antibody (green) and DAPI for nuclei (blue). Scale bars = 75 μm. (**b**) A significant increase in neurite length of neurons was found in groups treated with different concentrations of PEA at 24 h only compared to cells treated with Paclitaxel only. At 48 and 72-h post-treatment, 0.1 μM PEA or 1 μM PEA combined with 10 μM Paclitaxel demonstrated a significant increasing effect on the neurite lengths in comparison with 10 μM Paclitaxel alone (* $p < 0.05$, ** $p < 0.01$, *** $p < 0.001$, **** $p < 0.0001$). Data represented as (mean ± SEM), and the experiments were performed at least 3 independent times and 10–15 replicas. The asterisk denotes significant results regarding the respective measurement indicated with the bar graphs.

3.3.2. Soma Size

The four different concentrations of Paclitaxel led to an increase in the soma size of neurons 24 h post-treatment when compared to the control group ($p < 0.05$) (Figure 7a). At 48 and 72-h, all investigated groups treated with Paclitaxel showed apparent enlargements in areas of neuronal somata except for 0.01 µM of Paclitaxel when compared to the control group ($p < 0.05$) (Figure 7b,c). Effects of Paclitaxel on soma size of neuronal cell bodies were obviously time- and concentration-dependent. Treatment with PEA alone demonstrated no significant effects on the size of neuronal bodies at any time point when compared to control group ($p > 0.05$) (Figure S3d–f).

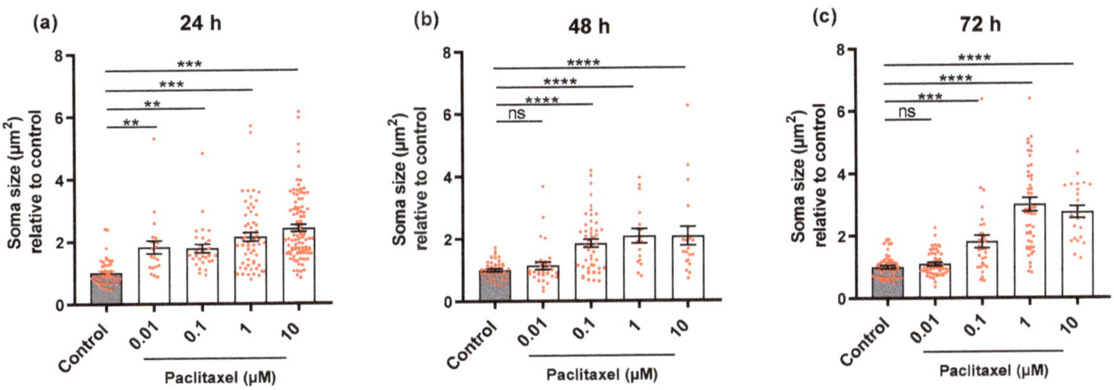

Figure 7. Effects of different Paclitaxel concentrations on soma size of DRG neurons at different time points. Bar charts show a significant increase in the soma size of DRG neurons after the application of different Paclitaxel concentrations at (**a**) 24 h, (**b**) 48 h, and (**c**) 72 h post-treatment compared to the control (** $p < 0.01$, *** $p < 0.001$, **** $p < 0.0001$). Asterisks denote significant results regarding the respective measurement indicated with the bar. Values are served as mean ± SEM of three independent experiments, n = 30–45 replicates. SEM: Standard error mean.

The effects of different combined groups of Paclitaxel plus PEA on soma size of DRG neurons were investigated in comparison to the cells treated with Paclitaxel alone. An increase in soma size was found for 0.01 µM Paclitaxel only at 24 h after treatment compared to control group. A total of 10 µM PEA combined with 0.01 µM Paclitaxel was the only group that demonstrated a significant decrease in neurons soma size in comparison to Paclitaxel group ($p < 0.05$) at 24 h after application, while 0.1 and 1 µM PEA did not show any protective effects against 0.01 µM Paclitaxel group ($p > 0.05$) at the same time point (Figure S6).

The 0.1 and 10 µM PEA co-applied with 0.1 µM Paclitaxel revealed an apparent decrease in neuronal somata sizes, whereas 1 µM PEA had no effect compared to neurons treated with Paclitaxel only at 24 h after treatment ($p < 0.05$) (Figure 8a). At 48 h post-treatment, 1 and 10 µM of PEA combined with 0.1 µM of Paclitaxel were the only groups with a significant neuroprotective effect on somata sizes when compared to 0.1 µM Pacli taxel ($p < 0.05$) (Figure 8a). At 72 h after application, all PEA concentrations combined with 0.1 µM Paclitaxel showed considerable protectant action except for the 10 µM PEA + 0.1 µM Paclitaxel group when compared to Paclitaxel only (Figure 8a).

Treatment of neurons with the three different concentrations of PEA co-applied with 1 µM Paclitaxel demonstrated a significant protective effect on somata sizes when compared to cells exposed to 1 µM Paclitaxel alone at 24- and 72-h time points ($p < 0.05$) (Figure 8b). At 48 h post-treatment, 1 µM of Paclitaxel in combination with 10 µM of PEA was the only group without any protective effects, whereas the other two combined groups led to a strong decrease when compared to individual Paclitaxel treated cells ($p < 0.05$) (Figure 8b).

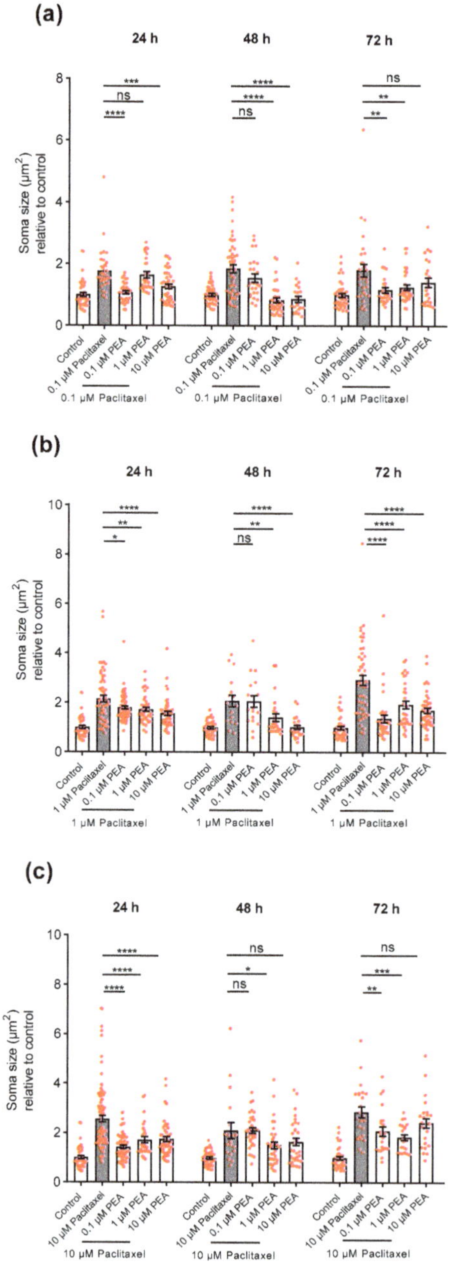

Figure 8. Effects of different PEA concentrations (0.1, 1, and 10 μM) in combination with different concentrations of Paclitaxel (**a**) 0. 1 μM, (**b**) 1 μM, and (**c**) 10 μM on soma sizes of DRG neurons at 24, 48, and 72-h post-treatment. The combined groups of Paclitaxel plus PEA demonstrated a varied significant decrease in DRG neuronal cell bodies in comparison with neurons treated with Paclitaxel only (* $p < 0.05$, ** $p < 0.01$, *** $p < 0.001$, **** $p < 0.0001$). Data represented as (mean ± SEM), and the experiments were performed at least 3 independent times with n = 30–45 replicas. The asterisk denotes significant results regarding the respective measurement indicated with the bar graphs.

The three concentrations of PEA co-applied with 10 μM Paclitaxel showed statistically significant protective effects on DRG neurons' cell bodies against the 10 μM Paclitaxel group at 24 h after treatment ($p < 0.05$) (Figure 8c). Similarly, both concentrations, 0.1 and 1 μM of PEA, combined with 10 μM of Paclitaxel showed a significant neuroprotective effect on soma sizes, while the combination of 10 μM Paclitaxel and 10 μM PEA did not have any protective effect at 72 h post-treatment compared to Paclitaxel alone ($p < 0.05$) (Figure 8c). The 10 μM Paclitaxel plus 1 μM PEA group demonstrated a significant protective effect; however, 0.1 and 10 μM of PEA combined with 10 μM Paclitaxel groups did not reveal any protective effect on the size of cell bodies of neurons when compared to neurons treated only with 10 μM Paclitaxel at 48 h after treatment ($p < 0.05$) (Figure 8c).

Overall, the neuroprotective actions of PEA against the induced toxicity of Paclitaxel on soma size of DRG neurons were time and concentration independent ($p < 0.05$).

4. Discussion

Peripheral neuropathy (PN) is one of the most common side effects of Paclitaxel, affecting up to 97% of all gynecological and urological cancer patients [46,47]. Paclitaxel causes cell death in cancer cells by interfering with mitosis via microtubule stabilization; however, Paclitaxel also affects the peripheral nervous system, causing PN [48]. The primary symptoms are hand and foot numbness besides pain caused by Paclitaxel accumulation in the DRG. DRG neurons are highly susceptible to Paclitaxel accumulation presumably due to a more permeable blood nerve barrier [49]. In the current study, the toxicity of Paclitaxel on viability of neurons was apparent at only 72 h post-treatment in comparison to the control group, while, at all-time windows studied, different Paclitaxel concentrations resulted in a significant reduction in neurite length of DRG neurons. These findings were in line with previous research on Paclitaxel-induced peripheral neuropathy with axonal sensory neuropathy that was length-dependent [50]. A significant reduction in neurite length was reported when DRG neurons were exposed to 10 μM Paclitaxel for 24 h [17]. In addition, Paclitaxel's toxic effects on neurons resulted in the enlargement of neuronal cell bodies obviously at 24, 48, and 72 h post-treatment. These findings are in agreement with previous findings, in which Paclitaxel treatment caused a significant increase in DRG neuron soma size after 24 h of treatment [15,16] and induced a significant enlargement of DRG nucleolus size [51].

Our data obviously demonstrated that Paclitaxel neurotoxicity on neurite outgrowth and soma size is time- and dose-dependent. Similar earlier studies reported on a dose- and infusion time-dependent-induced neurotoxicity that could be exacerbated by underlying conditions or co-application with other drugs [52,53]. The differences in toxic effects of Paclitaxel on the viability and morphology of neurons might possibly be due to higher susceptibility or vulnerability of neurites to toxins than neuronal somata [15,16]. As a result, the toxic effects of Paclitaxel were more rapid, with a significant reduction in neurite length of neurons after 24 h post-treatment. The data are in agreement with a previous work reporting on reduction of axon length after Paclitaxel treatment. Therefore, Paclitaxel seems to act directly on axons and causes axonal degeneration probably through local mechanisms [54]. Additionally, Paclitaxel disrupted intracellular microtubules and bindings with beta-tubulin inside cell soma, resulting in accumulation of non-functional beta-tubulin units of microtubules, vacuolization of mitochondria and cytoplasm in neuronal cell bodies, and cell enlargement [55]. Paclitaxel increased cell size, and, after 72 h of treatment, neurons may explode and die due to a non-apoptotic effect. Taken together, Paclitaxel targets the nerve fibers and causes local axonopathy in still viable neurons with increased soma sizes.

PEA is a bioactive lipid that is used as an anti-nociceptive agent in different animal models of neuropathic pain, including spinal cord injury [56] and diabetes-induced peripheral neuropathy [57]. In humans, PEA accumulates in painful tissues, as observed in the trapezius muscle of women suffering from chronic neck pain [58]. Moreover, PEA protected nerve tissue in neuropathic conditions [37] and prevented neurotoxicity and neurodegeneration [59,60]. Furthermore, PEA also alleviated painful diabetic neuropathy, chemotherapy neuropathy, idiopathic axonal neuropathy, nonspecific neuropathy, and sciatic and lumbosacral spine disease pain [61]. Here, we demonstrated that PEA partially counteracted the toxicity of Paclitaxel on DRG neurons. Regardless of PEA concentration, combining PEA with Paclitaxel significantly increased neuron cell viability compared to treatment with Paclitaxel at 72 h after treatment. These results were consistent with a previous study that showed a reduction of positive propidium iodide (PI) neuronal nuclei after the application of PEA to N-methyl D-aspartate (NMDA)-treated organotypic hippocampal slice cultures [36]. The positive effects on cell viability seem not to be confined to neurons. In astrocytes, different PEA concentrations increased the cell viability from 30 min to 18 h [62]. The data imply for Paclitaxel the need for a longer interaction with damaged cells. Paclitaxel's slow action might provide a good opportunity for PEA to exert protective effects and reverse the toxic effects of Paclitaxel, resulting in increasing the viability of DRG neurons.

In the present study, PEA plus Paclitaxel groups showed a significant increase in neurite length and a strongly decreased soma size of DRG neurons at all studied time points when compared to individual Paclitaxel treatment groups. Interestingly, at 24 h after treatment, PEA produced a protective effect on neurite length and size of cell bodies of neurons against toxicity of Paclitaxel, independent of PEA concentration. This phenomenon might be attributed to the short period of exposure of DRG neurons to Paclitaxel and PEA treatment, allowing PEA to mask and alleviate the toxicity of Paclitaxel. In line with previous evidence in a rat model of Oxaliplatin-induced neurotoxicity, acute intraperitoneal administration of PEA (30 mg kg-1) substantially relieved pain 30 min after administration [63].

Notably, 10 μM PEA did not show any significant protective effect on neurite outgrowth at 48 h post-treatment, although it enhanced neurite extension at 72 h after treatment. The results might be interpreted as an attempt of damaged neurons to develop a survival pressure to resist death caused by Paclitaxel toxicity. They retract and aggregate short neurites. Therefore, neurite extension might become a secondary process at 48 h post-treatment and, as a result, PEA remains unable to express any protective effects. These data are agreed with previous studies on PEA effects on preserving myelin sheet thickness and axonal diameter and preventing myelin degeneration [37]. PEA reduced myelin loss caused by sciatic nerve injury, maintained neuron cell diameters, reduced nerve edema, and restored nerve function, all of which were associated with decreased hypersensitivity [64].

In summary, PEA induced strong neuroprotective actions against Paclitaxel toxicity in DRG neurons and improved their viability and morphology.

5. Conclusions

Our findings showed the ability of PEA to attenuate the toxicity of Paclitaxel on DRG neurons. The effects of Paclitaxel on neuronal viability alone were apparent at 72 h post-treatment only. Furthermore, treatment with Paclitaxel led to a strong reduction in neurite length and enlargement of neuronal cell bodies at all investigated time windows. PEA showed neuroprotective effects by partially reversing the toxic effects of Paclitaxel, including increasing cell viability, enhancing DRG neuron neurite outgrowth, and decreasing swelling of neuronal soma. These findings contribute to our understanding of Paclitaxel's site and mode of action on the peripheral nervous system and highlight the critical need for novel peripheral neuropathy protective strategies. More research will be needed to elucidate the signaling pathways underlying PEA's neuroprotective effects against Paclitaxel neurotoxicity. With these results, PEA might be a promising therapeutic option for cancer patients suffering from CIPN.

Supplementary Materials: The following supporting information can be downloaded at: https://www.mdpi.com/article/10.3390/biom12121873/s1. Figure S1: Representative example for tracing neurites by ImageJ program; Figure S2: Effects of different PEA concentrations on neuronal cell viability (%) (mean ± SEM) at 72 h post-treatment; Figure S3: Effects of different concentrations of PEA on neurite lengths and soma sizes (mean ± SEM) of DRG neurons at 24, 48, and 72-h post-treatment; Figure S4: Effects of different PEA concentrations (0.1, 1, and 10 μM) combined with 0.01 μM Paclitaxel on neurite length of DRG neurons at 24, 48, and 72-h post-treatment; Figure S5: Effects of different PEA concentrations (0.1, 1, and 10 μM) combined with 0.1 μM Paclitaxel at 24, 48, and 72-h post-treatment; Figure S6: The effects of different PEA concentrations (0.1, 1, and 10 μM) co-applied with 0.01 μM Paclitaxel on soma sizes of DRG neurons at 24, 48, and 72-h post-treatment.

Author Contributions: Conceptualization, F.D. and A.E.; methodology, A.E.; formal analysis, A.E.; investigation, A.E.; resources, F.D.; data curation, A.E.; writing—original draft preparation, A.E. and F.D.; writing—review and editing, F.D. and A.E.; visualization, A.E.; supervision, F.D.; project administration, F.D. All authors have read and agreed to the published version of the manuscript.

Funding: This research received no external funding.

Institutional Review Board Statement: All animal experiments were carried out in accordance with the policy on ethics and the policy on the use of animals in neuroscience research, as specified in directive 2010/63/EU of the European Parliament and of the Council of the European union on the protection of animals used for scientific purposes and were approved by local authorities for laboratory animal care and use (State of Saxony-Anhalt, Germany, permission number: I11M27).

Informed Consent Statement: Not applicable.

Data Availability Statement: The data presented in this study are available on request from the authors.

Acknowledgments: We acknowledge Katholischer akademischer Ausländer-Dienst (KAAD) for the Ph.D. scholarship to A.E. Authors would like to thank Chalid Ghadban and Candy Rothgänger-Strube for excellent technical assistance and Tim Hohmann, Marc Richard Kolbe, Urszula Hohmann, Johanna Marie Dela Cruz, Derek C Molliver, and Steven Fagan for helpful discussions. We acknowledge the financial support of the Open Access Publication Fund of the Martin Luther University Halle-Wittenberg.

Conflicts of Interest: The authors declare no conflict of interest.

References

1. Grisold, W.; Cavaletti, G.; Windebank, A.J. Peripheral Neuropathies from Chemotherapeutics and Targeted Agents: Diagnosis, Treatment, and Prevention. *Neuro. Oncol.* **2012**, *14* (Suppl. 4), iv45–iv54. [CrossRef]
2. Masocha, W. Targeting the Endocannabinoid System for Prevention or Treatment of Chemotherapy-Induced Neuropathic Pain: Studies in Animal Models. *Pain Res. Manag.* **2018**, *2018*, 5234943. [CrossRef]
3. Rowinsky, E.K.; Donehower, R.C. Paclitaxel (Taxol). *N. Engl. J. Med.* **1995**, *332*, 1004–1014. [CrossRef]
4. Tankanow, R.M. Docetaxel: A Taxoid for the Treatment of Metastatic Breast Cancer. *Am. J. Health Pharm.* **1998**, *55*, 1777–1791. [CrossRef]
5. Vyas, D.M.; Kadow, J.F. 6 Paclitaxel: A Unique Tubulin Interacting Anticancer Agent. *Prog. Med. Chem.* **1995**, *32*, 289–337. [CrossRef]
6. Schiff, P.B.; Fant, J.; Horwitz, S.B. Promotion of Microtubule Assembly in Vitro by Taxol. *Nature* **1979**, *277*, 665–667. [CrossRef]
7. Rowinsky, E.K.; Cazenave, L.A.; Donehower, R.C. Taxol: A Novel Investigational Antimicrotubule Agent. *JNCI J. Natl. Cancer Inst.* **1990**, *82*, 1247–1259. [CrossRef]
8. Seretny, M.; Currie, G.L.; Sena, E.S.; Ramnarine, S.; Grant, R.; MacLeod, M.R.; Colvin, L.A.; Fallon, M. Incidence, Prevalence, and Predictors of Chemotherapy-Induced Peripheral Neuropathy: A Systematic Review and Meta-Analysis. *Pain* **2014**, *155*, 2461–2470. [CrossRef]
9. Farquhar-Smith, P. Chemotherapy-Induced Neuropathic Pain. *Curr. Opin. Support. Palliat. Care* **2011**, *5*, 1–7. [CrossRef]
10. Dougherty, P.M.; Cata, J.P.; Cordella, J.V.; Burton, A.; Weng, H.-R. Taxol-Induced Sensory Disturbance Is Characterized by Preferential Impairment of Myelinated Fiber Function in Cancer Patients. *Pain* **2004**, *109*, 132–142. [CrossRef]
11. Sharma, S.; Venkitaraman, R.; Vas, P.R.J.; Rayman, G. Assessment of Chemotherapy-Induced Peripheral Neuropathy Using the LDI FLARE Technique: A Novel Technique to Detect Neural Small Fiber Dysfunction. *Brain Behav.* **2015**, *5*, e00354. [CrossRef]
12. Windebank, A.J.; Smith, A.G.; Russell, J.W. The Effect of Nerve Growth Factor, Ciliary Neurotrophic Factor, and ACTH Analogs on Cisplatin Neurotoxicity in Vitro. *Neurology* **1994**, *44 Pt 1*, 488–494. [CrossRef]
13. Gill, J.S.; Windebank, A.J. Cisplatin-Induced Apoptosis in Rat Dorsal Root Ganglion Neurons Is Associated with Attempted Entry into the Cell Cycle. *J. Clin. Investig.* **1998**, *101*, 2842–2850. [CrossRef]

14. Fischer, S.J.; McDonald, E.S.; Gross, L.; Windebank, A.J. Alterations in Cell Cycle Regulation Underlie Cisplatin Induced Apoptosis of Dorsal Root Ganglion Neurons in Vivo. *Neurobiol. Dis.* **2001**, *8*, 1027–1035. [CrossRef]
15. Scuteri, A.; Nicolini, G.; Miloso, M.; Bossi, M.; Cavaletti, G.; Windebank, A.J.; Tredici, G. Paclitaxel Toxicity in Post-Mitotic Dorsal Root Ganglion (DRG) Cells. *Anticancer Res.* **2006**, *26*, 1065–1070.
16. Guo, L.; Hamre, J.; Eldridge, S.; Behrsing, H.P.; Cutuli, F.M.; Mussio, J.; Davis, M. Multiparametric Image Analysis of Rat Dorsal Root Ganglion Cultures to Evaluate Peripheral Neuropathy-Inducing Chemotherapeutics. *Toxicol. Sci.* **2017**, *156*, kfw254. [CrossRef]
17. Park, S.H.; Eber, M.R.; Fonseca, M.M.; Patel, C.M.; Cunnane, K.A.; Ding, H.; Hsu, F.-C.; Peters, C.M.; Ko, M.-C.; Strowd, R.E.; et al. Usefulness of the Measurement of Neurite Outgrowth of Primary Sensory Neurons to Study Cancer-Related Painful Complications. *Biochem. Pharmacol.* **2021**, *188*, 114520. [CrossRef]
18. Akin, E.J.; Alsaloum, M.; Higerd, G.P.; Liu, S.; Zhao, P.; Dib-Hajj, F.B.; Waxman, S.G.; Dib-Hajj, S.D. Paclitaxel Increases Axonal Localization and Vesicular Trafficking of Nav1.7. *Brain* **2021**, *144*, 1727–1737. [CrossRef]
19. Shin, G.J.; Pero, M.E.; Hammond, L.A.; Burgos, A.; Kumar, A.; Galindo, S.E.; Lucas, T.; Bartolini, F.; Grueber, W.B. Integrins Protect Sensory Neurons in Models of Paclitaxel-Induced Peripheral Sensory Neuropathy. *Proc. Natl. Acad. Sci. USA* **2021**, *118*, e2006050118. [CrossRef]
20. Rao, S.S.C.; Seaton, K.; Miller, M.; Brown, K.; Nygaard, I.; Stumbo, P.; Zimmerman, B.; Schulze, K. Randomized Controlled Trial of Biofeedback, Sham Feedback, and Standard Therapy for Dyssynergic Defecation. *Clin. Gastroenterol. Hepatol.* **2007**, *5*, 331–338. [CrossRef]
21. Kautio, A.-L.; Haanpää, M.; Leminen, A.; Kalso, E.; Kautiainen, H.; Saarto, T. Amitriptyline in the Prevention of Chemotherapy-Induced Neuropathic Symptoms. *Anticancer Res.* **2009**, *29*, 2601–2606.
22. Hershman, D.L.; Lacchetti, C.; Loprinzi, C.L. Prevention and Management of Chemotherapy-Induced Peripheral Neuropathy in Survivors of Adult Cancers: American Society of Clinical Oncology Clinical Practice Guideline Summary. *J. Oncol. Pract.* **2014**, *10*, e421–e424. [CrossRef]
23. Kim, J.H.; Dougherty, P.M.; Abdi, S. Basic Science and Clinical Management of Painful and Non-Painful Chemotherapy-Related Neuropathy. *Gynecol. Oncol.* **2015**, *136*, 453–459. [CrossRef]
24. Aizpurua-Olaizola, O.; Elezgarai, I.; Rico-Barrio, I.; Zarandona, I.; Etxebarria, N.; Usobiaga, A. Targeting the Endocannabinoid System: Future Therapeutic Strategies. Drug Discov. *Today* **2017**, *22*, 105–110. [CrossRef]
25. Pacher, P.; Kunos, G. Modulating the Endocannabinoid System in Human Health and Disease—Successes and Failures. *FEBS J.* **2013**, *280*, 1918–1943. [CrossRef]
26. Skaper, S.D.; Facci, L.; Barbierato, M.; Zusso, M.; Bruschetta, G.; Impellizzeri, D.; Cuzzocrea, S.; Giusti, P. N-Palmitoylethanolamine and Neuroinflammation: A Novel Therapeutic Strategy of Resolution. *Mol. Neurobiol.* **2015**, *52*, 1034–1042. [CrossRef]
27. di Marzo, V.D.; Skaper, S. Editorial (Hot Topic: Palmitoylethanolamide: Biochemistry, Pharmacology and Therapeutic Use of a Pleiotropic Anti-Inflammatory Lipid Mediator). *CNS Neurol. Disord. Drug Targets* **2013**, *12*, 4–6. [CrossRef]
28. Re, G.; Barbero, R.; Miolo, A.; Di Marzo, V. Palmitoylethanolamide, Endocannabinoids and Related Cannabimimetic Compounds in Protection against Tissue Inflammation and Pain: Potential Use in Companion Animals. *Vet. J.* **2007**, *173*, 21–30. [CrossRef]
29. Esposito, E.; Cuzzocrea, S. Palmitoylethanolamide in Homeostatic and Traumatic Central Nervous System Injuries. CNS Neurol. Disord. *Drug Targets* **2013**, *12*, 55–61. [CrossRef]
30. Petrosino, S.; Schiano Moriello, A.; Cerrato, S.; Fusco, M.; Puigdemont, A.; De Petrocellis, L.; Di Marzo, V. The Anti-Inflammatory Mediator Palmitoylethanolamide Enhances the Levels of 2-Arachidonoyl-Glycerol and Potentiates Its Actions at TRPV1 Cation Channels. *Br. J. Pharmacol.* **2016**, *173*, 1154–1162. [CrossRef]
31. Iannotti, F.A.; Vitale, R.M. The Endocannabinoid System and PPARs: Focus on Their Signalling Crosstalk, Action and Transcriptional Regulation. *Cells* **2021**, *10*, 586. [CrossRef]
32. Ho, W.-S.V.; Barrett, D.A.; Randall, M.D. "Entourage" Effects of N-Palmitoylethanolamide and N-Oleoylethanolamide on Vasorelaxation to Anandamide Occur through TRPV1 Receptors. *Br. J. Pharmacol.* **2008**, *155*, 837–846. [CrossRef]
33. Hohmann, U.; Pelzer, M.; Kleine, J.; Hohmann, T.; Ghadban, C.; Dehghani, F. Opposite Effects of Neuroprotective Cannabinoids, Palmitoylethanolamide, and 2-Arachidonoylglycerol on Function and Morphology of Microglia. *Front. Neurosci.* **2019**, *13*, 1180. [CrossRef]
34. Lo Verme, J.; Fu, J.; Astarita, G.; La Rana, G.; Russo, R.; Calignano, A.; Piomelli, D. The Nuclear Receptor Peroxisome Proliferator-Activated Receptor-Alpha Mediates the Anti-Inflammatory Actions of Palmitoylethanolamide. *Mol. Pharmacol.* **2005**, *67*, 15–19. [CrossRef]
35. Bougarne, N.; Weyers, B.; Desmet, S.J.; Deckers, J.; Ray, D.W.; Staels, B.; De Bosscher, K. Molecular Actions of PPARα in Lipid Metabolism and Inflammation. *Endocr. Rev.* **2018**, *39*, 760–802. [CrossRef]
36. Koch, M.; Kreutz, S.; Böttger, C.; Benz, A.; Maronde, E.; Ghadban, C.; Korf, H.-W.; Dehghani, F. Palmitoylethanolamide Protects Dentate Gyrus Granule Cells via Peroxisome Proliferator-Activated Receptor-Alpha. *Neurotox. Res.* **2011**, *19*, 330–340. [CrossRef]
37. Di Cesare Mannelli, L.; D'Agostino, G.; Pacini, A.; Russo, R.; Zanardelli, M.; Ghelardini, C.; Calignano, A. Palmitoylethanolamide Is a Disease-Modifying Agent in Peripheral Neuropathy: Pain Relief and Neuroprotection Share a PPAR-Alpha-Mediated Mechanism. *Mediat. Inflamm.* **2013**, *2013*, 328797. [CrossRef]
38. Malin, S.A.; Davis, B.M.; Molliver, D.C. Production of Dissociated Sensory Neuron Cultures and Considerations for Their Use in Studying Neuronal Function and Plasticity. *Nat. Protoc.* **2007**, *2*, 152–160. [CrossRef]

39. Tolkovsky, A.M.; Brelstaff, J. Sensory Neurons from Tau Transgenic Mice and Their Utility in Drug Screening. *Methods Mol. Biol.* **2018**, *1727*, 93–105. [CrossRef]
40. Liebmann, J.; Cook, J.; Lipschultz, C.; Teague, D.; Fisher, J.; Mitchell, J. Cytotoxic Studies of Paclitaxel (Taxol®) in Human Tumour Cell Lines. *Br. J. Cancer* **1993**, *68*, 1104–1109. [CrossRef]
41. Zasadil, L.M.; Andersen, K.A.; Yeum, D.; Rocque, G.B.; Wilke, L.G.; Tevaarwerk, A.J.; Raines, R.T.; Burkard, M.E.; Weaver, B.A. Cytotoxicity of Paclitaxel in Breast Cancer Is Due to Chromosome Missegregation on Multipolar Spindles. *Sci. Transl. Med.* **2014**, *6*, 229ra43. [CrossRef] [PubMed]
42. Musella, A.; Fresegna, D.; Rizzo, F.R.; Gentile, A.; Bullitta, S.; De Vito, F.; Guadalupi, L.; Centonze, D.; Mandolesi, G. A Novel Crosstalk within the Endocannabinoid System Controls GABA Transmission in the Striatum. *Sci. Rep.* **2017**, *7*, 7363. [CrossRef] [PubMed]
43. Scuderi, C.; Valenza, M.; Stecca, C.; Esposito, G.; Carratù, M.R.; Steardo, L. Palmitoylethanolamide Exerts Neuroprotective Effects in Mixed Neuroglial Cultures and Organotypic Hippocampal Slices via Peroxisome Proliferator-Activated Receptor-α. *J. Neuroinflammation* **2012**, *9*, 49. [CrossRef] [PubMed]
44. Beggiato, S.; Tomasini, M.C.; Cassano, T.; Ferraro, L. Chronic Oral Palmitoylethanolamide Administration Rescues Cognitive Deficit and Reduces Neuroinflammation, Oxidative Stress, and Glutamate Levels in A Transgenic Murine Model of Alzheimer's Disease. *J. Clin. Med.* **2020**, *9*, 428. [CrossRef] [PubMed]
45. Pool, M.; Thiemann, J.; Bar-Or, A.; Fournier, A.E. NeuriteTracer: A Novel ImageJ Plugin for Automated Quantification of Neurite Outgrowth. *J. Neurosci. Methods* **2008**, *168*, 134–139. [CrossRef] [PubMed]
46. Klein, I.; Lehmann, H. Pathomechanisms of Paclitaxel-Induced Peripheral Neuropathy. *Toxics* **2021**, *9*, 229. [CrossRef]
47. Mielke, S.; Sparreboom, A.; Mross, K. Peripheral Neuropathy: A Persisting Challenge in Paclitaxel-Based Regimes. *Eur. J. Cancer* **2006**, *42*, 24–30. [CrossRef]
48. Fukuda, Y.; Li, Y.; Segal, R.A. A Mechanistic Understanding of Axon Degeneration in Chemotherapy-Induced Peripheral Neuropathy. *Front. Neurosci.* **2017**, *11*, 481. [CrossRef]
49. Cavaletti, G.; Cavalletti, E.; Oggioni, N.; Sottani, C.; Minoia, C.; D'Incalci, M.; Zucchetti, M.; Marmiroli, P.; Tredici, G. Distribution of Paclitaxel within the Nervous System of the Rat after Repeated Intravenous Administration. *Neurotoxicology* **2000**, *21*, 389–393. [CrossRef]
50. Staff, N.P.; Grisold, A.; Grisold, W.; Windebank, A.J. Chemotherapy-Induced Peripheral Neuropathy: A Current Review. *Ann. Neurol.* **2017**, *81*, 772–781. [CrossRef]
51. Jamieson, S.M.F.; Liu, J.; Hsu, T.; Baguley, B.C.; McKeage, M.J. Paclitaxel Induces Nucleolar Enlargement in Dorsal Root Ganglion Neurons in vivo Reducing Oxaliplatin Toxicity. *Br. J. Cancer* **2003**, *88*, 1942–1947. [CrossRef] [PubMed]
52. Jaggi, A.S.; Singh, N. Mechanisms in Cancer-Chemotherapeutic Drugs-Induced Peripheral Neuropathy. *Toxicology* **2012**, *291*, 1–9. [CrossRef] [PubMed]
53. Carlson, K.; Ocean, A.J. Peripheral Neuropathy with Microtubule-Targeting Agents: Occurrence and Management Approach. *Clin. Breast Cancer* **2011**, *11*, 73–81. [CrossRef] [PubMed]
54. Yang, I.H.; Siddique, R.; Hosmane, S.; Thakor, N.; Höke, A. Compartmentalized Microfluidic Culture Platform to Study Mechanism of Paclitaxel-Induced Axonal Degeneration. *Exp. Neurol.* **2009**, *218*, 124–128. [CrossRef] [PubMed]
55. Barbuti, A.M.; Chen, Z.-S. Paclitaxel Through the Ages of Anticancer Therapy: Exploring Its Role in Chemoresistance and Radiation Therapy. *Cancers* **2015**, *7*, 2360–2371. [CrossRef]
56. Genovese, T.; Esposito, E.; Mazzon, E.; Paola, R.D.; Meli, R.; Bramanti, P.; Piomelli, D.; Calignano, A.; Cuzzocrea, S. Effects of Palmitoylethanolamide on Signaling Pathways Implicated in the Development of Spinal Cord Injury. *J. Pharmacol. Exp. Ther.* **2008**, *326*, 12–23. [CrossRef]
57. Donvito, G.; Bettoni, I.; Comelli, F.; Colombo, A.; Costa, B. Palmitoylethanolamide Relieves Pain and Preserves Pancreatic Islet Cells in a Murine Model of Diabetes. *CNS Neurol. Disord. Drug Targets* **2015**, *14*, 452–462. [CrossRef]
58. Ghafouri, N.; Ghafouri, B.; Larsson, B.; Stensson, N.; Fowler, C.J.; Gerdle, B. Palmitoylethanolamide and Stearoylethanolamide Levels in the Interstitium of the Trapezius Muscle of Women with Chronic Widespread Pain and Chronic Neck-Shoulder Pain Correlate with Pain Intensity and Sensitivity. *Pain* **2013**, *154*, 1649–1658. [CrossRef]
59. D'Agostino, G.; Russo, R.; Avagliano, C.; Cristiano, C.; Meli, R.; Calignano, A. Palmitoylethanolamide Protects Against the Amyloid-B25-35-Induced Learning and Memory Impairment in Mice, an Experimental Model of Alzheimer Disease. *Neuropsychopharmacology* **2012**, *37*, 1784–1792. [CrossRef]
60. Lambert, D.M.; Vandevoorde, S.; Diependaele, G.; Govaerts, S.J.; Robert, A.R. Anticonvulsant Activity of N-Palmitoylethanolamide, a Putative Endocannabinoid, in Mice. *Epilepsia* **2002**, *42*, 321–327. [CrossRef]
61. Andresen, S.R.; Bing, J.; Hansen, R.M.; Biering-Sørensen, F.; Johannesen, I.L.; Hagen, E.M.; Rice, A.S.C.; Nielsen, J.F.; Bach, F.W.; Finnerup, N.B. Ultramicronized Palmitoylethanolamide in Spinal Cord Injury Neuropathic Pain: A Randomized, Double-Blind, Placebo-Controlled Trial. *Pain* **2016**, *157*, 2097–2103. [CrossRef] [PubMed]
62. Morsanuto, V.; Galla, R.; Molinari, C.; Uberti, F. A New Palmitoylethanolamide Form Combined with Antioxidant Molecules to Improve Its Effectivess on Neuronal Aging. *Brain Sci.* **2020**, *10*, 457. [CrossRef] [PubMed]

63. Di Cesare Mannelli, L.; Pacini, A.; Corti, F.; Boccella, S.; Luongo, L.; Esposito, E.; Cuzzocrea, S.; Maione, S.; Calignano, A.; Ghelardini, C. Antineuropathic Profile of N-Palmitoylethanolamine in a Rat Model of Oxaliplatin-Induced Neurotoxicity. *PLoS ONE* **2015**, *10*, e0128080. [CrossRef] [PubMed]
64. Truini, A.; Biasiotta, A.; Di Stefano, G.; La Cesa, S.; Leone, C.; Cartoni, C.; Federico, V.T.; Petrucci, M.; Cruccu, G. Palmitoylethanolamide Restores Myelinated-Fibre Function in Patients with Chemotherapy-Induced Painful Neuropathy. *CNS Neurol. Disord. Drug Targets* **2011**, *10*, 916–920. [CrossRef]

biomolecules

MDPI

Article

Aerosol-Administered Adelmidrol Attenuates Lung Inflammation in a Murine Model of Acute Lung Injury

Livia Interdonato [1,†], Ramona D'amico [1,†], Marika Cordaro [2], Rosalba Siracusa [1], Roberta Fusco [1], Alessio Filippo Peritore [1], Enrico Gugliandolo [3], Rosalia Crupi [3], Stefano Coaccioli [4], Tiziana Genovese [1], Daniela Impellizzeri [1,*], Rosanna Di Paola [3,‡] and Salvatore Cuzzocrea [1,‡]

1 Department of Chemical, Biological, Pharmaceutical and Environmental Sciences, University of Messina, 98166 Messina, Italy
2 Department of Biomedical, Dental and Morphological and Functional Imaging, University of Messina, Via Consolare Valeria, 98125 Messina, Italy
3 Department of Veterinary Science, University of Messina, 98168 Messina, Italy
4 General Medical Clinic and Medical Therapy, Rheumatology and Medical Therapy of the Pain, University of Perugia, "Polo di Terni", "AO Santa Maria" of Terni, 06129 Perugia, Italy
* Correspondence: dimpellizzeri@unime.it; Tel.: +39-090-676-5208
† These authors equally contributed to the work.
‡ These authors shared senior authorship.

Abstract: Acute lung injury (ALI) is a common and devastating clinical disorder with a high mortality rate and no specific therapy. The pathophysiology of ALI is characterized by increased alveolar/capillary permeability, lung inflammation, oxidative stress and structural damage to lung tissues, which can progress to acute respiratory distress syndrome (ARDS). Adelmidrol (ADM), an analogue of palmitoylethanolamide (PEA), is known for its anti-inflammatory and antioxidant functions, which are mainly due to down-modulating mast cells (MCs) and promoting endogenous antioxidant defense. The aim of this study is to evaluate the protective effects of ADM in a mice model of ALI, induced by intratracheal administration of lipopolysaccharide (LPS) at the dose of 5 mg/kg. ADM 2% was administered by aerosol 1 and 6 h after LPS instillation. In this study, we clearly demonstrated that ADM reduced lung damage and airway infiltration induced by LPS instillation. At the same time, ADM counteracted the increase in MC number and the expression of specific markers of MC activation, i.e., chymase and tryptase. Moreover, ADM reduced oxidative stress by upregulating antioxidant enzymes as well as modulating the Nf-kB pathway and the resulting pro-inflammatory cytokine release. These results suggest that ADM could be a potential candidate in the management of ALI.

Keywords: acute lung injury; Adelmidrol; mast cells; inflammation; oxidative stress

Citation: Interdonato, L.; D'amico, R.; Cordaro, M.; Siracusa, R.; Fusco, R.; Peritore, A.F.; Gugliandolo, E.; Crupi, R.; Coaccioli, S.; Genovese, T.; et al. Aerosol-Administered Adelmidrol Attenuates Lung Inflammation in a Murine Model of Acute Lung Injury. *Biomolecules* **2022**, *12*, 1308. https://doi.org/10.3390/biom12091308

Academic Editor: Vladimir N. Uversky

Received: 25 July 2022
Accepted: 14 September 2022
Published: 16 September 2022

1. Introduction

Acute lung injury (ALI) and acute respiratory distress syndrome (ARDS), two acute inflammatory conditions, are a major cause of respiratory failure and one of the most challenging clinical conditions with significant morbidity and mortality [1]. ALI is characterized by alteration of the endothelium and alveolar epithelial barrier, resulting in increased microvascular permeability, pulmonary edema, and polymorphonuclear neutrophil infiltration, all of which contribute to decreased respiratory function [2]. Evidence has proposed that several pathophysiological pathways are activated during ALI, especially during the early phase of the disease [3,4], in which inflammatory response plays a key role [1]. ALI can be modeled in rodents by the administration of LPS through tracheal instillation [5–8]. Local administration of LPS causes an acute and vigorous migration of inflammatory cells into the lung tissue, leading to the overproduction of pro-inflammatory cytokines, including interleukin (IL)-6, IL-1β and tumor necrosis factor (TNF)-α [9,10].

Among the inflammatory cells, mast cells (MCs) stand out for their involvement in the pathophysiology of ALI [11]. In particular, MCs activation induce the release of the contents of their granules, including specific proteases such as chymase and tryptase that contribute to the progression of inflammatory diseases on the respiratory system [12,13]. Additionally, several studies support the role of oxidants and oxidative stress in the pathogenesis of ALI [14–16]. In the context of ALI/ARDS, there are many potential sources of reactive oxygen species (ROS), including leukocytes (neutrophils, monocytes, and macrophages), parenchymal cells (endothelial and epithelial cells, fibroblasts, and myocytes) and circulating oxidant-generating enzymes [16]. Excessive ROS production generated by the injured endothelium/epithelium, as well as recruited leukocytes, amplifies the tissue damage and pulmonary edema [17,18]. Thus, a cross-link between inflammatory response and oxidative stress is involved in the development of ALI [19,20]. Therefore, new approaches are needed to improve the clinical outcomes of the patients affected with the disease. In this regard, we investigated the properties of Adelmidrol (ADM), a palmitoylethanolamide (PEA) analogue that belongs to the ALIAmide family (Autacoid Local Injury Antagonist Amides) [21]. It is well known that ADM has important anti-inflammatory properties due to the regulation of MC activation [22–24]. Recently, it has also been shown that ADM is able to boost endogenous antioxidant defense [25], indirectly enhancing its protective function. Therefore, the aim of this study is to evaluate the beneficial effects of ADM in an LPS-induced ALI model, through the modulation of inflammatory and oxidative pathways.

2. Materials and Methods

2.1. Animals

Male CD1 mice (25–30 g, Envigo, Milan, Italy) were housed in a controlled environment, with food and water ad libitum. The University of Messina Review Board for animal care (OPBA) approved the study (ethical protocol code: 266/2021-PR). All in vivo experiments followed the new directives of the USA, Europe, Italy, and the ARRIVE guidelines.

2.2. Induction of Acute Lung Injury

For intratracheal (i.t.) instillation, animals were anesthetized with isoflurane (2%), and LPS was instilled as previously described [26,27]. Briefly, a 1 cm long ventral midline cervical incision was used to expose the trachea, and LPS was injected using a bent 27-gauge tuberculin needle. Escherichia coli LPS (026: B6L3755, Sigma Aldrich, St. Louis, MO, USA) was administered by a single i.t. instillation at the dose of 5 mg/kg suspended in saline solution (total volume = 0.05 mL per animal) [1]. Sham animals were subjected to the same procedure but received saline instead of LPS. ADM 2% in isotonic solution was administered by aerosol, with a Lovelace nebulizer (In-Tox Products, Albuquerque, NM, USA) being used to create an atmosphere in an exposure chamber (Research and Consulting Co., AG, Basel, Switzerland), as previously described by D'Amico et al. [28].

2.3. Experimental Groups

Mice were randomized into the following experimental groups (n = 12/group):

- LPS group: Mice received LPS i.t. and were treated with the vehicle (saline);
- LPS + ADM group: Mice received LPS i.t. and were treated with ADM 2% aerosol 1 h and 6 h after LPS instillation;
- Sham group: Similar to LPS group, but mice received saline i.t. instead of LPS;
- Sham + ADM: Mice received saline i.t. and were treated with ADM 2% aerosol 1 h and 6 h after saline instillation (data not shown, as no significant difference was ever observed between Sham and Sham + ADM).

At 24 h after induction, all animals were sacrificed, and bronchoalveolar lavage fluid (BALF) as well as lung tissues were collected for further analysis.

2.4. Proteins Concentration and Cell Counts in BALF

The cell count in BALF was carried out as previously described [1]. Briefly, BALF was collected by cannulating the trachea and lavaging the lung twice with 0.7 mL of phosphate-buffered saline (PBS) [1]. The washing solution were removed by aspiration and BALF was centrifugated at 800 rpm [29]. The supernatant was stored at −20 °C, while the pelleted cells were resuspended in PBS. Then, the total cells in BALF were enumerated by counting with a hemocytometer in the presence of the trypan blue stain. For differential cell counting, Wright's Giemsa stain was performed, and the leukocyte and macrophage populations present in BALF were counted. After staining, the differential count was carried out by the standard morphological protocol under a light microscope [30]. To determine the protein concentration and to measure the pro-inflammatory cytokines, the supernatants in BALF were analyzed by a BCA Protein Assay Kit (ThermoFisher, 00161, Rome, Italy. while the levels of IL-6 (#DKW12-2060; Dakewe Biotech Co., Ltd., Bensheim, Germany), IL-1β (#MBS8800273; Biosource International, Camarillo, CA, USA) and TNF-α (#30907; BioLegend, San Diego, CA, USA)) were detected using ELISA [1,31].

2.5. Measurement of Lung Edema

At the end of experiment, wet lung weights were recorded. The lungs were subsequently dried for 48 h at 80 °C and weighed again. The water content in the lung tissues was calculated as the ratio of wet/dry weight of the lung [32].

2.6. Histological Examination

Lung sections were stained with Hematoxylin and Eosin (H&E) for histological analysis [33–36] and with toluidine blue to determine MC degranulation [37]. Every section was examined using a Leica DM6 microscope; (Leica Microsystems SpA, Milan, Italy) associated with Leica LAS X Navigator software (Leica Microsystems SpA, Milan, Italy). Every slide was viewed at a magnification of 10× and morphological changes were evaluated by two blinded investigators [38–41]. Lung injury score was measured according to the methods reported previously [42,43]. The criteria are as follows: 0 = no damage, 1 = mild damage, 2 = moderate damage, 3 = severe damage, 4 = very severe histologic changes.

2.7. Myeloperoxidase (MPO) Assay

The MPO activity was measured as previously described [44–46] and represented in units per gram of wet tissue weight, defined as the amount of enzyme capable of decomposing 1 μmol of peroxide per minute at 37 °C.

2.8. Immunohistochemical Localization of Chymase and Tryptase

Immunohistochemical analysis was performed as previously described [47–49]. Primary antibodies anti-MC chymase (1:100, Santa Cruz Biotechnology (SCB) Heidelberg, Germany, #sc59586) and anti-MC tryptase (1:100, SCB, #sc59587) were incubated overnight on the lung tissue sections. Images were collected using a Leica DM6 microscope; a 10× magnification is shown (Leica Microsystems SpA, Milan, Italy) following a typical procedure [50–53]. The positive pixel intensity value obtained was connected to the histogram profile [54,55].

2.9. Measurement of Oxidative Stress

The malondialdehyde (MDA; #A003-1-2, Nanjing, China), glutathione (GSH; #A006-2-1, Nanjing, China) and catalase (CAT; #A007-1-1, Nanjing, China) levels in the lung tissues were measured using activity assay kits (Nanjing Jiancheng Bioengineering Institute) [52,56–58].

2.10. Analysis of Western Blots

Western blots were performed on lung samples as described in our previous studies [59–61]. The following antibodies were used: anti-IkBα (1:1000, SCB, #sc1643), anti-NF-kB p65 (1:1000; SCB, #sc8414), anti-Nrf2 (1:5000; SCB, #sc365949), anti-HO-1 (1:5000; SCB,

#sc136960), MnSOD (1:5000 SCB #sc137254), anti-β-actin (1:5000; SCB, #sc8432) and anti-lamin A/C antibody (1:5000; Sigma-Aldrich, St. Louis, MO, USA). The membranes were then incubated with IgG peroxidase-conjugated secondary antibody-conjugated bovine mouse IgG or IgG peroxidase-conjugated goat anti-rabbit (1:2000, Jackson ImmunoResearch, Baltimore, MD, USA) [58,62–64]. Protein expression was quantified by densitometry with BIORAD ChemiDocTM XRS + software and normalized to housekeeping genes β-actin and lamin A/C as previously reported [64,65]. Images of blot signals were imported to analysis software (Image Quant TL, v2003, Rome, Italy.) [41,60].

2.11. *Materials*

Unless, otherwise stated, all compounds used in this study were purchased from Sigma-Aldrich Company Ltd. (Milan, Italy). ADM was obtained from Epitech Group SpA.

2.12. *Statistical Evaluation*

All values are expressed as mean ± standard error of the mean (SEM) of N observations. The images shown are representative of the last three experiments performed on diverse experimental days on tissue sections collected from all animals in each group. For in vivo studies, N represents the number of animals used. The results were analyzed by one-way ANOVA followed by a Bonferroni post hoc test for multiple comparisons. A p value less than 0.05 was considered significant.

3. Results

3.1. *ADM 2% Aerosol on Histopathological Analysis and Neutrophil Activity*

First, we analyzed the ADM effects on histopathological damage, including alveolar congestion, bleeding, neutrophil infiltration and thickness of alveolar wall/hyaline membrane formation. H&E exhibited extensive tissue damage and extracellular matrix deposition in the lungs of LPS-treated animals (Figure 1B,D) compared to the sham groups (Figure 1A,D). Aerosol treatment with ADM 2% significantly minimized lung damage (Figure 1C,D). We also evaluated the presence of lung edema by the ratio of wet/dry weight of the lung and neutrophil infiltration by the MPO assay. The ratio of wet/dry weight of the lung and MPO activity were increased by i.t. injection of LPS, while ADM 2% significantly reduced both parameters. (Figure 1E,F).

3.2. *ADM 2% Aerosol on Inflammatory Cells and Pro-Inflammatory Cytokines in BALF*

To determine whether ADM was able to reduce cell infiltration, we measured inflammatory cell counts in the BALF 24 h after LPS i.t. instillation. We found a substantial increase in total cell counts (Figure 2A), macrophages (Figure 2B) and neutrophils (Figure 2C) in BALF taken from LPS-treated animals compared to Sham mice. The number of inflammatory cells in BALF was significantly reduced after ADM 2% aerosol (Figure 2A–C). Additionally, we examined BALF levels of the pro-inflammatory cytokines. TNF-α (Figure 2D), IL-1β (Figure 2E) and IL-6 (Figure 2F) levels were significantly increased in the LPS group compared to Sham mice. On the contrary, cytokines release in BALF was markedly reduced in mice treated with ADM (Figure 2D–F).

3.3. *ADM 2% Aerosol on MC Number*

Toluidine blue staining of lung sections was used to assess the MC number. We detected a higher number of MCs in the LPS group (Figure 3B,B1,D), compared to Sham animals (Figure 3A,A1,D). ADM 2% aerosol reduced in a significant manner MC hyperplasia in lung tissues (Figure 3C,C1,D).

3.4. *ADM 2% Aerosol on Chymase and Tryptase Expression*

To confirm the activity of MCs and their activation, we evaluated the chymase and tryptase expressions by immunohistochemical analysis. LPS instillation enhanced chymase activity in the lungs (Figure 4B,B1,D), compared to Sham mice (Figure 4A,A1,D). At the same way, LPS increased tryptase expression in the lungs (Figure 5B,B1,D) compared to Sham mice (Figure 5A,A1,D). ADM 2% aerosol was able to reduce both preformed mediators expression (Figure 4C,C1,D for chymase; Figure 5C,C1,D for tryptase).

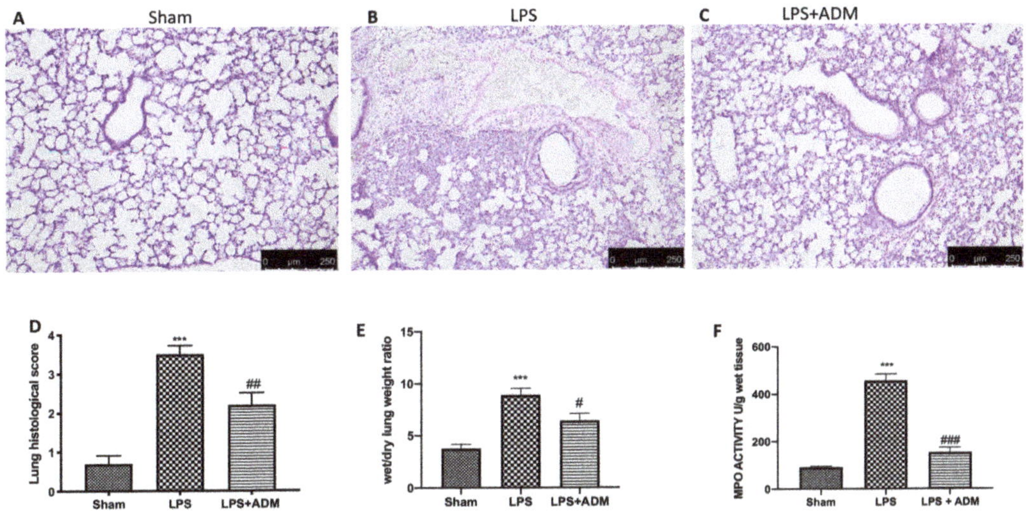

Figure 1. Histological analysis: sham (**A**), LPS (**B**), LPS + ADM 2% (**C**). Histological score (**D**). Wet/dry lung weight ratio (**E**). MPO activity (**F**). A 10× magnification is shown. Data are expressed as the mean ± SEM of N = 6 mice/group. *** $p < 0.001$ vs. sham; # $p < 0.05$ vs. LPS; ## $p < 0.01$ vs. LPS; ### $p < 0.001$ vs. LPS.

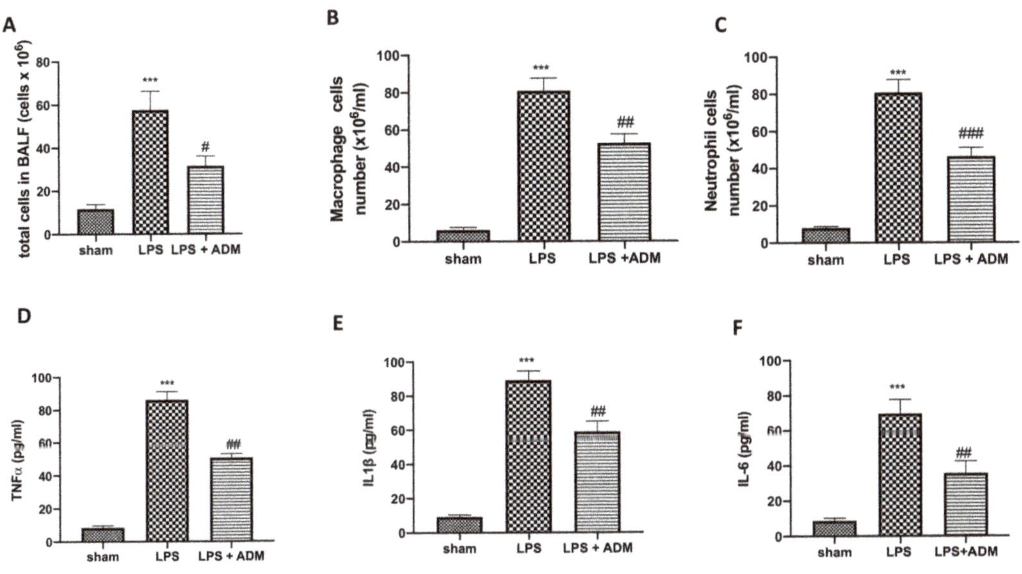

Figure 2. Cell infiltration expression in BALF. Total cell number (**A**), Macrophages (**B**), Neutrophils (**C**). Expression of proinflammatory cytokine: TNF-α (**D**), IL-1β (**E**), IL-6 (**F**). Data are expressed as the mean ± SEM of N = 6 mice/group. *** $p < 0.001$ vs. sham; # $p < 0.05$ vs. LPS; ## $p < 0.01$ vs. LPS; ### $p < 0.001$ vs. LPS.

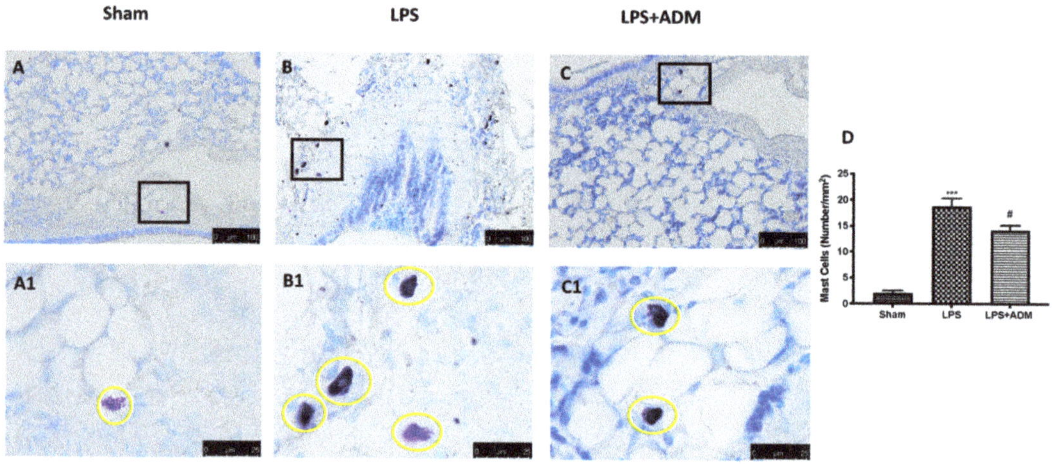

Figure 3. Mast cells indicated by toluidine blue staining: Sham (**A**), LPS (**B**), LPS + ADM 2% (**C**), mast cell count (**D**). A 20× and 100× magnification is shown. Data are expressed as the mean ± SEM of N = 6 mice/group. *** $p < 0.001$ vs. sham; # $p < 0.05$ vs. LPS.

3.5. ADM 2% Aerosol on Oxidative Stress

To evaluate the effect of ADM on oxidative stress, we performed MDA activity as an indicator of lipid peroxidation. LPS-treated mice showed increased MDA levels, while the LPS + ADM group showed lower levels of MDA (Figure 6A). Additionally, we investigated CAT and GSH levels for the oxidative response. LPS induced an important decrease in CAT (Figure 6B) and GSH (Figure 6C) levels, compared to the Sham groups. Both levels of antioxidant indicators were markedly increased by ADM 2% aerosol treatment.

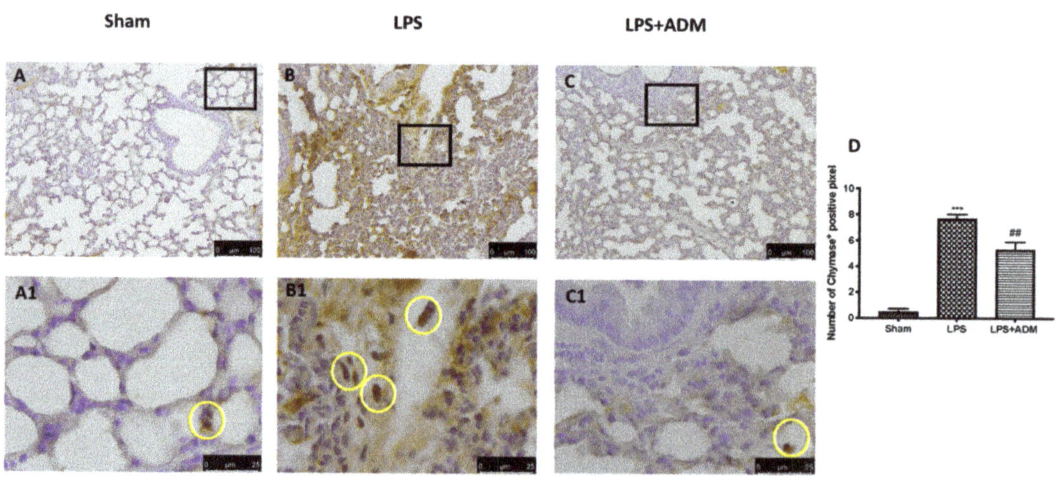

Figure 4. Immunohistochemical analysis for chymase: Sham (**A,A1**), LPS (**B,B1**), LPS + ADM 2% (**C,C1**), graphical quantification (**D**). A 20× and 100× magnification is shown. Data are expressed as the mean ± SEM of N = 6 mice/group. *** $p < 0.001$ vs. sham; ## $p < 0.01$ vs. LPS.

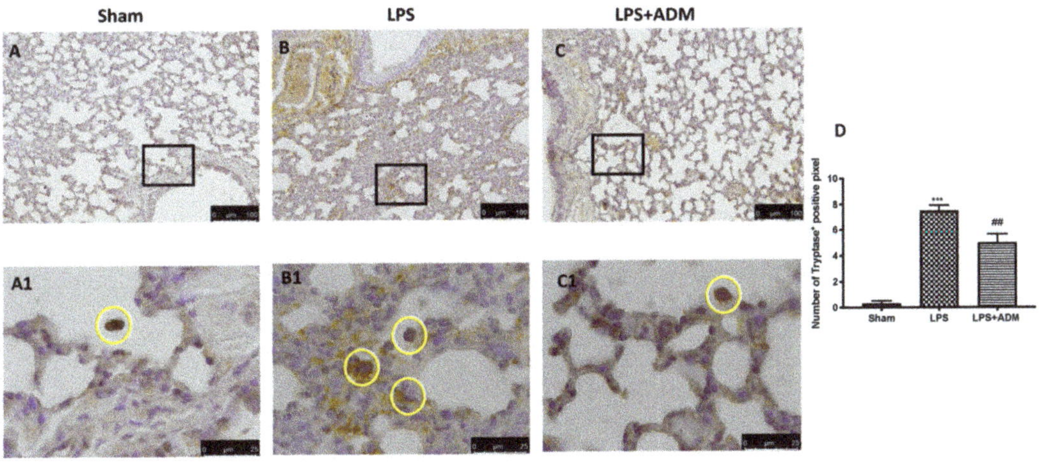

Figure 5. Immunohistochemical analysis for tryptase: Sham (**A**,**A1**), LPS (**B**,**B1**), LPS + ADM 2% (**C**,**C1**), graphical quantification (**D**). A 20× and 100× magnification is shown. Data are expressed as the mean ± SEM of N = 6 mice/group. *** $p < 0.001$ vs. sham; ## $p < 0.01$ vs. LPS.

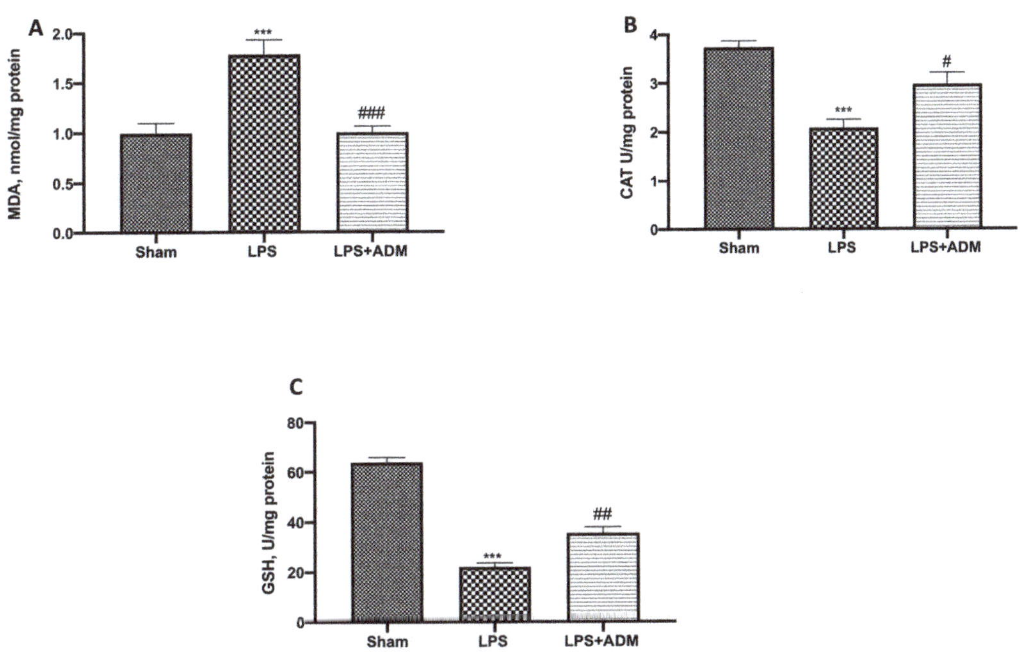

Figure 6. Markers of oxidative stress: MDA (**A**), CAT (**B**), and GSH (**C**). Data are expressed as the mean ± SEM of N = 6 mice/group. *** $p < 0.001$ vs. sham; # $p < 0.05$ vs. LPS; ## $p < 0.01$ vs. LPS; ### $p < 0.001$ vs. LPS.

3.6. ADM 2% Aerosol on Nrf2 Pathway

To confirm the antioxidant effect of ADM, we evaluated the Nrf2 pathway by Western blot analysis. Our results showed an important reduction in Nrf2 expression in the vehicle group compared to the sham group, while ADM was able to upregulate Nrf2 expression

(Figure 7A). Consequently, we evaluated HO-1 (Figure 7B) and MnSOD (Figure 7C) expression, which are regulated by the Nrf-2 pathway. Our results showed an important decrease in HO-1 and MnSOD expression in the LPS group, compared to sham mice; on the contrary, ADM partially restored the expression of both endogenous enzymes (Figure 7B,C).

3.7. ADM 2% Aerosol on Inflammatory Pathway

Additionally, we investigated one of the key inflammatory pathways involved in LPS-induced ALI, the NF-κB pathway. Our Western blot analysis showed a basal expression of IκB-α in Sham mice, while LPS i.t. instillation significantly decreased IκB-α expression in lung samples (Figure 8A). At the same time, nuclear NF-κB expression was significantly higher in LPS-treated animals compared to the sham group (Figure 8B). ADM treatment reduced IKB-α degradation and, consequently, nuclear translocation of NF-κB induced by LPS (Figure 8A,B). Additionally, to confirm the anti-inflammatory effect of ADM, we measured the levels of pro-inflammatory cytokines in lung tissues. We found that IL-1β (Figure 8C), IL-6 (Figure 8D) and TNF-α (Figure 8E) levels were markedly increased in the LPS group, compared to the sham mice. On the contrary, ADM was able to decrease the lung levels of these pro-inflammatory cytokines (Figure 8C–E).

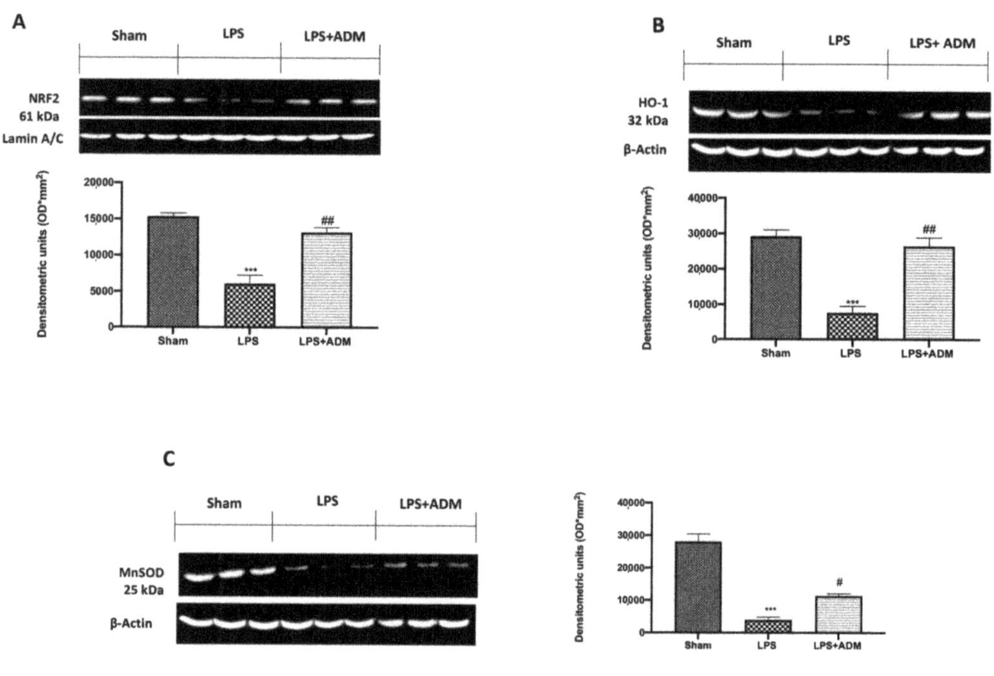

Figure 7. Western blot analysis for: Nrf2 (**A**); HO-1 (**B**); MnSOD (**C**). A demonstrative blot of lysates with a densitometric analysis for all animals is shown. Data are expressed as the mean ± SEM of N = 6 mice/group. *** $p < 0.001$ vs. sham; # $p < 0.05$ vs. LPS; ## $p < 0.01$ vs. LPS.

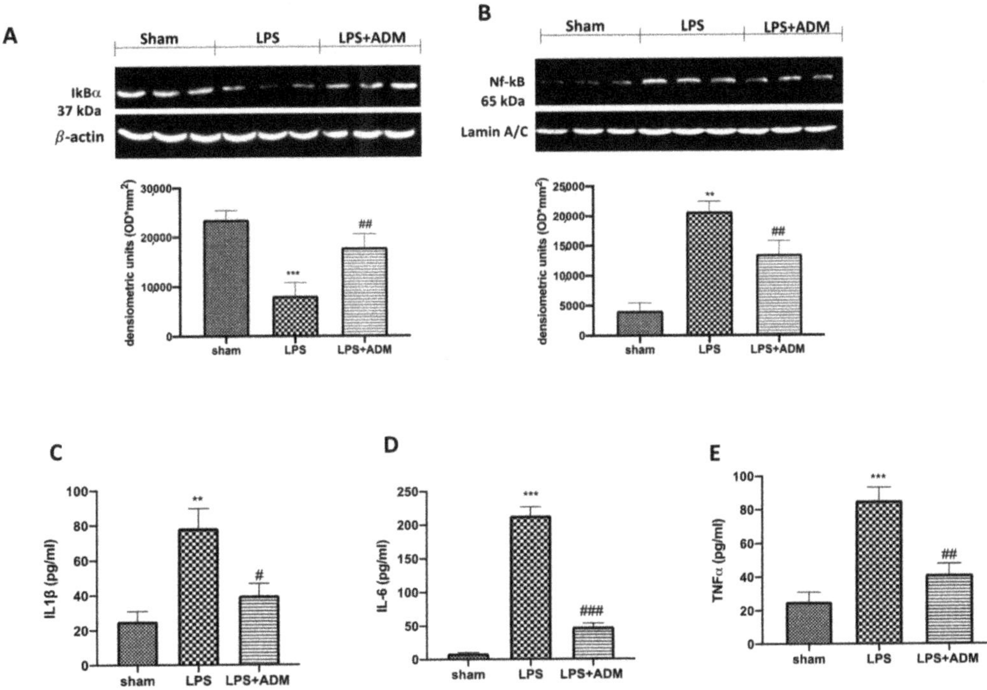

Figure 8. Western blot analysis for NF-κB (**B**) and IκB-α (**A**). Levels of inflammatory cytokine in lung tissues: IL-1β (**C**) IL-6 (**D**), TNF-α (**E**). A demonstrative blot of lysates for NF-kB and IκB-α with a densitometric analysis for all animals is shown. Data are expressed as the mean ± SEM of N = 6 mice/group. ** $p < 0.01$ vs. sham; *** $p < 0.001$ vs. sham; # $p < 0.05$ vs. LPS; ## $p < 0.01$ vs. LPS; ### $p < 0.001$ vs. LPS.

4. Discussion

ALI is an acute inflammatory illness that can advance to a severe stage known as ARDS, which is marked by a high death rate. These clinical syndromes are characterized by a loss of barrier functionality by alveolar epithelial and pulmonary capillary endothelial cells, resulting in respiratory failure in critically ill individuals. To study the molecular mechanisms underlying ALI, the experimental endotoxin (bacterial LPS) model by intratracheal instillation in mice was used [7,66]. In experimental ALI, the lung parenchyma is damaged by the generation of a complex network of inflammatory cytokines and chemokine, including IL-1β, IL-6, and TNF-α [9]. Moreover, activation of oxidative stress with excessive release of ROS produced by activated pulmonary macrophages and transmigrated neutrophils in the interstitial and alveolar compartments has been demonstrated [66,67]. Then, an imbalance is created between the oxidant/antioxidant system, which, combined with the activated inflammatory response, causes diffuse alveolar damage with intrapulmonary hemorrhage, edema and fibrin deposition. Therefore, this study was designed to evaluate the effects of ADM in controlling the inflammatory and oxidative response in LPS-induced ALI. The anti-inflammatory and antioxidant properties of ADM, a member of the ALIAmide family, have been extensively demonstrated in previous studies [23,25,34,50,68,69]. First, histopathological investigation showed that ADM 2% aerosol administration significantly repaired the morphological and histological alterations in lung tissue, induced by LPS instillation. Extensive neutrophil infiltration, the large release of inflammatory mediators, an increase in capillary permeability, and severe interstitial edema are all thought to play important roles in the pathogenesis of ALI [70–73]. ADM 2% treatment was able to reduce all these parameters, as demonstrated by a reduction in MPO activity and levels of cell infiltration in BALF, as well as significant decrease in pro-inflammatory cytokines and lung edema. ALI and ARDS are often characterized by inappropriately/chronically activated MCs [74]. Many proteases, such as chymase and tryptase, are substances secreted by MCs activation, and contribute to inflammatory cells infiltration, cytokine production, and increased vascular permeability, exacerbating inflammation [75]. At this regard, the anti-inflammatory properties of ADM are mainly due to the control of MC activation and, as expected, our results confirmed a reduced number of MCs after ADM treatment. Consequently, we observed an important reduction in chymase and tryptase expression after ADM 2% administration, confirming the control of ADM on MCs activity. Additionally, it has also been demonstrated that ADM increased endogenous levels of antioxidant enzymes [25], indirectly modulating the NF-κB pathway. Indeed, ALI is characterized by excessive ROS production, causing imbalance to antioxidant system, and resulting in the release of substances modulating the endothelial dysfunction and disruption responsible for the principal clinical manifestations of the syndrome. ADM 2% aerosol administration also had positive results on endogenous levels of enzymes involved in oxidative stress; in fact, the treatment significantly counteracted the LPS-induced down-regulation of antioxidant indicators, as shown by the effect of CAT and GSH levels, as well as HO-1 and MnSOD expressions. These antioxidant enzymes are regulated by the Nrf2 pathway. Moreover, the functional crosstalk between Nrf2 and NF-κB is well known. The absence of Nrf2 is associated with increased oxidative stress, leading to an increase in cytokine production, as NF-κB is more readily activated in oxidative conditions [76,77]. Our Western blot analysis showed that ADM 2% was able to upregulate Nrf2 expression, responsible for antioxidant response, as well as to modulate the NF-κB pathway. To confirm the protective function of ADM, we also investigated the levels of proinflammatory cytokines in lung tissue. Again, ADM treatment significantly counteracted the LPS-induced increase in inflammatory mediator levels.

5. Conclusions

In conclusion, our data demonstrated that ADM 2% aerosol was able to reduce lung damage and cell infiltration, as well as the overexpression of proinflammatory cytokines. The protective effects of ADM 2% aerosol, probably due to the control of MC degranulation and the upregulation of endogenous antioxidant enzymes, modulate the inflammatory and oxidative response. Therefore, we suggest that ADM 2% aerosol can be considered as a potential candidate in the management of ALI.

Author Contributions: Conceptualization, R.D.; methodology, T.G. and M.C.; software, R.F.; validation, A.F.P. and E.G.; formal analysis, R.C. and D.I.; investigation, A.F.P.; resources, R.S. and M.C.; data curation, R.D.; writing original draft preparation, L.I. and M.C.; writing—review and editing, R.D.P.; visualization, S.C. (Stefano Coaccioli); supervision, D.I.; project administration, S.C. (Salvatore Cuzzocrea); funding acquisition, S.C. (Salvatore Cuzzocrea). All authors have read and agreed to the published version of the manuscript.

Funding: This research received no external funding.

Institutional Review Board Statement: This study was approved by the University of Messina Review Board for the care of animals. Animal care conformed to Italian regulations on the use of animals for experimental and scientific purposes (D.Lgs 2014/26 and EU Directive 2010/63).

Informed Consent Statement: Not applicable.

Data Availability Statement: For a rule of our laboratory the datasets used in the current study are available from the corresponding author on reasonable request.

Acknowledgments: We would like to thank Valentina Malvagni for editorial support with the manuscript.

Conflicts of Interest: Salvatore Cuzzocrea is a coinventor on patent WO2013121449 A8 (Epitech Group Srl), which deals with methods and compositions for the modulation of amidases capable of hydrolyzing N-acylethanolamines employable in the treatment of inflammatory diseases. This invention is wholly unrelated to the present study. Moreover, Cuzzocrea is also, with Epitech Group, a coinventor on the patents EP 2 821 083, MI2014 A001495, and 102015000067344, which are unrelated to the study. The remaining authors report no conflict of interest.

References

1. Peritore, A.F.; D'Amico, R.; Siracusa, R.; Cordaro, M.; Fusco, R.; Gugliandolo, E.; Genovese, T.; Crupi, R.; Di Paola, R.; Cuzzocrea, S.; et al. Management of Acute Lung Injury: Palmitoylethanolamide as a New Approach. *Int. J. Mol. Sci.* **2021**, *22*, 5533. [CrossRef] [PubMed]
2. Leung, P.O.; Lee, H.H.; Kung, Y.C.; Tsai, M.F.; Chou, T.C. Therapeutic effect of C-phycocyanin extracted from blue green algae in a rat model of acute lung injury induced by lipopolysaccharide. *Evid.-Based Complement. Altern. Med.* **2013**, *2013*, 916590. [CrossRef]
3. Domscheit, H.; Hegeman, M.A.; Carvalho, N.; Spieth, P.M. Molecular Dynamics of Lipopolysaccharide-Induced Lung Injury in Rodents. *Front. Physiol.* **2020**, *11*, 36. [CrossRef] [PubMed]
4. Menezes, S.L.; Bozza, P.T.; Neto, H.C.; Laranjeira, A.P.; Negri, E.M.; Capelozzi, V.L.; Zin, W.A.; Rocco, P.R. Pulmonary and extrapulmonary acute lung injury: Inflammatory and ultrastructural analyses. *J. Appl. Physiol.* **2005**, *98*, 1777–1783. [CrossRef] [PubMed]
5. Wyns, H.; Plessers, E.; De Backer, P.; Meyer, E.; Croubels, S. In Vivo porcine lipopolysaccharide inflammation models to study immunomodulation of drugs. *Vet Immunol. Immunopathol.* **2015**, *166*, 58–69. [CrossRef]
6. Matute-Bello, G.; Frevert, C.W.; Martin, T.R. Animal models of acute lung injury. *Am. J. Physiol. Lung Cell Mol. Physiol.* **2008**, *295*, L379–L399. [CrossRef]
7. Chen, H.; Bai, C.; Wang, X. The value of the lipopolysaccharide-induced acute lung injury model in respiratory medicine. *Expert Rev. Respir. Med.* **2010**, *4*, 773–783. [CrossRef]
8. Lee, J.-W.; Seo, K.-H.; Ryu, H.W.; Yuk, H.J.; Park, H.A.; Lim, Y.; Ahn, K.-S.; Oh, S.-R. Anti-inflammatory effect of stem bark of *Paulownia tomentosa* Steud. in lipopolysaccharide (LPS)-stimulated RAW264.7 macrophages and LPS-induced murine model of acute lung injury. *J. Ethnopharmacol.* **2018**, *210*, 23–30. [CrossRef]
9. Goodman, R.B.; Pugin, J.; Lee, J.S.; Matthay, M.A. Cytokine-mediated inflammation in acute lung injury. *Cytokine Growth Factor Rev.* **2003**, *14*, 523–535. [CrossRef]
10. Cuzzocrea, S.; Nocentini, G.; Di Paola, R.; Agostini, M.; Mazzon, E.; Ronchetti, S.; Crisafulli, C.; Esposito, E.; Caputi, A.P.; Riccardi, C. Proinflammatory role of glucocorticoid-induced TNF receptor-related gene in acute lung inflammation. *J. Immunol.* **2006**, *177*, 631–641. [CrossRef]
11. Hu, Y.; Jin, Y.; Han, D.; Zhang, G.; Cao, S.; Xie, J.; Xue, J.; Li, Y.; Meng, D.; Fan, X. Mast cell-induced lung injury in mice infected with H5N1 influenza virus. *J. Virol.* **2012**, *86*, 3347–3356. [CrossRef] [PubMed]
12. Caughey, G.H. Mast cell proteases as protective and inflammatory mediators. *Adv. Exp. Med. Biol.* **2011**, *716*, 212–234. [CrossRef] [PubMed]
13. Lundequist, A.; Pejler, G. Biological implications of preformed mast cell mediators. *Cell. Mol. Life Sci.* **2011**, *68*, 965–975. [CrossRef] [PubMed]
14. Kellner, M.; Noonepalle, S.; Lu, Q.; Srivastava, A.; Zemskov, E.; Black, S.M. ROS Signaling in the Pathogenesis of Acute Lung Injury (ALI) and Acute Respiratory Distress Syndrome (ARDS). In *Pulmonary Vasculature Redox Signaling in Health and Disease*; Springer: Cham, Switzerland, 2017; Volume 967, pp. 105–137. [CrossRef]
15. Yang, H.; Lv, H.; Li, H.; Ci, X.; Peng, L. Oridonin protects LPS-induced acute lung injury by modulating Nrf2-mediated oxidative stress and Nrf2-independent NLRP3 and NF-kappaB pathways. *Cell Commun. Signal.* **2019**, *17*, 98. [CrossRef]
16. Chow, C.W.; Herrera Abreu, M.T.; Suzuki, T.; Downey, G.P. Oxidative stress and acute lung injury. *Am. J. Respir. Cell Mol. Biol.* **2003**, *29*, 427–431. [CrossRef]
17. Han, Y.K.; Kim, J.S.; Lee, G.B.; Lim, J.H.; Park, K.M. Oxidative stress following acute kidney injury causes disruption of lung cell cilia and their release into the bronchoaveolar lavage fluid and lung injury, which are exacerbated by Idh2 deletion. *Redox Biol.* **2021**, *46*, 102077. [CrossRef]
18. Fan, D.; Wang, D.; Zhu, L. Protective role of scutellarin on LPS induced—Acute lung injury and regulation of apoptosis, oxidative stress and reduction of mitochondrial dysfunction. *Saudi. J. Biol. Sci.* **2022**, *29*, 371–378. [CrossRef]
19. Cui, Y.R.; Qu, F.; Zhong, W.J.; Yang, H.H.; Zeng, J.; Huang, J.H.; Liu, J.; Zhang, M.Y.; Zhou, Y.; Guan, C.X. Beneficial effects of aloperine on inflammation and oxidative stress by suppressing necroptosis in lipopolysaccharide-induced acute lung injury mouse model. *Phytomedicine* **2022**, *100*, 154074. [CrossRef]
20. Cui, X.F.; Lin, P.; Yu, J.; Liu, L.; Wang, Z.Y.; Tang, X.J. Dimethyl fumarate attenuates lipopolysaccharide-induced acute lung injury by inhibiting inflammation and oxidative stress. *J. Biol. Regul. Homeost Agents* **2021**, *35*. [CrossRef]
21. D'Amico, R.; Impellizzeri, D.; Cuzzocrea, S.; Di Paola, R. ALIAmides Update: Palmitoylethanolamide and Its Formulations on Management of Peripheral Neuropathic Pain. *Int. J. Mol. Sci.* **2020**, *21*, 5330. [CrossRef]
22. Impellizzeri, D.; Di Paola, R.; Cordaro, M.; Gugliandolo, E.; Casili, G.; Morittu, V.M.; Britti, D.; Esposito, E.; Cuzzocrea, S. Adelmidrol, a palmitoylethanolamide analogue, as a new pharmacological treatment for the management of acute and chronic inflammation. *Biochem. Pharmacol.* **2016**, *119*, 27–41. [CrossRef] [PubMed]

23. Cordaro, M.; Impellizzeri, D.; Gugliandolo, E.; Siracusa, R.; Crupi, R.; Esposito, E.; Cuzzocrea, S. Adelmidrol, a Palmitoylethanolamide Analogue, as a New Pharmacological Treatment for the Management of Inflammatory Bowel Disease. *Mol. Pharmacol.* **2016**, *90*, 549–561. [CrossRef]
24. D'Amico, R.; Siracusa, R.; Fusco, R.; Cordaro, M.; Genovese, T.; Peritore, A.F.; Gugliandolo, E.; Crupi, R.; Impellizzeri, D.; Cuzzocrea, S.; et al. Protective effects of Colomast®, A New Formulation of Adelmidrol and Sodium Hyaluronate, in A Mouse Model of Acute Restraint Stress. *Int. J. Mol. Sci.* **2020**, *21*, 8136. [CrossRef] [PubMed]
25. Fusco, R.; Cordaro, M.; Genovese, T.; Impellizzeri, D.; Siracusa, R.; Gugliandolo, E.; Peritore, A.F.; D'Amico, R.; Crupi, R.; Cuzzocrea, S. Adelmidrol: A new promising antioxidant and anti-inflammatory therapeutic tool in pulmonary fibrosis. *Antioxidants* **2020**, *9*, 601. [CrossRef] [PubMed]
26. Ribeiro, A.; Mayer, C.; Wilson, C.; Martin, R.; MacFarlane, P. Intratracheal LPS administration attenuates the acute hypoxic ventilatory response: Role of brainstem IL-1β receptors. *Respir. Physiol. Neurobiol.* **2017**, *242*, 45–51. [CrossRef]
27. Gugliandolo, E.; Fusco, R.; Ginestra, G.; D'amico, R.; Bisignano, C.; Mandalari, G.; Cuzzocrea, S.; Di Paola, R. Involvement of TLR4 and PPAR-α receptors in host response and NLRP3 inflammasome activation, against pulmonary infection with pseudomonas aeruginosa. *Shock* **2019**, *51*, 221–227. [CrossRef]
28. D'Amico, R.; Monaco, F.; Fusco, R.; Peritore, A.F.; Genovese, T.; Impellizzeri, D.; Crupi, R.; Interdonato, L.; Sforza, A.M.; Gugliandolo, E.; et al. Exposure to Atrazine Induces Lung Inflammation through Nrf2-HO1 and Beclin 1/LC3 Pathways. *Cell Physiol. Biochem.* **2021**, *55*, 413–427. [CrossRef]
29. Conte, E.; Iemmolo, M.; Fagone, E.; Gili, E.; Fruciano, M.; Genovese, T.; Esposito, E.; Cuzzocrea, S.; Vancheri, C. Thymosin beta4 reduces IL-17-producing cells and IL-17 expression, and protects lungs from damage in bleomycin-treated mice. *Immunobiology* **2014**, *219*, 425–431. [CrossRef]
30. Saadat, S.; Beheshti, F.; Askari, V.R.; Hosseini, M.; Mohamadian Roshan, N.; Boskabady, M.H. Aminoguanidine affects systemic and lung inflammation induced by lipopolysaccharide in rats. *Respir. Res.* **2019**, *20*, 96. [CrossRef]
31. Impellizzeri, D.; D'Amico, R.; Fusco, R.; Genovese, T.; Peritore, A.F.; Gugliandolo, E.; Crupi, R.; Interdonato, L.; Di Paola, D.; Di Paola, R.; et al. Acai Berry Mitigates Vascular Dementia-Induced Neuropathological Alterations Modulating Nrf-2/Beclin1 Pathways. *Cells* **2022**, 11. [CrossRef] [PubMed]
32. Genovese, T.; Cuzzocrea, S.; Di Paola, R.; Failla, M.; Mazzon, E.; Sortino, M.A.; Frasca, G.; Gili, E.; Crimi, N.; Caputi, A.P.; et al. Inhibition or knock out of inducible nitric oxide synthase result in resistance to bleomycin-induced lung injury. *Respir. Res.* **2005**, *6*, 58. [CrossRef] [PubMed]
33. Mazzon, E.; Esposito, E.; Impellizzeri, D.; Di Paola, R.; Melani, A.; Bramanti, P.; Pedata, F.; Cuzzocrea, S. CGS 21680, an agonist of the adenosine (A2A) receptor, reduces progression of murine type II collagen-induced arthritis. *J. Rheumatol.* **2011**, *38*, 2119–2129. [CrossRef] [PubMed]
34. Di Paola, D.; Natale, S.; Iaria, C.; Cordaro, M.; Crupi, R.; Siracusa, R.; D'Amico, R.; Fusco, R.; Impellizzeri, D.; Cuzzocrea, S.; et al. Intestinal Disorder in Zebrafish Larvae (*Danio rerio*): The Protective Action of N-Palmitoylethanolamide-oxazoline. *Life* **2022**, *12*, 125. [CrossRef] [PubMed]
35. Di Paola, D.; Natale, S.; Iaria, C.; Crupi, R.; Cuzzocrea, S.; Spano, N.; Gugliandolo, E.; Peritore, A.F. Environmental Co-Exposure to Potassium Perchlorate and Cd Caused Toxicity and Thyroid Endocrine Disruption in Zebrafish Embryos and Larvae (*Danio rerio*). *Toxics* **2022**, *10*, 198. [CrossRef]
36. Genovese, T.; Impellizzeri, D.; D'Amico, R.; Fusco, R.; Peritore, A.F.; Di Paola, D.; Interdonato, L.; Gugliandolo, E.; Crupi, R.; Di Paola, R.; et al. Role of Bevacizumab on Vascular Endothelial Growth Factor in Apolipoprotein E Deficient Mice after Traumatic Brain Injury. *Int. J. Mol. Sci.* **2022**, *23*, 4162. [CrossRef]
37. Cordaro, M.; Fusco, R.; D'Amico, R.; Siracusa, R.; Peritore, A.F.; Gugliandolo, E.; Genovese, T.; Crupi, R.; Mandalari, G.; Cuzzocrea, S.; et al. Cashew (*Anacardium occidentale* L.) Nuts Modulate the Nrf2 and NLRP3 Pathways in Pancreas and Lung after Induction of Acute Pancreatitis by Cerulein. *Antioxidants* **2020**, *9*, 992. [CrossRef]
38. Di Paola, D.; Iaria, C.; Lanteri, G.; Cordaro, M.; Crupi, R.; Siracusa, R.; D'Amico, R.; Fusco, R.; Impellizzeri, D.; Cuzzocrea, S. Sensitivity of Zebrafish Embryogenesis to Risk of Fotemustine Exposure. *Fishes* **2022**, *7*, 67. [CrossRef]
39. D'Iglio, C.; Albano, M.; Famulari, S.; Savoca, S.; Panarello, G.; Di Paola, D.; Perdichizzi, A.; Rinelli, P.; Lanteri, G.; Spano, N.; et al. Intra- and interspecific variability among congeneric *Pagellus* otoliths. *Sci. Rep.* **2021**, *11*, 16315. [CrossRef]
40. Di Paola, D.; Abbate, J.M.; Iaria, C.; Cordaro, M.; Crupi, R.; Siracusa, R.; D'Amico, R.; Fusco, R.; Impellizzeri, D.; Cuzzocrea, S.; et al. Environmental Risk Assessment of Dexamethasone Sodium Phosphate and Tocilizumab Mixture in Zebrafish Early Life Stage (*Danio rerio*). *Toxics* **2022**, *10*, 279. [CrossRef]
41. Di Paola, D.; Capparucci, F.; Abbate, J.M.; Cordaro, M.; Crupi, R.; Siracusa, R.; D'Amico, R.; Fusco, R.; Genovese, T.; Impellizzeri, D.; et al. Environmental Risk Assessment of Oxaliplatin Exposure on Early Life Stages of Zebrafish (*Danio rerio*). *Toxics* **2022**, *10*, 81. [CrossRef]
42. Wang, M.; Zhong, H.; Zhang, X.; Huang, X.; Wang, J.; Li, Z.; Chen, M.; Xiao, Z. EGCG promotes PRKCA expression to alleviate LPS-induced acute lung injury and inflammatory response. *Sci. Rep.* **2021**, *11*, 11014. [CrossRef] [PubMed]
43. Su, X.; Wang, L.; Song, Y.; Bai, C. Inhibition of inflammatory responses by ambroxol, a mucolytic agent, in a murine model of acute lung injury induced by lipopolysaccharide. *Intensive Care Med.* **2004**, *30*, 133–140. [CrossRef] [PubMed]

44. Cordaro, M.; Siracusa, R.; Fusco, R.; D'Amico, R.; Peritore, A.F.; Gugliandolo, E.; Genovese, T.; Scuto, M.; Crupi, R.; Mandalari, G. Cashew (*Anacardium occidentale* L.) nuts counteract oxidative stress and inflammation in an acute experimental model of Carrageenan-induced Paw edema. *Antioxidants* **2020**, *9*, 660. [CrossRef] [PubMed]
45. Mariotto, S.; Esposito, E.; Di Paola, R.; Ciampa, A.; Mazzon, E.; de Prati, A.C.; Darra, E.; Vincenzi, S.; Cucinotta, G.; Caminiti, R.; et al. Protective effect of Arbutus unedo aqueous extract in carrageenan-induced lung inflammation in mice. *Pharmacol. Res.* **2008**, *57*, 110–124. [CrossRef] [PubMed]
46. Fusco, R.; Cordaro, M.; Siracusa, R.; Peritore, A.F.; Gugliandolo, E.; Genovese, T.; D'Amico, R.; Crupi, R.; Smeriglio, A.; Mandalari, G.; et al. Consumption of *Anacardium Occidentale* L. (Cashew Nuts) Inhibits Oxidative Stress through Modulation of the Nrf2/HO-1 and NF-kB Pathways. *Molecules* **2020**, *25*, 4426. [CrossRef]
47. Cordaro, M.; Siracusa, R.; Impellizzeri, D.; D'Amico, R.; Peritore, A.F.; Crupi, R.; Gugliandolo, E.; Fusco, R.; Di Paola, R.; Schievano, C. Safety and efficacy of a new micronized formulation of the ALIAmide palmitoylglucosamine in preclinical models of inflammation and osteoarthritis pain. *Arthritis Res. Ther.* **2019**, *21*, 1–17. [CrossRef]
48. Galuppo, M.; Esposito, E.; Mazzon, E.; Di Paola, R.; Paterniti, I.; Impellizzeri, D.; Cuzzocrea, S. MEK inhibition suppresses the development of lung fibrosis in the bleomycin model. *Naunyn-Schmiedeberg's Arch. Pharmacol.* **2011**, *384*, 21–37. [CrossRef]
49. D'Amico, R.; Gugliandolo, E.; Cordaro, M.; Fusco, R.; Genovese, T.; Peritore, A.F.; Crupi, R.; Interdonato, L.; Di Paola, D.; Cuzzocrea, S.; et al. Toxic Effects of Endocrine Disruptor Exposure on Collagen-Induced Arthritis. *Biomolecules* **2022**, *12*, 564. [CrossRef]
50. Siracusa, R.; Impellizzeri, D.; Cordaro, M.; Gugliandolo, E.; Peritore, A.F.; Di Paola, R.; Cuzzocrea, S. Topical application of adelmidrol+ trans-traumatic acid enhances skin wound healing in a streptozotocin-induced diabetic mouse model. *Front. Pharmacol.* **2018**, *9*, 871. [CrossRef]
51. D'Amico, R.; Fusco, R.; Cordaro, M.; Siracusa, R.; Peritore, A.F.; Gugliandolo, E.; Crupi, R.; Scuto, M.; Cuzzocrea, S.; Di Paola, R.; et al. Modulation of NLRP3 Inflammasome through Formyl Peptide Receptor 1 (Fpr-1) Pathway as a New Therapeutic Target in Bronchiolitis Obliterans Syndrome. *Int. J. Mol. Sci.* **2020**, *21*, 2144. [CrossRef]
52. Di Paola, D.; Capparucci, F.; Lanteri, G.; Cordaro, M.; Crupi, R.; Siracusa, R.; D'Amico, R.; Fusco, R.; Impellizzeri, D.; Cuzzocrea, S.; et al. Combined Toxicity of Xenobiotics Bisphenol A and Heavy Metals on Zebrafish Embryos (*Danio rerio*). *Toxics* **2021**, *9*, 344. [CrossRef] [PubMed]
53. Di Paola, D.; Capparucci, F.; Lanteri, G.; Crupi, R.; Marino, Y.; Franco, G.A.; Cuzzocrea, S.; Spano, N.; Gugliandolo, E.; Peritore, A.F. Environmental Toxicity Assessment of Sodium Fluoride and Platinum-Derived Drugs Co-Exposure on Aquatic Organisms. *Toxics* **2022**, *10*, 272. [CrossRef] [PubMed]
54. Varghese, F.; Bukhari, A.B.; Malhotra, R.; De, A. IHC Profiler: An open source plugin for the quantitative evaluation and automated scoring of immunohistochemistry images of human tissue samples. *PLoS ONE* **2014**, *9*, e96801. [CrossRef] [PubMed]
55. D'Amico, R.; Gugliandolo, E.; Siracusa, R.; Cordaro, M.; Genovese, T.; Peritore, A.F.; Crupi, R.; Interdonato, L.; Di Paola, D.; Cuzzocrea, S.; et al. Toxic Exposure to Endocrine Disruptors Worsens Parkinson's Disease Progression through NRF2/HO-1 Alteration. *Biomedicines* **2022**, *10*, 1073. [CrossRef]
56. Zhang, J.; Cui, R.; Feng, Y.; Gao, W.; Bi, J.; Li, Z.; Liu, C. Serotonin exhibits accelerated bleomycin-induced pulmonary fibrosis through TPH1 knockout mouse experiments. *Mediat. Inflamm.* **2018**, *2018*, 7967868. [CrossRef]
57. Saini, R.; Verma, S.; Singh, A.; Gupta, M.L. Role of Active Principles of Podophyllum hexandrum in Amelioration of Radiation Mediated Lung Injuries by Reactive Oxygen/Nitrogen Species Reduction. *CellBio* **2013**, *2*, 36989. [CrossRef]
58. Di Paola, D.; Iaria, C.; Capparucci, F.; Cordaro, M.; Crupi, R.; Siracusa, R.; D'Amico, R.; Fusco, R.; Impellizzeri, D.; Cuzzocrea, S.; et al. Aflatoxin B1 Toxicity in Zebrafish Larva (*Danio rerio*): Protective Role of *Hericium erinaceus*. *Toxins* **2021**, *13*, 710. [CrossRef]
59. Peritore, A.F.; Crupi, R.; Scuto, M.; Gugliandolo, E.; Siracusa, R.; Impellizzeri, D.; Cordaro, M.; D'Amico, R.; Fusco, R.; Di Paola, R.; et al. The Role of Annexin A1 and Formyl Peptide Receptor 2/3 Signaling in Chronic Corticosterone-Induced Depression-Like behaviors and Impairment in Hippocampal-Dependent Memory. *CNS Neurol. Disord.-Drug Targets* **2020**, *19*, 27–43. [CrossRef]
60. Fusco, R.; Gugliandolo, E.; Siracusa, R.; Scuto, M.; Cordaro, M.; D'Amico, R.; Evangelista, M.; Peli, A.; Peritore, A.F.; Impellizzeri, D.; et al. Formyl Peptide Receptor 1 Signaling in Acute Inflammation and Neural Differentiation Induced by Traumatic Brain Injury. *Biology* **2020**, *9*, 238. [CrossRef]
61. Marfella, R.; Esposito, K.; Nappo, F.; Siniscalchi, M.; Sasso, F.C.; Portoghese, M.; Di Marino, M.P.; Baldi, A.; Cuzzocrea, S.; Di Filippo, C.; et al. Expression of angiogenic factors during acute coronary syndromes in human type 2 diabetes. *Diabetes* **2004**, *53*, 2383–2391. [CrossRef]
62. Cordaro, M.; Siracusa, R.; D'Amico, R.; Genovese, T.; Franco, G.; Marino, Y.; Di Paola, D.; Cuzzocrea, S.; Impellizzeri, D.; Di Paola, R.; et al. Role of Etanercept and Infliximab on Nociceptive Changes Induced by the Experimental Model of Fibromyalgia. *Int. J. Mol. Sci.* **2022**, *23*, 6139. [CrossRef] [PubMed]
63. Di Paola, D.; Iaria, C.; Capparucci, F.; Arangia, A.; Crupi, R.; Cuzzocrea, S.; Spano, N.; Gugliandolo, E.; Peritore, A.F. Impact of Mycotoxin Contaminations on Aquatic Organisms: Toxic Effect of Aflatoxin B1 and Fumonisin B1 Mixture. *Toxins* **2022**, *14*, 518. [CrossRef] [PubMed]
64. Di Paola, D.; Natale, S.; Gugliandolo, E.; Cordaro, M.; Crupi, R.; Siracusa, R.; D'Amico, R.; Fusco, R.; Impellizzeri, D.; Cuzzocrea, S.; et al. Assessment of 2-Pentadecyl-2-oxazoline Role on Lipopolysaccharide-Induced Inflammation on Early Stage Development of Zebrafish (*Danio rerio*). *Life* **2022**, *12*, 128. [CrossRef] [PubMed]

65. Siracusa, R.; Fusco, R.; Peritore, A.F.; Cordaro, M.; D'Amico, R.; Genovese, T.; Gugliandolo, E.; Crupi, R.; Smeriglio, A.; Mandalari, G. The antioxidant and anti-inflammatory properties of *Anacardium occidentale* L. cashew nuts in a mouse model of colitis. *Nutrients* **2020**, *12*, 834. [CrossRef]
66. Rittirsch, D.; Flierl, M.A.; Day, D.E.; Nadeau, B.A.; McGuire, S.R.; Hoesel, L.M.; Ipaktchi, K.; Zetoune, F.S.; Sarma, J.V.; Leng, L.; et al. Acute lung injury induced by lipopolysaccharide is independent of complement activation. *J. Immunol.* **2008**, *180*, 7664–7672. [CrossRef]
67. Kabir, K.; Gelinas, J.P.; Chen, M.; Chen, D.; Zhang, D.; Luo, X.; Yang, J.H.; Carter, D.; Rabinovici, R. Characterization of a murine model of endotoxin-induced acute lung injury. *Shock* **2002**, *17*, 300–303. [CrossRef]
68. Di Paola, R.; Fusco, R.; Impellizzeri, D.; Cordaro, M.; Britti, D.; Morittu, V.M.; Evangelista, M.; Cuzzocrea, S. Adelmidrol, in combination with hyaluronic acid, displays increased anti-inflammatory and analgesic effects against monosodium iodoacetate-induced osteoarthritis in rats. *Arthritis Res. Ther.* **2016**, *18*, 291. [CrossRef]
69. Campolo, M.; Siracusa, R.; Cordaro, M.; Filippone, A.; Gugliandolo, E.; Peritore, A.F.; Impellizzeri, D.; Crupi, R.; Paterniti, I.; Cuzzocrea, S. The association of adelmidrol with sodium hyaluronate displays beneficial properties against bladder changes following spinal cord injury in mice. *PLoS ONE* **2019**, *14*, e0208730. [CrossRef]
70. Parekh, D.; Dancer, R.C.; Thickett, D.R. Acute lung injury. *Clin. Med.* **2011**, *11*, 615. [CrossRef]
71. Ma, C.; Dong, L.; Li, M.; Cai, W. Qidonghuoxue Decoction Ameliorates Pulmonary Edema in Acute Lung Injury Mice through the Upregulation of Epithelial Sodium Channel and Aquaporin-1. *Evid.-Based Complement. Altern. Med.* **2020**, *2020*, 2492304. [CrossRef]
72. He, Y.Q.; Zhou, C.C.; Deng, J.L.; Wang, L.; Chen, W.S. Tanreqing Inhibits LPS-Induced Acute Lung Injury In Vivo and In Vitro Through Downregulating STING Signaling Pathway. *Front. Pharmacol.* **2021**, *12*, 746964. [CrossRef] [PubMed]
73. Li, J.; Lu, K.; Sun, F.; Tan, S.; Zhang, X.; Sheng, W.; Hao, W.; Liu, M.; Lv, W.; Han, W. Panaxydol attenuates ferroptosis against LPS-induced acute lung injury in mice by Keap1-Nrf2/HO-1 pathway. *J. Transl. Med.* **2021**, *19*, 96. [CrossRef] [PubMed]
74. Virk, H.; Arthur, G.; Bradding, P. Mast cells and their activation in lung disease. *Transl. Res.* **2016**, *174*, 60–76. [CrossRef] [PubMed]
75. Huang, P.; Liu, D.; Gan, X.; Zhang, R.; Gao, W.; Xia, Z.; Hei, Z. Mast cells activation contribute to small intestinal ischemia reperfusion induced acute lung injury in rats. *Injury* **2012**, *43*, 1250–1256. [CrossRef]
76. Genovese, T.; D'Amico, R.; Fusco, R.; Impellizzeri, D.; Peritore, A.F.; Crupi, R.; Interdonato, L.; Gugliandolo, E.; Cuzzocrea, S.; Paola, R.D.; et al. Acai (Euterpe Oleraceae Mart.) Seeds Regulate NF-kappaB and Nrf2/ARE Pathways Protecting Lung against Acute and Chronic Inflammation. *Cell Physiol. Biochem.* **2022**, *56*, 1–20. [CrossRef]
77. Cordaro, M.; D'Amico, R.; Morabito, R.; Fusco, R.; Siracusa, R.; Peritore, A.F.; Impellizzeri, D.; Genovese, T.; Crupi, R.; Gugliandolo, E.; et al. Physiological and Biochemical Changes in NRF2 Pathway in Aged Animals Subjected to Brain Injury. *Cell Physiol. Biochem.* **2021**, *55*, 160–179. [CrossRef]

biomolecules

MDPI

Article

N-Palmitoyl-D-Glucosamine Inhibits TLR-4/NLRP3 and Improves DNBS-Induced Colon Inflammation through a PPAR-α-Dependent Mechanism

Irene Palenca [1], Luisa Seguella [1], Alessandro Del Re [1], Silvia Basili Franzin [1], Chiara Corpetti [1], Marcella Pesce [2], Sara Rurgo [2], Luca Steardo [3], Giovanni Sarnelli [2] and Giuseppe Esposito [1,*]

1 Department of Physiology and Pharmacology "V. Erspamer", Sapienza University of Rome, Piazzale Aldo Moro 5, 00185 Rome, Italy
2 Department of Clinical Medicine and Surgery, University of Naples "Federico II", 80131 Naples, Italy
3 Department of Psychiatry, Giustino Fortunato University, 82100 Benevento, Italy
* Correspondence: giuseppe.esposito@uniroma1.it

Abstract: Similar to canine inflammatory enteropathy, inflammatory bowel disease (IBD) is a chronic idiopathic condition characterized by remission periods and recurrent flares in which diarrhea, visceral pain, rectal bleeding/bloody stools, and weight loss are the main clinical symptoms. Intestinal barrier function alterations often persist in the remission phase of the disease without ongoing inflammatory processes. However, current therapies include mainly anti-inflammatory compounds that fail to promote functional symptoms-free disease remission, urging new drug discoveries to handle patients during this step of the disease. ALIAmides (ALIA, autacoid local injury antagonism) are bioactive fatty acid amides that recently gained attention because of their involvement in the control of inflammatory response, prompting the use of these molecules as plausible therapeutic strategies in the treatment of several chronic inflammatory conditions. N-palmitoyl-D-glucosamine (PGA), an under-researched ALIAmide, resulted in being safe and effective in preclinical models of inflammation and pain, suggesting its potential engagement in the treatment of IBD. In our study, we demonstrated that micronized PGA significantly and dose-dependently reduces colitis severity, improves intestinal mucosa integrity by increasing the tight junction proteins expression, and downregulates the TLR-4/NLRP3/iNOS pathway via PPAR-α receptors signaling in DNBS-treated mice. The possibility of clinically exploiting micronized PGA as support for the treatment and prevention of inflammation-related changes in IBD patients would represent an innovative, effective, and safe strategy.

Keywords: micronized N-palmitoyl-D-glucosamine; ulcerative colitis; IBD; intestinal inflammation; toll-like receptors; PPARs; NLRP3; intestinal barrier

Citation: Palenca, I.; Seguella, L.; Del Re, A.; Franzin, S.B.; Corpetti, C.; Pesce, M.; Rurgo, S.; Steardo, L.; Sarnelli, G.; Esposito, G. N-Palmitoyl-D-Glucosamine Inhibits TLR-4/NLRP3 and Improves DNBS-Induced Colon Inflammation through a PPAR-α-Dependent Mechanism. *Biomolecules* **2022**, *12*, 1163. https://doi.org/10.3390/biom12081163

Academic Editors: Salvatore Cuzzocrea and Rosalia Crupi

Received: 22 July 2022
Accepted: 20 August 2022
Published: 22 August 2022

1. Introduction

Inflammatory Bowel Disease (IBD) is a complex of chronic and relapsing diseases of the gastrointestinal (GI) tract that converges environmental, microbial, immunological, and genetic factors both in humans [1] and dogs [2]. In humans, these heterogeneous GI disorders present as two major clinical phenotypes, ulcerative colitis (UC) and Crohn's disease (CD), which are characterized by periods of remission and flare-ups of the disease, in which diarrhea, visceral hypersensitivity, and fever are commonly referred [3]. Acute inflammation markedly impairs intestinal physiology and function, and persistent alterations are often observed after the resolution of intestinal inflammation, consistently suggesting a role for inflammatory effects in generating the symptoms that occur during remission in patients with IBD. These long-term changes involve motility, abnormal secretion, and altered visceral sensation, and no effective treatments are currently available to handle this phase of the disease. In recent decades, IBD has alarming emerged in Western countries, predominantly

UC, suggesting that this epidemiological evolution is likely related to the westernization of lifestyle associated with changes in diet, antibiotic use, hygiene status, and microbial exposures [4]. UC primarily involves confluent inflammation of the colonic mucosa and impairs the epithelial barrier integrity, and intestinal homeostasis. The consequent abnormal translocation of luminal microbes and their products across the impaired intestinal barrier leads to robust activation of resident macrophages and antigen-presenting cells (APCs), and the massive release of Tumor Necrosis Factor- α (TNF-α), Interleukin-1β (IL-1β), IL-13, IL-9, IL-23, IL-36, and other pro-inflammatory mediators [5]. This acute inflammation is responsible for most alterations in intestinal functions that often persist following the resolution of the acute inflammation process and are frequently observed during remission in patients with IBD. Currently, there are no effective treatments for UC that prevent the inflammation-related acute and long-term intestinal dysfunctions, but only combined pharmacological and nutritional therapies, which encompass assessment of daily caloric intake and periodic measurement of functional capacities. 5-aminosalicylic acid is the first therapeutic approach with corticosteroids and thiopurines. However, the high incidence of side effects, including hypersensitivity reactions, liver toxicity, immunosuppression, and pancreatitis [5] associated with the increasing number of corticosteroids/thiopurines-resistant UC led to the development of alternative therapeutical strategies that include anti-TNF-α, anti-IL-12/23 p40, and anti-integrin or JAK inhibitors [6]. Biological therapies, such as infliximab and adalimumab, also have limitations mainly related to the reduced compliance for parenteral administration, immunogenicity, and high costs that reduce their clinical application. N-palmitoyl-D-glucosamine (PGA) is a natural amide of palmitic acid and glucosamine that shares the anti-inflammatory properties with its endogenous analog palmitoylethanolamide (PEA) and those of glucosamine [7]. PGA belongs to the ALIAmide family (ALIA, autacoid local injury antagonism), a class of both synthetic and endogenous fatty acid amides that display a wide range of homeostatic effects in response to increased oxidative stress and cell damage, including anti-inflammatory and pain-relieving effects [8]. In particular, PEA has repeatedly been shown to improve clinical and histological signs of colitis in different murine models [9,10]. In line, micronized PGA resulted in being safe (LD50 $\geq$ 2000 mg/kg) and effective in preclinical models of inflammation and osteoarthritis (OA) pain, taking advantage of particle size reduction that enhances its anti-inflammatory activity [7]. To note, the anti-inflammatory action of PGA might derive from both the amide and monosaccharide portions [11]. In fact, glucosamine improved colitis symptoms in DSS-treated mice by preventing intestinal epithelial cell activation and tight junction proteins expression decrease, with a parallel decrease in the nuclear factor-kappa B (NF-kB) activity and reduced TNF-α and IL-1β release [11,12]. Moreover, oral glucosamine and chondroitin compositions positively impact microbiota composition in the intestine of healthy adults, supporting PGA similar acitivity [13]. Iannotta et al. also reported that micronized PGA acts as a toll-like receptor (TLR)-4 antagonist, thanks to its structural similarity to the lipid A component of lipopolysaccharide (LPS). TLR-4 is involved in activating the nucleotide-binding oligomerization domain leucine-rich repeat and pyrin domain-containing protein 3 (NLRP3) inflammasome complex, which is overactivated in several inflammatory syndromes, including intestinal inflammation, neurodegenerative, and metabolic diseases [14]. NLRP3 inflammasome plays a critical role in inflammatory response as a major component of innate immunity and is involved in exacerbating the mucosal immune response and intestinal epithelial barrier damage during colitis [15]. ALIAmide-mediated activation of the peroxisome proliferator-activated receptors-α (PPAR-α) counters the NLRP3 activity [16], although whether PGA exerts its anti-inflammatory effects via PPAR-α receptors is still unknown.

In this study, we investigated the effectiveness and the mechanisms of action of micronized PGA in a murine model of colitis induced by the 2,4-dinitrobenzene sulfonic acid (DNBS) and assessed the in vivo effects of orally administered micronized PGA on (i) colitis severity, (ii) mucosal inflammation and immune cells infiltration, (iii) release of pro-inflammatory cytokines, and (iv) NLRP3 and TLR-4 intestinal activation in a resolution phase of intestinal inflammation.

2. Materials and Methods

2.1. Animals and Experimental Design

Eight-week-old male C57BL/6J mice (Charles River, Lecco, Italy) (n = 50) have been used for the experiments. The procedures included in the experimental plan have been approved by Sapienza University's Ethics Committee. Animal care followed the (International Association for the Study of Pain) IASP and European Community (EC L358/1 18/12/86) guidelines on the use and protection of animals in experimental research. Colitis groups received a single intracolonic administration of 4 mg DNBS (Sigma Aldrich, St. Louis, MO, USA) in 100 µL of 50% ethanol (Sigma Aldrich, St. Louis, MO, USA) and saline (Thermo Fisher Scientific, Waltham, MA, USA), whereas the vehicle group received a single intracolonic administration of saline and ethanol as described below. Overnight-fasted mice were treated with DNBS on day 0 through a soft cannula (Hugo-Sachs Elektronik, March, Germany) quickly inserted around 3cm away from the anus without anesthetization. DNBS solution was slowly administrated into the colon-rectal tract, and animals were maintained slightly sloped for the entire procedure. Thus, mice were placed back into their cages and kept overnight on a heating pad to aid recovery. Disease activity index (DAI) parameters were daily recorded from day 0 to day 7 to assess colitis severity. Micronized PGA was suspended in carboxymethylcellulose (CMC) (Thermo Fisher Scientific, Waltham, MA, USA) and 1X PBS (Sigma Aldrich, St. Louis, MO, USA), and 200 µL of 30 mg/kg and 100 mg/kg PGA suspension were daily given by a single gavage from days 1 to 6 based on the experimental design [7]. PGA was used in the micronized formulation kindly provided by Epitech Group S.p.A (Saccolongo, Italy). Micronized formulation of PGA with a smaller particle size overcomes the low water-solubility issue of this compound, increasing its oral absorption and bioavailability. In fact, drug solubility is strongly related to particle size and its reduction leads to an increase in the specific surface area with enhanced solubility and potentially higher bioavailability. PPAR-α antagonist MK886 (Selleck Chemicals, Houston, TX) was dissolved in 1X PBS and given daily at the dose of 10 mg/kg by a single intraperitoneal (IP) administration (150 µL) from days 1 to Mice were randomly divided into the following groups (n = 10 each): (1) vehicle group receiving single intracolonic administration of saline; (2) colitis group; (3) colitis group receiving a daily gavage with 30 mg/kg micronized PGA; (4) colitis group receiving a daily gavage with 100 mg/kg micronized PGA; (5) colitis group receiving a daily gavage with 100 mg/kg micronized PGA associated with daily intraperitoneal administration of 10 mg/kg PPAR-α antagonist MK886. Animals in the vehicle and colitis groups also received a daily gavage of CMC solution in 1X PBS (200 µL) and a daily IP with 1X PBS (150 µL) from days 1 to day 6. Mice were euthanized on day 7 by cervical dislocation; thus, spleen weight and colon length were measured, and blood samples, as well as colon tissues, were collected to conduct histochemical and biochemical analyses as described below. Each experimental group included n = 10 mice. All the experiments were performed in triplicate on the distal colon by randomly using N = 5 colon for histological staining and N = 5 colon for immunofluorescence analysis.

2.2. Disease Activity Index (DAI)

The DAI score was used to evaluate the colitis severity and progression during the 7 days of the experimental protocol, according to the criteria developed by Cooper et al. [17]. The scored parameters were: (i) changes in body weight; (ii) stool consistency; (iii) rectal bleeding. DAI score was recorded daily (from day 0 to 7), scores were given depending on the severity of the symptoms, and the results were expressed as cumulative average scores in each experimental group.

2.3. Histopathological Analyses

Colonic tissues were fixed in 4% paraformaldehyde (PFA) (Thermo Fisher Scientific, Waltham, MA, USA) and cryo-sectioned in 15 μm slices. Slices were stained with hematoxylin and eosin (H&E) (Sigma Aldrich, St. Louis, MO, USA) to evaluate the histopathological damage score according to Li et al. [18]. Criteria included: (i) distortion and loss of crypt architecture; (ii) inflammatory cells infiltration; (iii) muscle thickening; (iv) goblet cells depletion; (v) crypts absence.Slices were analyzed with a microscope Nikon Eclipse 80i by Nikon Instruments Europe (Nikon Corporation, Tokyo, Japan), and images were captured at 10× magnification by a high-resolution digital camera (Nikon Digital Sight DS-U1). Cumulative damage scores obtained from each experimental group were expressed as average scores.

2.4. Immunofluorescence Analysis on Colonic Sections

Immunofluorescence analyses were performed on 15 μm colonic slices fixed in ice-cold 4% PFA. Sections were blocked with a solution composed of 1X PBS, 4% Normal Donkey Serum, 0.4% (Merk Millipore, St. Louis, MO, USA) TRITON-100 (Sigma Aldrich, St. Louis, MO, USA), and 1% Bovine Serum Albumin (BSA) (Sigma Aldrich, St. Louis, MO, USA) for 45 min and subsequently incubated at +4 °C overnight with primary antibody (Table 1). Slices were then washed with 1X PBS and incubated in the dark at +4 °C with fluorescein isothiocyanate-conjugated anti-rabbit (1:1000 dilution *v*/*v*; Abcam, Cambridge, UK) or Texas Red-conjugated anti-mouse (1:500 dilution *v*/*v*, mouse; Abcam, Cambridge, USA). Sections were analyzed with a microscope Nikon Eclipse 80i, and images were captured at 20× and 40× magnification by a high-resolution digital camera (Nikon Digital Sight DS-U1). Results were expressed as relative fluorescence units (RFU) and fluorescence intensity percentage (FI%).

Table 1. Primary antibodies used in immunofluorescence analyses on cryo-sectioned colon slides.

Antibody	Host	Clonality	Dilution	Brand
ZO-1	Mouse	Monoclonal	6 microgram *w*/*v*	Invitrogen, Thermo Fisher, Waltham, MA, USA
Occludin	Rabbit	Polyclonal	1:100 *v*/*v*	Bioss Antibodies, Boston, MA, USA
NLRP3	Rabbit	Polyclonal	1:1000 *v*/*v*	Invitrogen, Thermo Fisher, Waltham, MA, USA
TLR-4	Rabbit	Polyclonal	1:150 *v*/*v*	Bioss Antibodies, Boston, MA, USA
iNOS	Mouse	Monoclonal	1:1000 *v*/*v*	Novusbio, Centennial, CO, USA

Zonula occludens (ZO-1); nucleotide-binding oligomerization domain leucine-rich repeat and pyrine domain-containing protein 3 (NLRP3); toll-like receptor (TLR)-4; inducible nitric oxide synthase (iNOS).

2.5. Enzyme-Linked Immunosorbent Assay (ELISA) for IL-1β and PGE_2

Enzyme-linked immunosorbent assay (ELISA) for PGE_2 and IL-1β (Thermo Fisher Scientific, Waltham, MA, USA) was carried out on mouse plasma isolated from blood samples according to the manufacturer's protocol. Absorbance was measured on a microtiter plate reader. PGE_2 and IL-1β levels were determined using standard curve methods.

2.6. Statistical Analyses

Results were expressed as the mean ± SD. Statistical analysis was performed using parametric one-way analysis of variance (ANOVA) and multiple comparisons were performed by Bonferroni's post hoc test. p-values < 0.05 were considered statistically significant. Data were analyzed by using Graphpad Prism and ImageJ software.

3. Results

3.1. Micronized PGA Improves the Disease Spectrum and Macroscopic Signs of Colitis in a Dose-Dependent Manner through PPAR-α Involvement

The DAI score was significantly increased in the colitis group during the 7 days that followed DNBS administration (7.9 ± 0.141, $p < 0.0001$; Figure 1A), with a parallel colonic shortening (3.83 ± 0.543 cm, $p < 0.0001$; Figure 1B,D) and spleen weight increase (0.153 ± 0.0269 g, $p < 0.05$; Figure 1C) in comparison to the vehicle group. PGA resulted in a dose-dependent improvement of overall colitis hallmarks, leading to a significant decrease in DAI score (4.65 ± 0.354, $p < 0.001$ and 3.07 ± 0.305, $p < 0.0001$ for the 30 mg/kg and 100 mg/kg dose, respectively; Figure 1A), increase in colon length (5.82 ± 0.676 cm and 7.55 ± 0.572 cm, $p < 0.0001$ for 30 mg/kg and 100 mg/kg dose, respectively; Figure 1B), and reduction of spleen weight (0.0897 ± 0.011 g, $p < 0.05$ for 100 mg/kg m-PGA, Figure 1C) as compared to DNBS-treated mice. DAI score, colon length, and spleen weight were comparable to those in DNBS mice treated with 100 mg/Kg um-PGA and PPAR-α antagonist MK886 (10 mg/kg), suggesting that PGA exerts its beneficial effects on colitis through the selective involvement of PPAR-α receptors.

3.2. Micronized PGA Ameliorates Mucosal Integrity and Prevents Colonic Histological Damage in DNBS-Treated Mice

DNBS-treated mice maintained a significant impairment of colonic mucosal integrity on day 7 following colitis induction, as demonstrated by the decreased expression of the two tight junction proteins, zonula occludens-1 (ZO-1) and occludin, compared to the vehicle group (25.08 ± 3.196 FI%, $p < 0.0001$ for ZO-1 and 18.23 ± 1.322 FI%, $p < 0.0001$ for occludin; Figure 2A–C). The loss of ZO-1 and occludin was partially prevented by PGA at 30 mg/kg dose (37.4 ± 2.507 FI%, $p < 0.0001$ for ZO-1 and 33.97 ± 2.217 FI%, $p < 0.0001$ for occludin; Figure 2A–C), whereas a marked restoration was observed in colitis mice treated with 100 mg/kg PGA in comparison to the DNBS group (87.8 ± 2.579 FI% and 94.75 ± 3.988 FI%, $p < 0.0001$ for ZO-1 and occludin, respectively; Figure 2A–C). The effects of PGA (100 mg/kg) were reverted by MK886 (10 mg/kg) in colitis mice, further demonstrating the involvement of PPAR-α receptors. Histopathological scores revealed extensive damage in the colonic mucosal barrier of DNBS-treated mice with marked neutrophil infiltration and damaged mucosal integrity compared to vehicle (8.2 ± 0.9189, $p < 0.0001$; Figure 2D,E). Micronized PGA (30 mg/kg) preserved mucosal integrity and counteracted the neutrophil infiltration within the mucosa, although the number of crypts was lower in comparison to the DNBS group (5.5 ± 0.9718, $p < 0.0001$; Figure 2D,E). At the higher dose (100 mg/kg), PGA widely restored mucosal integrity, significantly reducing the neutrophil infiltration, and preserving the architecture and number of the crypts in comparison to the DNBS group (2.5 ± 0.8498, $p < 0.0001$; Figure 2D,E). Co-administration of PGA and MK886 resulted in colonic histological damage comparable to DNBS-group, further supporting that PGA preserves the intestinal mucosal integrity by PPAR-α engagement.

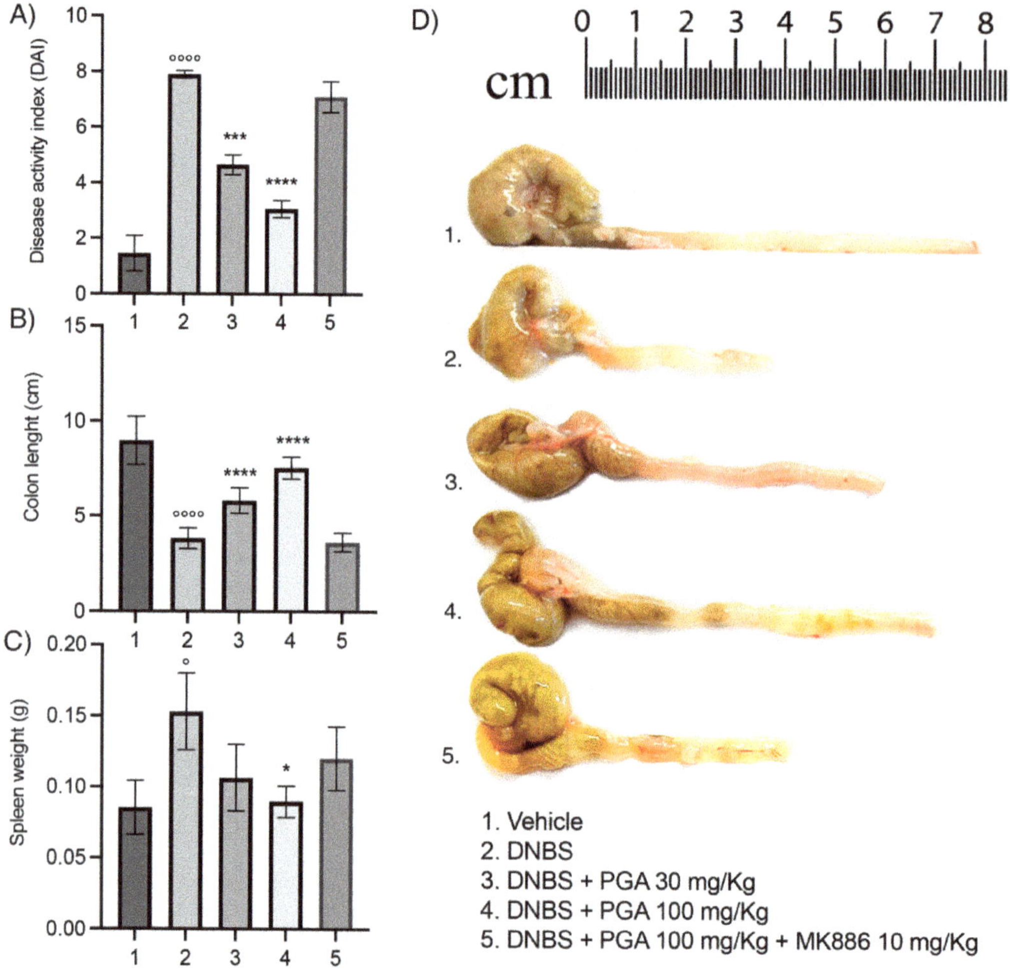

Figure 1. Micronized PGA significantly improves colitis hallmarks in PPAR-α-dependent manner. The effects of oral administration of micronized PGA on (**A**) DAI score, (**B**,**D**) colonic length, and (**C**) spleen weight in DNBS-treated mice. Results are expressed as the mean ± SD of n = 5 experiments ° $p < 0.05$ vs. vehicle; °°°° $p < 0.0001$ vs. vehicle; * $p < 0.05$ vs. DNBS; *** $p < 0.001$ vs. DNBS; **** $p < 0.0001$ vs. DNBS. Peroxisome proliferator-activated receptors-α (PPAR-α); Disease Activity Index (DAI); 2,4-dinitrobenzene sulfonic acid (DNBS).

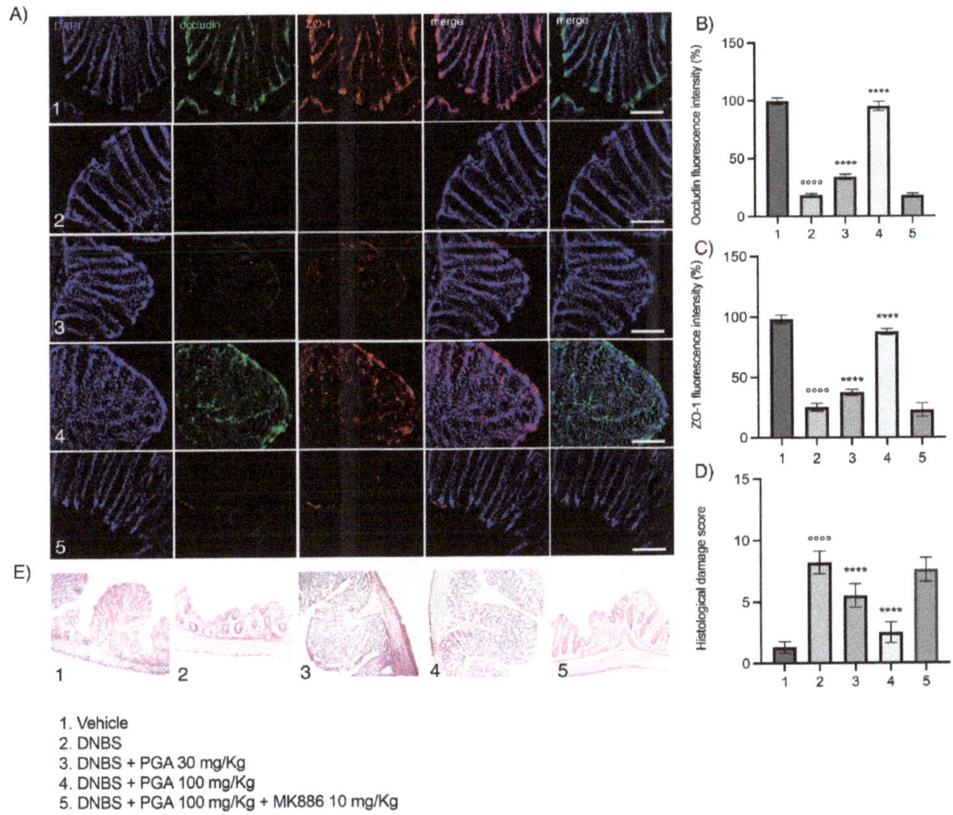

Figure 2. Micronized PGA prevents the loss of tight junction proteins ZO-1 and occludin, and colonic barrier disruption through PPAR-α involvement. (**A**) Representative images show double-label immunohistochemistry for occludin (green) and ZO-1 (red) in the colon with (**B**,**C**) relative quantification. Nuclei were also labeled by DAPI. (**E**) Representative images of H&E-stained on distal colon sections, and (**D**) relative histological damage score. Results are expressed as the mean ± SD of n = 5 experiments (25 slices for each animal). °°°° $p < 0.0001$ vs. vehicle; **** $p < 0.0001$ vs. DNBS. Scale bar = 100 μm; magnification 20×. Zonula occludens (ZO-1); peroxisome proliferator-activated receptors-α (PPAR-α); hematoxylin and eosin (H&E), 2,4-dinitrobenzene sulfonic acid (DNBS).

3.3. Micronized PGA Downregulates the TLR-4/NLRP3/iNOS Expression and Decreases the Release of Inflammatory Mediators in DNBS-Treated Mice via PPAR-α Receptors

Our results show that TLR-4, NLRP3, and inducible nitric oxide synthase (iNOS) expression was markedly increased within the mucosa following DNBS-induced colitis compared to the vehicle (31.38 ± 1.169 RFU, 14.54 ± 1.282 RFU, and 33.93 ± 1.665 RFU for TLR-4, NLRP3 and iNOS, respectively; $p < 0.0001$ for all comparisons; Figure 3A–D). In DNBS mice receiving 30 mg/kg PGA, the expression of TLR-4, NLRP3, and iNOS was significantly reduced (21.08 ± 1.904 RFU, 10.40 ± 1.235 RFU, and 18.26 ± 1.668 RFU for TLR-4, NLRP3 and iNOS, respectively; $p < 0.0001$ for all comparisons; Figure 3A–D) and completely downregulated by PGA 100 mg/kg (14.98 ± 0.712 RFU, 7.54 ± 0.537 RFU, and 12.56 ± 1.105 RFU for TLR-4, NLRP3, and iNOS, respectively; $p < 0.0001$ for all comparisons; Figure 3A–D). According to previous results, 100 mg/kg PGA did not show any effect in DNBS mice co-administered with PPAR-α antagonist MK866 (10 mg/kg), supporting that PGA prevents the activation of the TLR4/NLRP3/iNOS pathway through the PPAR-α-mediated mechanism. Further, the increased plasma levels of IL-1β and

PGE_2 detected in DNBS mice at day 7 after colitis induction (113.1 ± 32.71 pg/mL and 348.5 ± 45.22 pg/mL for IL-1β and PGE_2, respectively; $p < 0.0001$ vs. vehicle for both comparisons; Figure 3E,F) were significantly and dose-dependently decreased by PGA (62.1 ± 15.09 pg/mL and 36.7 ± 13.12 pg/mL for IL-1β in the lower and higher dose group, respectively; 217.5 ± 61.07 pg/mL and 169.7 ± 37.12 pg/mL for PGE_2 in the lower and higher dose group, respectively; $p < 0.0001$ vs. DNBS for all comparisons; Figure 3E,F). The anti-inflammatory effect of PGA was significantly inhibited in the presence of PPAR-α antagonist MK886, displaying comparable plasma cytokines levels with the DNBS group.

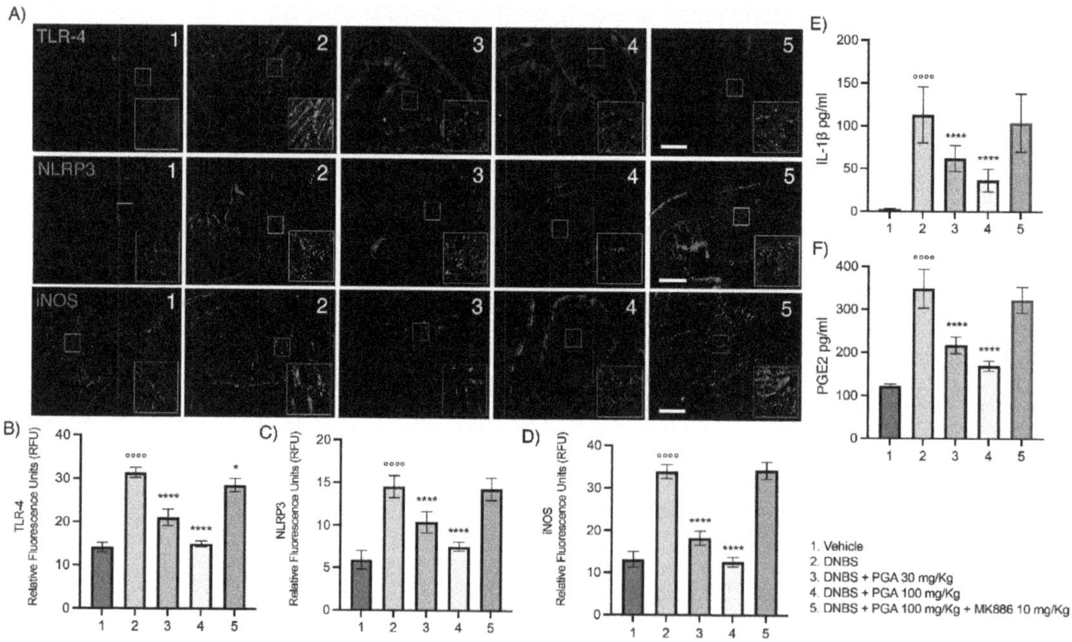

Figure 3. Micronized PGA decreases TLR-4/NLRP3/iNOS expression in mice colon and pro-inflammatory cytokines release in plasma samples throughout selective PPAR-α involvement in DNBS-treated mice. (**A**) Immunofluorescence images display TLR-4 (green), NLRP3 (magenta), and iNOS (red) staining, and the relative quantification for (**B**) TLR-4, (**C**) NLRP3, and (**D**) iNOS. (**E**) IL-1β and (**F**) PGE_2 release in mice plasma. Results are expressed as the mean ± SD of n = 5 experiments (25 slices for each animal). °°°° $p < 0.0001$ vs. vehicle; * $p < 0.05$ vs. DNBS; **** $p < 0.0001$ vs. DNBS. Scale bar = 100 μm; magnification 20× 40×. Toll-like receptor (TLR)-4; nucleotide-binding oligomerization domain leucine-rich repeat and pyrine domain-containing protein 3 (NLRP3); inducible nitric oxide synthase (iNOS); peroxisome proliferator-activated receptors-α (PPAR-α); interleukin-1β (IL-1β); prostaglandin E2 (PGE_2.), 2,4-dinitrobenzene sulfonic acid (DNBS).

4. Discussion

IBD showed an increasing trend of incidence worldwide, and current therapies mainly consist of the chronic administration of immunosuppressive drugs [19]. However, these drugs display a short-term efficacy, and they are not suitable as a maintenance therapy due to related systemic adverse reactions [20]. The possibility of exploiting new compounds to target different stages and pathways of the inflammatory response represents an outstanding challenge for IBD therapy. In the present study, we provided evidence that micronized PGA significantly and dose-dependently counteracts the occasional inflammatory-mediated intestinal functions alteration during the colitis resolution phase that characterizes the remission state of the disease.

Oral administration of micronized PGA improved the DAI score and resulted in macroscopic amelioration of intestinal inflammation, as shown by the increased expression of tight junction proteins (ZO-1 and occludin) in the colonic mucosa and lower histological damage score. Similar to the homologous ALIAmide PEA, which was successful in several colitis models treatment in mice and humans [21–25], PGA revealed an anti-inflammatory effect through a PPAR-α-dependent mechanism since in the presence of PPAR-α antagonist MK886, PGA-mediated effects were abolished. We demonstrated that PGA caused a significant decrease of pro-inflammatory mediators, such as iNOS and NLRP3 protein expression in colonic tissues as well as PGE_2 and IL-1β cytokines release in mice plasma. This anti-inflammatory effect of PGA might be related to its monosaccharide portion, glucosamine. In support, glucosamine was able to reduce colitis-associated symptoms in DSS-treated mice by reducing TNF-α, IL-1β, and NF-kB expression in the colonic mucosa, and increasing the expression of ZO-1 and occludin tight junction proteins [12]. Patients recovering from an exacerbation of ulcerative/membranous colitis display increased levels of glucosamine synthetase. This enzyme synthesizes N-acetyl-glucosamine (NAG), which was recently implicated in the tissue regeneration phase in pediatric IBD [26–28]. In addition, glucosamine and/or chondroitin are used as dietary supplements to reduce colonrectal cancer risk and systemic inflammation [29–32]. This evidence supports the hypothesis that glucosamine might be responsible, at least in part, for PGA anti-inflammatory activity once hydrolyzed by the degradative enzymes (FAAH, NAAA), although further confirmation is required.

In agreement with previous studies demonstrating that PGA negatively regulates TLR-4 signaling enhanced by intestinal inflammation and neuropathic pain [8,14], our results show that micronized PGA dose-dependently reduced TLR-4 expression in the colonic mucosa of colitis mice, suggesting that this may contribute to its potent anti-inflammatory activity. The sequel of events triggered by TLR-4 activation, including NF-kB and NLRP3 activation, and the release of IL-1β, IL-6, and TNF-α, are considered the most involved in persistent intestinal inflammation triggering and maintenance during colitis [8,14], and HIV-1 Tat-induced diarrhea [33]. In this context, PPAR-α agonists, such as PEA, may mediate TLR-4 down-regulation and efficiently suppress the inflammatory process similar to the evidence observed in vitro [15], clinical studies [34], endotoxin induced-uveitis rat model, and DSS-induced colitis mice model [21,35].

In addition to PPAR-α agonism, PGA displays structural similarity with the lipopolysaccharide (LPS) component Lipid A, which results in a direct antagonism on TLR-4 [14]. Thus, the anti-inflammatory properties of PGA might be related to two distinct mechanisms, although this hypothesis requires further investigation.

Previous studies have shown that PPAR-α activation inhibits immune cell infiltration in colonic mucosa and decreases the expression and release of pro-inflammatory markers in mice and humans [36,37]. In line, we have shown that micronized PGA reduces immune cell infiltration within the colonic mucosa and downregulates NLRP3 inflammasome expression via PPAR-α activation. This anti-inflammatory activity probably allows PGA to preserve the intestinal epithelial barrier integrity. Actually, NLRP3 plays a key role in exacerbating the mucosal immune response and intestinal epithelial barrier damage [38]. The activation of PPAR-α receptors prevents NLRP3 hyper-activation, which is linked to several inflammatory syndromes, including intestinal inflammation [36]. Here, we demonstrated the efficient and dose-dependent counteraction of NLRP3 overexpression by oral administration of micronized PGA. NLRP3 inhibition was associated with reduced colitis severity and histological damage of colonic mucosa, supporting that PGA exerts an anti-inflammatory effect by suppressing NLRP3 activity through PPAR-α receptor involvement.

To gain more mechanistic insights, we evaluated the effect of micronized PGA on TLR-4 expression, which plays an essential role in innate immunity activation by recognizing microbial antigens, such as LPS [39]. The increased translocation of luminal microbe-derived products that follows the epithelial barrier breakdown during intestinal inflammation leads to higher TLR-4 activation. The downstream sequel events triggered by the TLR-4 activation, including NLRP3 overexpression, iNOS upregulation, and increased release of pro-inflammatory cytokines such as IL-1β and PGE_2 [40,41], are strictly related to mucosa damage expansion [42], and visceral hypersensitivity associated with the inflammatory process [43,44]. Oral administration of micronized PGA resulted in decreased expression of TLR-4 and related downstream pro-inflammatory pathways in DNBS-treated mice, as demonstrated by the parallel downregulation of iNOS and NLRP3 expression, as well as by the reduced release of IL-1β and PGE_2. This suggests that PGA might act as a safe "multitarget" modulator of intestinal inflammation and prevents the long-term intestinal dysfunction that generally follows the acute phase of colitis.

To note, TLR-4 activation is also associated with visceral hypersensitivity and related behavioral disorders, resulting in episodes of anxiety and recurrent flares of abdominal pain [45–47]. This latter was also associated with long-term changes in the intestinal microbiota composition, pointing out that compounds able to restore and maintain microbiota homeostasis might provide benefit in IBD patients with recurrent visceral allodynia and hyperalgesia [48]. Interestingly, a bacteria strain derived from legumes is able to produce PGA (*rhizobium leguminosarum*) [49], and successfully colonize the intestinal microenvironment by acting as a xenobiotic metabolizer [47]. Taking the evidence that oral glucosamine and chondroitin compositions modulate positively the intestinal microbiota in healthy adults [13], micronized PGA might have the capability to target distinct pathological aspects of intestinal inflammation by downregulating pro-inflammatory mediators, decreasing mucosal damage and visceral hypersensitivity, and even restoring microbiota homeostasis.

In addition, antagonism on TRL-4 could potentially determine the regulation of mucosal blood flow since a reduced expression of TRL-4 and consequent down-regulation of NF-kB leads to inactivity of iNOS and reduced release of nitric oxide (NO). Adequate organ blood flow also demonstrates a protective and healing-promoting role in the gastrointestinal tract [50–52]. Despite further studies being needed to confirm this hypothesis, our results provide the first evidence on the ability of micronized PGA to target colitis through a double mechanism of action: PPAR-α agonism and TLR-4 antagonism [14]. In consideration of the obtained results and the safety profile of micronized PGA, more studies are advised to explore the protective effects of micronized PGA in IBD management.

Author Contributions: Conceptualization, G.E.; methodology, I.P. and L.S. (Luisa Seguella).; software, I.P. and L.S. (Luisa Seguella), C.C.; validation, I.P., A.D.R. and S.B.F.; formal analysis, I.P., A.D.R., C.C., and S.R.; investigation, I.P.; resources, S.B.F.; data curation, I.P.; writing—original draft preparation, I.P. and L.S. (Luisa Seguella) and M.P.; supervision, G.S., G.E., and L.S. (Luca Steardo); project administration, I.P. All authors have read and agreed to the published version of the manuscript.

Funding: This research received no external funding.

Institutional Review Board Statement: All experimental procedures were approved by Sapienza University's Ethics Committee. Animal care was in compliance with the IASP and European Community (EC L358/1 18/12/86, 30/09/2016) guidelines on the use and protection of animals in experimental research.

Informed Consent Statement: Not applicable.

Data Availability Statement: The data presented in this study are available on request from the corresponding author. The data are not publicly available due to the policy of our research group, but we will share data unreservedly on online data sharing platforms upon request.

Acknowledgments: In Memoriam of Francesco Della Valle for his profound passion for science and his unparalleled contribution to scientific progress that has given an impulse to pharmaceutical research.

Conflicts of Interest: The authors declare no conflict of interest.

References

1. Zhang, Y.-Z.; Li, Y.-Y. Inflammatory bowel disease: Pathogenesis. *World J. Gastroenterol.* **2014**, *20*, 91–99. [CrossRef] [PubMed]
2. Siel, D.; Beltrán, C.J.; Martínez, E.; Pino, M.; Vargas, N.; Salinas, A.; Pérez, O.; Pereira, I.; Ramírez-Toloza, G. Elucidating the Role of Innate and Adaptive Immune Responses in the Pathogenesis of Canine Chronic Inflammatory Enteropathy—A Search for Potential Biomarkers. *Animals* **2022**, *12*, 1645. [CrossRef] [PubMed]
3. Kucharzik, T.; Koletzko, S.; Kannengießer, K.; Dignaß, A. Ulcerative Colitis—Diagnostic and Therapeutic Algorithms. *Dtsch. Ärzteblatt Int.* **2020**, *117*, 564–574. [CrossRef] [PubMed]
4. Du, L.; Ha, C. Epidemiology and Pathogenesis of Ulcerative Colitis. *Gastroenterol. Clin. N. Am.* **2020**, *49*, 643–654. [CrossRef]
5. Gearry, R.B.; Barclay, M.L.; Burt, M.J.; Collett, J.A.; Chapman, B.A. Thiopurine drug adverse effects in a population of New Zealand patients with inflammatory bowel disease. *Pharmacoepidemiol. Drug Saf.* **2004**, *13*, 563–567. [CrossRef] [PubMed]
6. Kobayashi, T.; Siegmund, B.; Le Berre, C.; Wei, S.C.; Ferrante, M.; Shen, B.; Bernstein, C.N.; Danese, S.; Peyrin-Biroulet, L.; Hibi, T. Ulcerative colitis. *Nat. Rev. Dis. Prim.* **2020**, *6*, 1–20. [CrossRef] [PubMed]
7. Cordaro, M.; Siracusa, R.; Impellizzeri, D.; D'Amico, R.; Peritore, A.F.; Crupi, R.; Gugliandolo, E.; Fusco, R.; Di Paola, R.; Schievano, C.; et al. Safety and efficacy of a new micronized formulation of the ALIAmide palmitoylglucosamine in preclinical models of inflammation and osteoarthritis pain. *Arthritis Res. Ther.* **2019**, *21*, 254. [CrossRef]
8. Gugliandolo, E.; Peritore, A.F.; Piras, C.; Cuzzocrea, S.; Crupi, R. Palmitoylethanolamide and Related ALIAmides: Prohomeostatic Lipid Compounds for Animal Health and Wellbeing. *Veter.-Sci.* **2020**, *7*, 78. [CrossRef]
9. Capasso, R.; Orlando, P.; Pagano, E.; Aveta, T.; Buono, L.; Borrelli, F.; Di Marzo, V.; Izzo, A.A. Palmitoylethanolamide normalizes intestinal motility in a model of post-inflammatory accelerated transit: Involvement of CB_1 receptors and TRPV1 channels. *J. Cereb. Blood Flow Metab.* **2014**, *171*, 4026–4037. [CrossRef]
10. Borrelli, F.; Romano, B.; Petrosino, S.; Pagano, E.; Capasso, R.; Coppola, D.; Battista, G.; Orlando, P.; Di Marzo, V.; Izzo, A.A. Palmitoylethanolamide, a naturally occurring lipid, is an orally effective intestinal anti-inflammatory agent. *Br. J. Pharmacol.* **2014**, *172*, 142–158. [CrossRef]
11. Yomogida, S.; Kojima, Y.; Tsutsumi-Ishii, Y.; Hua, J.; Sakamoto, K.; Nagaoka, I. Glucosamine, a naturally oc-curring amino monosaccharide, suppresses dextran sulfate sodium-induced colitis in rats. *Int. J. Mol. Med.* **2008**, *22*, 317–323. [PubMed]
12. Bak, Y.-K.; Lampe, J.W.; Sung, M.-K. Effects of dietary supplementation of glucosamine sulfate on intestinal inflammation in a mouse model of experimental colitis. *J. Gastroenterol. Hepatol.* **2014**, *29*, 957–963. [CrossRef]
13. Navarro, S.L.; Levy, L.; Curtis, K.R.; Lampe, J.W.; Hullar, M.A.J. Modulation of Gut Microbiota by Glucosamine and Chondroitin in a Randomized, Double-Blind Pilot Trial in Humans. *Microorganisms* **2019**, *7*, 610. [CrossRef] [PubMed]
14. Iannotta, M.; Belardo, C.; Trotta, M.C.; Iannotti, F.A.; Vitale, R.M.; Maisto, R.; Boccella, S.; Infantino, R.; Ricciardi, F.; Mirto, B.F.; et al. N-palmitoyl-D-glucosamine, a Natural Monosaccharide-Based Glycolipid, Inhibits TLR4 and Prevents LPS-Induced Inflammation and Neuropathic Pain in Mice. *Int. J. Mol. Sci.* **2021**, *22*, 1491. [CrossRef] [PubMed]
15. Del Re, A.; Corpetti, C.; Pesce, M.; Seguella, L.; Steardo, L.; Palenca, I.; Rurgo, S.; De Conno, B.; Sarnelli, G.; Esposito, G. Ultramicronized Palmitoylethanolamide Inhibits NLRP3 Inflammasome Expression and Pro-Inflammatory Response Activated by SARS-CoV-2 Spike Protein in Cultured Murine Alveolar Macrophages. *Metabolites* **2021**, *11*, 592. [CrossRef] [PubMed]
16. Impellizzeri, D.; Di Paola, R.; Cordaro, M.; Gugliandolo, E.; Casili, G.; Morittu, V.M.; Britti, D.; Esposito, E.; Cuzzocrea, S. Adelmidrol, a palmitoylethanolamide analogue, as a new pharmacological treatment for the management of acute and chronic inflammation. *Biochem. Pharmacol.* **2016**, *119*, 27–41. [CrossRef]
17. Cooper, H.S.; Murthy, S.N.; Shah, R.S.; Sedergran, D.J. Clinicopathologic study of dextran sulfate sodium experimental murine colitis. Laboratory investigation. *J. Tech. Methods Pathol.* **1993**, *69*, 238–249.
18. Li, R.; Kim, M.-H.; Sandhu, A.K.; Gao, C.; Gu, L. Muscadine Grape (*Vitis rotundifolia*) or Wine Phytochemicals Reduce Intestinal Inflammation in Mice with Dextran Sulfate Sodium-Induced Colitis. *J. Agric. Food Chem.* **2017**, *65*, 769–776. [CrossRef]
19. Cushing, K.; Higgins, P.D.R. Management of Crohn Disease. *JAMA* **2021**, *325*, 69–80. [CrossRef]
20. Bruscoli, S.; Febo, M.; Riccardi, C.; Migliorati, G. Glucocorticoid Therapy in Inflammatory Bowel Disease: Mechanisms and Clinical Practice. *Front. Immunol.* **2021**, *12*. [CrossRef]
21. Shen, W.; Gao, Y.; Lu, B.; Zhang, Q.; Hu, Y.; Chen, Y. Negatively regulating TLR4/NF-κB signaling via PPARα in endotoxin-induced uveitis. *Biochim. Et Biophys. Acta (BBA)—Mol. Basis Dis.* **2014**, *1842*, 1109–1120. [CrossRef] [PubMed]
22. Couch, D.G.; Cook, H.; Ortori, C.; Barrett, D.; Lund, J.N.; O'Sullivan, S.E. Palmitoylethanolamide and Cannabidiol Prevent Inflammation-induced Hyperpermeability of the Human Gut In Vitro and In Vivo—A Randomized, Placebo-controlled, Double-blind Controlled Trial. *Inflamm. Bowel Dis.* **2019**, *25*, 1006–1018. [CrossRef] [PubMed]
23. Matalon, S.T.; Azar, S.; Meiri, D.; Hadar, R.; Nemirovski, A.; Abu Jabal, N.; Konikoff, F.M.; Drucker, L.; Tam, J.; Naftali, T. Endocannabinoid Levels in Ulcerative Colitis Patients Correlate with Clinical Parameters and Are Affected by Cannabis Consumption. *Front. Endocrinol.* **2021**, *12*. [CrossRef]
24. Esposito, G.; Corpetti, C.; Pesce, M.; Seguella, L.; Annunziata, G.; Del Re, A.; Vincenzi, M.; Lattanzi, R.; Lu, J.; Sanseverino, W.; et al. A Palmitoylethanolamide Producing Lactobacillus paracasei Improves Clostridium difficile Toxin A-Induced Colitis. *Front. Pharmacol.* **2021**, *12*. [CrossRef] [PubMed]
25. Esposito, G.; Pesce, M.; Seguella, L.; Lu, J.; Corpetti, C.; Del Re, A.; De Palma, F.; Esposito, G.; Sanseverino, W.; Sarnelli, G. Engineered *Lactobacillus paracasei* Producing Palmitoylethanolamide (PEA) Prevents Colitis in Mice. *Int. J. Mol. Sci.* **2021**, *22*, 2945. [CrossRef]

26. Goodman, M.J.; Kent, P.W.; Truelove, S.C. Glucosamine synthetase activity of the colonic mucosa in ulcerative colitis and Crohn's disease. *Gut* **1977**, *18*, 219–228. [CrossRef]
27. Goodman, M.J.; Kent, P.W.; Truelove, S.C. Glucosamine synthetase activity of the colonic mucosa in membranous colitis. *Gut* **1977**, *18*, 229–231. [CrossRef]
28. Salvatore, S.; Heuschkel, R.; Tomlin, S.; Davies, S.E.; Edwards, S.; Walker-Smith, J.A.; French, I.; Murch, S.H. A pilot study of N-acetyl glucosamine, a nutritional substrate for glycosaminoglycan synthesis, in paediatric chronic inflammatory bowel disease. *Aliment. Pharmacol. Ther.* **2000**, *14*, 1567–1579. [CrossRef]
29. Kantor, E.D.; Zhang, X.; Wu, K.; Signorello, L.B.; Chan, A.T.; Fuchs, C.S.; Giovannucci, E.L. Use of glucosamine and chondroitin supplements in relation to risk of colorectal cancer: Results from the Nurses' Health Study and Health Professionals follow-up study. *Int. J. Cancer* **2016**, *139*, 1949–1957. [CrossRef]
30. Kantor, E.D.; Newton, C.C.; Giovannucci, E.L.; McCullough, M.L.; Campbell, P.T.; Jacobs, E.J. Glucosamine use and risk of colorectal cancer: Results from the Cancer Prevention Study II Nutrition Cohort. *Cancer Causes Control* **2018**, *29*, 389–397. [CrossRef]
31. Lee, D.H.; Cao, C.; Zong, X.; Zhang, X.; O'Connell, K.; Song, M.; Wu, K.; Du, M.; Cao, Y.; Giovannucci, E.L.; et al. Glucosamine and Chondroitin Supplements and Risk of Colorectal Adenoma and Serrated Polyp. *Cancer Epidemiol. Biomark. Prev.* **2020**, *29*, 2693–2701. [CrossRef] [PubMed]
32. Navarro, S.L.; White, E.; Kantor, E.; Zhang, Y.; Rho, J.; Song, X.; Milne, G.; Lampe, P.D.; Lampe, J.W. Randomized Trial of Glucosamine and Chondroitin Supplementation on Inflammation and Oxidative Stress Biomarkers and Plasma Proteomics Profiles in Healthy Humans. *PLoS ONE* **2015**, *10*, e0117534. [CrossRef] [PubMed]
33. Sarnelli, G.; Seguella, L.; Pesce, M.; Lu, J.; Gigli, S.; Bruzzese, E.; Lattanzi, R.; D'Alessandro, A.; Cuomo, R.; Steardo, L.; et al. HIV-1 Tat-induced diarrhea is improved by the PPARalpha agonist, palmitoylethanolamide, by suppressing the activation of enteric glia. *J. Neuroinflamm.* **2018**, *15*, 94. [CrossRef]
34. Darmani, N.A.; Izzo, A.A.; Degenhardt, B.; Valenti, M.; Scaglione, G.; Capasso, R.; Sorrentini, I.; Di Marzo, V. Involvement of the cannabimimetic compound, N-palmitoyl-ethanolamine, in inflammatory and neuropathic conditions: Review of the available pre-clinical data, and first human studies. *Neuropharmacology* **2005**, *48*, 1154–1163. [CrossRef] [PubMed]
35. Grabacka, M.; Pierzchalska, M.; Płonka, P.M.; Pierzchalski, P. The Role of PPAR Alpha in the Modulation of Innate Immunity. *Int. J. Mol. Sci.* **2021**, *22*, 10545. [CrossRef] [PubMed]
36. Dana, N.; Vaseghi, G.; Haghjooy-Javanmard, S. Crosstalk between Peroxisome Proliferator-Activated Receptors andToll-Like Receptors: A Systematic Review. *Adv. Pharm. Bull.* **2019**, *9*, 12–21. [CrossRef]
37. Takeda, K.; Akira, S. Microbial recognition by Toll-like receptors. *J. Dermatol. Sci.* **2004**, *34*, 73–82. [CrossRef]
38. Zhen, Y.; Zhang, H. NLRP3 Inflammasome and Inflammatory Bowel Disease. *Front. Immunol.* **2019**, *10*, 276. [CrossRef]
39. Chen, M.-Y.; Ye, X.-J.; He, X.-H.; Ouyang, D.-Y. The Signaling Pathways Regulating NLRP3 Inflammasome Activation. *Inflammation* **2021**, *44*, 1229–1245. [CrossRef]
40. Zhuang, Y.; Zhao, F.; Liang, J.; Deng, X.; Zhang, Y.; Ding, G.; Zhang, A.; Jia, Z.; Huang, S. Activation of COX-2/mPGES-1/PGE2 Cascade via NLRP3 Inflammasome Contributes to Albumin-Induced Proximal Tubule Cell Injury. *Cell. Physiol. Biochem.* **2017**, *42*, 797–807. [CrossRef]
41. Luo, H.; Guo, P.; Zhou, Q. Role of TLR4/NF-κB in Damage to Intestinal Mucosa Barrier Function and Bacterial Translocation in Rats Exposed to Hypoxia. *PLoS ONE* **2012**, *7*, e46291. [CrossRef] [PubMed]
42. Morales-Soto, W.; Gulbransen, B.D. Enteric Glia: A New Player in Abdominal Pain. *Cell. Mol. Gastroenterol. Hepatol.* **2019**, *7*, 433–445. [CrossRef] [PubMed]
43. Yuan, B.; Tang, W.-H.; Lu, L.-J.; Zhou, Y.; Zhu, H.-Y.; Zhou, Y.-L.; Zhang, H.-H.; Hu, C.-Y.; Xu, G.-Y. TLR4 upregulates CBS expression through NF-κB activation in a rat model of irritable bowel syndrome with chronic visceral hypersensitivity. *World J. Gastroenterol.* **2015**, *21*, 8615–8628. [CrossRef]
44. Bettoni, I.; Comelli, F.; Rossini, C.; Granucci, F.; Giagnoni, G.; Peri, F.; Costa, B. Glial TLR4 receptor as new target to treat neuropathic pain: Efficacy of a new receptor antagonist in a model of peripheral nerve injury in mice. *Glia* **2008**, *56*, 1312–1319. [CrossRef]
45. Hutchinson, M.R.; Zhang, Y.; Brown, K.; Coats, B.D.; Shridhar, M.; Sholar, P.W.; Patel, S.J.; Crysdale, N.Y.; Harrison, J.A.; Maier, S.F. Non-stereoselective reversal of neuropathic pain by naloxone and naltrexone: Involvement of toll-like receptor 4 (TLR4). *Eur. J. Neurosci.* **2008**, *28*, 20–29. [CrossRef] [PubMed]
46. Lönnfors, S.; Vermeire, S.; Greco, M.; Hommes, D.; Bell, C.; Avedano, L. IBD and health-related quality of life—Discovering the true impact. *J. Crohn's Colitis* **2014**, *8*, 1281–1286. [CrossRef] [PubMed]
47. Cuesta, C.M.; Pascual, M.; Pérez-Moraga, R.; Rodríguez-Navarro, I.; García-García, F.; Ureña-Peralta, J.R.; Guerri, C. TLR4 Deficiency Affects the Microbiome and Reduces Intestinal Dysfunctions and Inflammation in Chronic Alcohol-Fed Mice. *Int. J. Mol. Sci.* **2021**, *22*, 12830. [CrossRef] [PubMed]
48. Philip-Hollingsworth, S.; Dazzo, F.B.; I Hollingsworth, R. Structural requirements of Rhizobium chitolipooligosaccharides for uptake and bioactivity in legume roots as revealed by synthetic analogs and fluorescent probes. *J. Lipid Res.* **1997**, *38*, 1229–1241. [CrossRef]
49. Das, A.; Srinivasan, M.; Ghosh, T.S.; Mande, S.S. Xenobiotic metabolism and gut microbiomes. *PLoS ONE* **2016**, *11*. [CrossRef]

50. Matuszyk, A.; Ceranowicz, D.; Warzecha, Z.; Ceranowicz, P.; Fyderek, K.; Gałązka, K.; Cieszkowski, J.; Bonior, J.; Jaworek, J.; Pihut, M.; et al. The Influence of Ghrelin on the Development of Dextran Sodium Sul-fate-Induced Colitis in Rats. *BioMed Res. Int.* **2015**, *2015*, 718314. [CrossRef]
51. Dembiński, A.; Warzecha, Z.; Ceranowicz, P.; Dembiński, M.; Cieszkowski, J.; Gosiewski, T.; Bulanda, M.; Kuśnierz-Cabala, B.; Gałązka, K.; Konturek, P.C. Synergic Interaction of Rifaximin and Mutaflor (*Escherichia coli* Nissle 1917) in the Treatment of Acetic Acid-Induced Colitis in Rats. *Gastroenterol. Res. Pract.* **2016**, *2016*, 3126280. [CrossRef] [PubMed]
52. Matuszyk, A.; Ceranowicz, P.; Warzecha, Z.; Cieszkowski, J.; Gałązka, K.; Bonior, J.; Jaworek, J.; Konturek, P.C.; Gil, K.; Dembiński, A. Pretreatment with obestatin inhibits the development of acetic acid-induced colitis in rats. *Arch. Med. Sci.* **2018**, *14*, 920–929. [CrossRef] [PubMed]

biomolecules

MDPI

Article

The Beneficial Effects of Ultramicronized Palmitoylethanolamide in the Management of Neuropathic Pain and Associated Mood Disorders Induced by Paclitaxel in Mice

Claudia Cristiano, Carmen Avagliano, Mariarosaria Cuozzo, Fabrizio Maria Liguori, Antonio Calignano and Roberto Russo *

Department of Pharmacy, University of Naples Federico II, Via D. Montesano 49, 80131 Naples, Italy
* Correspondence: roberto.russo@unina.it; Tel.: +39-081678465

Abstract: Chemotherapy-induced peripheral neuropathy (CIPN) is a common complication of antineoplastic drugs, particularly paclitaxel (PTX). It can affect the quality of patients' lives and increase the risk of developing mood disorders. Although several drugs are recommended, they yielded inconclusive results in clinical trials. The aim of the present work is to investigate whether the palmitoylethanolamide (PEA) would reduce PTX-induced CIPN and associated mood disorders. Moreover, the role PPAR-α and the endocannabinoid system will also be investigated. CIPN was induced by intraperitoneally injection of PTX (8 mg/kg) every other day for a week. PEA, 30 mg/kg, was orally administrated in a bioavailable form (i.e., ultramicronized PEA, um-PEA) one hour after the last PTX injection, for 7 days. In the antagonism experiments, AM281 (1 mg/kg) and GW6471 (2 mg/kg) were administrated 30 min before um-PEA. Our results demonstrated that um-PEA reduced the development of hypersensitivity with the effect being associated with the reduction in spinal and hippocampal pro-inflammatory cytokines, as well as antidepressive and anxiolytic effects. Moreover, the PPAR-α and CB1 receptor antagonists blocked the behavioral and antinociceptive effects of um-PEA. Our findings suggest that um-PEA is a promising adjunct in CIPN and associated mood disorders through the activation of PPAR-α, which influences the endocannabinoid system.

Keywords: paclitaxel; um-PEA; inflammation; neuropathic pain; behavior; cannabinoids; PPAR-α

Citation: Cristiano, C.; Avagliano, C.; Cuozzo, M.; Liguori, F.M.; Calignano, A.; Russo, R. The Beneficial Effects of Ultramicronized Palmitoylethanolamide in the Management of Neuropathic Pain and Associated Mood Disorders Induced by Paclitaxel in Mice. *Biomolecules* **2022**, *12*, 1155. https://doi.org/10.3390/biom12081155

Academic Editors: Salvatore Cuzzocrea and Rosalia Crupi

Received: 29 July 2022
Accepted: 18 August 2022
Published: 20 August 2022

1. Introduction

Cancer is one of the leading causes of death worldwide. In the last few years, the survival rates for most cancers have been increasing, probably due, at least in part, to new and improved treatments [1]. Despite the large interest in new drugs to fight cancer, old drugs still have an important use due to their high efficacy, despite their marked toxicity.

Paclitaxel (PTX) is a taxane chemotherapeutic agent used in the treatment of several cancer types. Although it shows beneficial antitumoral effects, it also produces important secondary effects both in the central and peripheral nervous system, resulting in emotional deficits and peripheral neuropathy, respectively [1–3]. Specifically, PTX causes chemotherapy-induced neuropathic pain (CIPN), a condition characterized by thermal and mechanical allodynia and hyperalgesia, which often persists for a long time [4,5]. In this case, the severity of symptoms can greatly increase the risk of developing mood disorders, including anxiety and depression, reinforcing the conception of CIPN as a chronic disease [6–8].

Unfortunately, there are currently no clinically effective interventions for CIPN, the efficacy of the existing therapies being only moderate [9].

PTX-induced neuropathy is initially characterized by oxidative stress followed by mitochondrial dysfunction [1,10–13]. Moreover, chemotherapeutical agents such as PTX have been shown to act similarly to LPS, i.e., increasing the production and release of pro-inflammatory interleukin (IL)-6 and IL-8 [14,15]. Notably, PTX-induced inflammation often interferes with its typical clinical efficacy, increasing tumor proliferation or chemoresistance [16–18] and contributing to the development of toxicity. Indeed, inflammation and neuroinflammation are a prominent characteristic of pain [19], as well as mood disorders [20]. Although acute inflammation can be considered a protective mechanism, chronic inflammation creates an array of detrimental effects. Despite the increasing understating of these mechanisms, novel analgesic strategies for treating PTX-induced toxicity are still lacking, and very few studies have investigated new possible effective treatments for CIPN and associated emotional components.

Palmitoylethanolamide (PEA) is a bioactive lipid mediator [21,22] belonging to the *N*-acyl-ethanolamine (NAE) family [23]. Although it is recognized that PEA primarily targets the nuclear receptor peroxisome proliferator-activated receptor alpha (PPAR-α), other receptors have been shown to mediate, at least in part, the effects of PEA, including G protein-coupled receptor 55 (GPR55), G protein-coupled receptor 119 (GPR119), and the transient receptor potential vanilloid receptor 1 (TRPV1) channels. Additionally, PEA indirectly activates cannabinoid receptors through the increase in the level of endocannabinoid mediators, such as anandamide (AEA), and 2-arachidonoylglycerol (2-AG) through the so-called called 'entourage effect' [24]. The well-known anti-inflammatory and antinociceptive effects exerted by PEA are considered to be an important non-pharmacological strategy in the management of neuropathic pain conditions, shown by several studies performed in preclinical models of inflammatory [25] and neuropathic pain [26–31], as well as clinical studies on human patients affected by osteoarthritis [32,33], neuropathic pain [34–36], fibromyalgia [37], and endometriosis [38]. Importantly, pain insensitivity in a human patient has been recently shown to be associated to a genetic deficiency in PEA-degrading enzymes, resulting in significantly higher PEA plasma levels compared to age-matched control patients [39].

In fact, PEA efficacy in chemotherapy-induced neuropathy has already been evaluated by previous studies via the oxaliplatin- [40] and PTX-induced CIPN model [41]. In particular, the last study has shown the effect to be mediated by PPAR-α, whose antagonism (i.e., through fenofibrate) is indeed known to reduce neuroinflammation in PTX-induced neuropathy [42].

PEA and its receptors are present in the central nervous system (CNS) [43,44], and exogenously administrated PEA is known to cross the blood–brain barrier [45], thereby exerting neuroprotective actions and promoting the resolution of neuroinflammation [46], and the regulation of behavior, mood, and cognition [47]. The therapeutic use of PEA in CNS disorders has produced promising results in several conditions, ranging from depression [48] to post-traumatic stress disorder [49] and autism spectrum disorders [50]. The ability of PEA to modulate neuroinflammation and associated symptoms (i.e., pain, depression, and anxiety) makes it a valid therapeutic tool in the treatment of several disorders.

The goal of our study was to investigate the activity of PEA on two PTX-induced side effects, i.e., mood disorders and peripheral neuropathy (CIPN), and to understand whether PPAR-α and the endocannabinoid system play a role. Moreover, we evaluated central and peripheral inflammatory modulation processes using this bioactive lipid. For our study, the ultramicronized formulation (um-PEA) was used, due to its higher solubility and bioavailability after systemic administration [51].

2. Materials and Methods

2.1. Animals

CD1 male mice (3 months old, 25–30 g, Charles Rivers, Calco, Lecco, Italy) were placed in a controlled area (room maintained at 22 $\pm$ 1 °C with 12 h light/dark cycle) and supplied with water and food ad libitum, in the animal care facility at the Department of Pharmacy of the University of Naples Federico II, Italy.

2.2. Drug Treatment

Paclitaxel (PTX, Cat#S1150 Selleckchem, Houston, TX, USA) at the dose of 8 mg/kg was dissolved in a solution made up of 5% DMSO, 40% PEG 300, 5% Tween 80, and ddH_2O according to the manufacture's guidelines. Ultramicronized palmitoylethanolamide (um-PEA), kindly provided by Epitech Group SpA (Saccolongo, Italy), was dissolved in 1% carboxymethylcellulose. The CB1 antagonist, AM281, and the PPAR- α antagonist GW6471, obtained from Tocris Cookson (Bristol, UK), were dissolved in 4% DMSO and saline.

2.3. Experimental Groups and Procedures

Mice were divided into five groups of $n = 8$ mice each, as follows:

- **Vehicle:** mice receiving saline intraperitoneally (IP).
- **PTX:** mice receiving PTX (8 mg/kg, 100 μL/mouse) IP at day 1, 3, 5, and 7.
- **PTX + um - PEA:** mice receiving PTX and then um-PEA (30 mg/kg for 7 days) by oral gavage.
- **PTX + um - PEA+AM281**: mice receiving PTX, then um-PEA, and on the last day AM281 (1 mg/kg) IP.
- **PTX+um-PEA+GW6471:** mice receiving PTX, then um-PEA, and on the last day GW6471 (2 mg/kg) IP.

Briefly, animals received PTX or vehicle treatments every other day for a week (days 1, 3, 5, and 7), as described previously [42,52]. One hour after the last day of PTX injection, mice started to receive oral um-PEA administration for 7 days. For antagonism experiments, on the last day of um-PEA injection, a different set of mice received AM281 or GW6471 IP 1 h before the um-PEA injection. The doses of AM281 and GW6471 selected in this study were based on the results of published data [41]. On the last day, mice were subjected to behavioral tests 1 h after um-PEA administration, and after euthanasia, the hippocampus and spinal cord were collected for ex vivo analysis.

2.4. Behavioral Tests

2.4.1. Depressive-like Behavior

Tail suspension test (TST). Mice were individually suspended by the tail 30–40 cm above the floor, using adhesive tape. The duration of immobility, recorded in seconds using a timer, was recorded during the 6-min test. Mice were considered immobile when they did not show any body movement, hung passively, with the absence of escape-oriented behavior.

Forced swimming test (FST). Mice were gently placed into individual glass cylinders (30 cm × 45 cm) filled with water maintained at 27 °C for 10 min, and their immobility times were recorded using a timer. Mice were considered immobile when floating in an upright position and only making small movements to keep their head above water, but without displacement. After the test, mice were allowed to dry and return to their home cage.

2.4.2. Anxiety-like Behavior

Elevated plus-maze (EPM). The maze was composed of a central square, two open arms and two closed arms enclosed by vertical walls, placed 50 cm above the floor. Mice were individually placed on the central area of one of the open arms and allowed to move freely. The number of entries into the open arms during a 5 min exploration period were recorded and analyzed by video tracking software (Any-maze, Stoelting, Wood Dale, IL, USA). An entry was counted only if all four paws were inside the arm. At the end of the test, the apparatus was wiped with 70% ethanol.

Open field (OF) test. Mice were placed in an OF arena (25 cm × 25 cm), and locomotor activity (total distance travelled in meters) was recorded for 30 min and analyzed by video tracking software (Any-maze, Stoelting, Wood Dale, IL, USA). The apparatus was cleaned before and after each behavioral session with a solution of 70% ethanol.

2.4.3. Pain Behavior

Mechanical allodynia (von Frey test). To assess changes in sensation or in the development of mechanical allodynia, sensitivity to tactile stimulation was measured using a dynamic plantar aesthesiometer (DPA, Ugo Basile, Italy). Animals were placed in a chamber with a mesh metal floor covered by a plastic dome that enabled the animals to walk freely, but not to jump. The mechanical stimulus (paw withdrawal threshold) was then delivered to the mid-plantar skin of the hind paw. The cutoff was fixed at 5 g, while the increasing force rate (ramp duration) was settled at 20 s. The DPA automatically records the force at which the foot is withdrawn and the withdrawal latency. Each paw was tested twice per session. This test did not require any special pretraining, just an acclimation period to the environment and testing procedure.

Mechanical hyperalgesia (Randall–Selitto test). Mechanical hyperalgesia was assessed using a Randall–Selitto algesimeter (Ugo Basile). Before the test, each animal received 5 min of handling to get used to manipulation; then it was placed into a soft cotton cloth and carefully immobilized with the same hand used to hold the tested paw. The test consisted of the application of an increasing mechanical force, in which the tip of the device was applied onto the medial portion of the plantar surfaces until a withdrawal response resulted. The maximum force applied was limited to 200 g to avoid skin damage.

Thermal allodynia (cold test). Cold sensitivity was measured as the number of foot withdrawal responses after application of acetone to the dorsal surface of the paw. A drop of acetone was applied to the dorsal surface of the paw with a syringe connected to a thin polyethylene tube while the mice were standing on a metal mesh. A brisk foot withdrawal response after the spread of acetone over the dorsal surface of the paw was considered as a sign of cold allodynia (n° paw withdrawal).

Thermal hyperalgesia (plantar test). Heat hypersensitivity was assessed using the mice plantar test apparatus (Ugo Basile, Italy). The plantar test consisted of three Perspex boxes (22 × 19 × 25 cm) on an elevated glass table. Mice were housed in each box and left to acclimatize for at least 10 min. A mobile infrared heat source was applied to the plantar surface of the hind paws. The paw withdrawal latency was defined as the time (expressed in seconds) taken by the mice to remove its hind paw from the heat source. The heat source was calibrated to 15 IR intensity, and a cutoff point of 60 s was applied to prevent tissue damage.

2.5. *Ex Vivo Experiments*

2.5.1. Determination of Brain and Spinal Cord Markers of Inflammation

The hippocampus and spinal cord samples were collected and TNF-α, IL-1β, IL-6, and IL-10, COX-2, and iNOS levels were measured using real time (RT)-PCR. For RT-PCR, total RNA was extracted from brain areas using TRIzol Reagent (Bio-Rad Laboratories, Hercules, California, USA) according to the manufacturer's instructions. cDNA from 4 μg total RNA was retrotranscribed using a reverse transcription kit (NucleoSpin®, MACHEREY-NAGEL GmbH & Co, Düren, Germany). RT-PCR reactions were performed using Bio-Rad CFX96 PCR System and relative software (Bio-Rad Laboratories). Mouse primers for TNF-α, IL-1β, IL-6, IL-10, COX-2, and iNOS were purchased from Qiagen (Hilden, Germany). Glyceraldehyde 3-phosphate dehydrogenase (GAPDH) was used as a housekeeping gene for normalization. Data are expressed using the ΔΔCT method.

2.5.2. Western Blotting

Spinal cord samples were homogenized on ice-cold lysis buffer (20 mM Tris–HCl (pH 7.5), 10 mM NaF, 150 mM NaCl, 1% Nonidet P-40, 1 mM phenylmethylsulfonyl fluoride, 1 mM Na3VO4, leupeptin and trypsin inhibitor 10 μg/mL; 0.25/50 mg tissue). After 1 h, tissue lysates were obtained by centrifugation at 2.0 × 10^4 g for 15 min at 4 °C. Protein concentrations were estimated with the Bio-Rad protein assay (Bio-Rad Laboratories, Milan, Italy) using bovine serum albumin as standard. Lysate proteins were dissolved in Laemmli sample buffer, boiled for 5 min, and separated by SDS-polyacrylamide gel electrophoresis and transferred to a nitrocellulose membrane (240 mA for 40 min at room temperature). The filter was then blocked with 1×phosphate buffer saline (PBS) and 3% non-fat dried milk for 40 min at room temperature and probed with anti-peroxisome proliferator-activated receptor (PPAR)-α (dilution 1:1000, cat. no. P0369, Sigma-Aldrich, Milan, Italy), or anti-cannabinoid (CB) receptor 1 (dilution 1:1000, cat. no. NB120-23703, Novus Biologicals, Cambridge, UK) in 1×PBS, 3% non-fat dried milk, and 0.1% Tween 20 at 4 °C overnight. The secondary antibodies were incubated for 1 h at room temperature. Subsequently, the blots were extensively washed with PBS, developed using enhanced chemiluminescence detection reagents (Amersham Pharmacia Biotech, Piscataway, NJ, USA) according to the manufacturer's instructions. The immune complex was visualized by the ChemiDoc Imaging System (Bio-Rad Laboratories). These blots were loaded with equal amounts of protein lysates, they were also incubated in the presence of the antibody against β-actin (cat. no. A5441, Sigma-Aldrich).

2.6. Statistical Analysis

Statistical analyses were performed using Prism 9 GraphPad software (GraphPad Software Inc., San Diego, CA, USA). All in vivo data are presented as mean ± SEM. For all experimental data, the significances of the differences between groups were determined by one-way repeated measures ANOVA, followed by post hoc Bonferroni's multiple comparison test. A value of $p < 0.05$ was considered statistically significant for all tests.

3. Results

3.1. Effect of um-PEA on PTX-Induced Depressive- and Anxiety-like Behaviors

We investigated the possible effect of um-PEA in reducing PTX-induced depressive-like behaviors as assessed in the TST and FST. As expected, both tests showed that PTX increased immobility time compared to vehicle-treated mice (** $p < 0.01$; Figure 1A,B). Um-PEA administration for 7 days after the last PTX injection significantly reduced the time of immobility compared to vehicle-treated mice (° $p < 0.05$; Figure 1A,B).

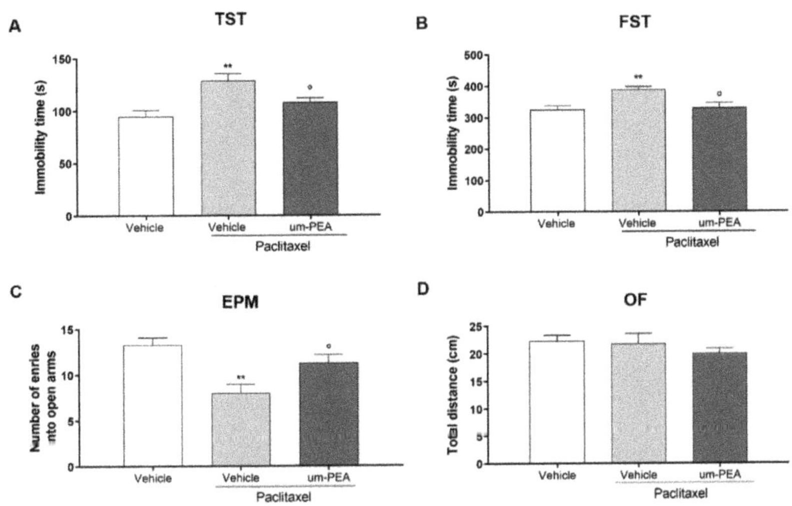

Figure 1. Effect of um-PEA on depressive- and anxiety-like behavior. (**A**) Time (in seconds) spent immobile in the TST; (**B**) time (in seconds) spent immobile in the FST test; (**C**) number of entries in the open arms of EPM test; (**D**) distance travelled (in cm) in the OF test; ** $p < 0.01$ versus vehicle group; ° $p < 0.05$ versus PTX group. Data are presented as mean ± SEM ($n = 8$). Differences were evaluated by ANOVA followed by Bonferroni's post hoc test for multiple comparisons.

The anxiety-like behavior was assessed in the EPM test. PTX-treated mice showed a significant decrease in the time spent in the open arms of the apparatus (** $p < 0.001$; Figure 1C) compared to vehicle-treated mice, which was significantly counteracted by repeated oral administration of um-PEA (° $p < 0.05$: Figure 1C). On the contrary, neither PTX nor- um-PEA had any effect on motor activity in the OF test (Figure 1D).

3.2. *CB1 and PPAR-α Are Involved in um-PEA Central Activity*

Since it was reported that um-PEA acts as both a CB1 and a PPAR-α receptor agonist [24], and these receptors are localized in the brain [53,54], we further characterized um-PEA anxiolytic and antidepressant effects by performing behavioral tests in the presence of CB1 and PPAR-α antagonists, AM281 and GW6471, respectively. Both of the receptor antagonists significantly inhibited the antidepressant effect of um-PEA ($^{\#}$ $p < 0.05$, $^{\#\#}$ $p < 0.01$; Figure 2A–B). Similar results were observed for the um-PEA anxiolytic effect compared to um-PEA+PTX-treated mice ($^{\#}$ $p < 0.05$, $^{\#\#}$ $p < 0.01$; Figure 2C). Additionally, in this case, neither AM281 nor GW6471 changed the motor activity of the mice (Figure 2D).

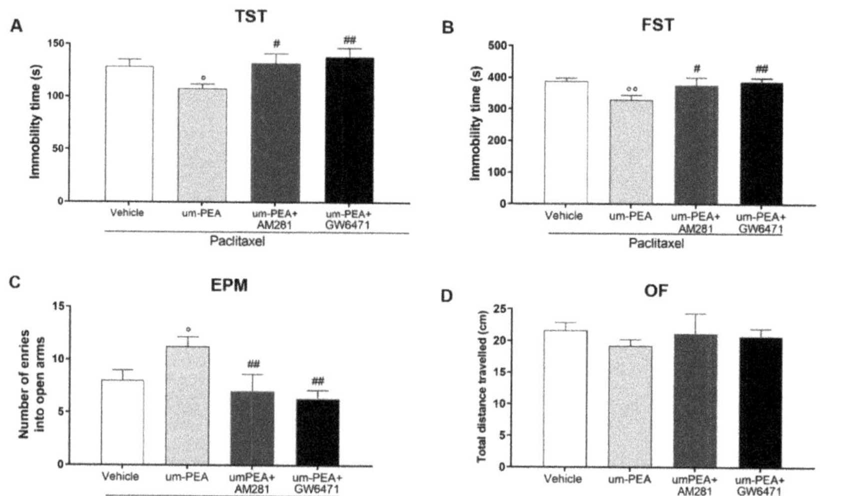

Figure 2. Effect of um-PEA on depressive- and anxiety-like behavior in presence of CB1 or PPAR-α antagonists. (**A**) Time (in seconds) spent immobile in the tail suspension test; (**B**) time (in seconds) spent immobile in the forced swim test; (**C**) number of entries in the open arms of elevated plus-maze test; (**D**) distance travelled (in cm) in the open field test; ° $p < 0.05$ and °° $p < 0.01$ versus PTX group; $^{\#}$ $p < 0.05$ and $^{\#\#}$ $p < 0.01$ versus PTX+ um-PEA. Data are presented as mean ± SEM ($n = 8$). Differences were evaluated by ANOVA followed by Bonferroni's post hoc test for multiple comparisons.

3.3. *Effect of um-PEA on Spinal and Sovraspinal Inflammatory Mediators in PTX Mice*

The mRNA levels of pro-inflammatory cytokines in hippocampus and spinal cord tissues were analyzed using RT-PCR. A significant (* $p < 0.05$, ** $p < 0.01$, and *** $p < 0.001$ versus vehicle) induction in the expression of these cytokines was observed in the PTX-treated group at both the central (Figure 3A–D) and the spinal level (Figure 4A,B). Um-PEA treatment significantly decreased all of the pro-inflammatory cytokine gene levels analyzed (TNF-α, IL-1β, and IL-6), while increasing the levels of IL-10 (* $p < 0.05$ versus PTX; Figure 3A–D). In the spinal cord, um-PEA treatment significantly decreased pro-inflammatory cytokine gene levels of TNF-α and IL-1β (° $p < 0.05$ and °° $p < 0.01$ versus PTX; Figure 4A-B). To further determine the inhibitory effect of um-PEA on inflammatory mediators in the spinal cord, we also evaluated the mRNA expression level of iNOS and COX-2. There was an upregulation of iNOS and COX-2 mRNA expression after PTX treatment, which was significantly (* $p < 0.05$ and *** $p < 0.001$ versus vehicle) reduced by um-PEA treatment (° $p < 0.05$ and °° $p < 0.01$ versus PTX; Figure 4C, D).

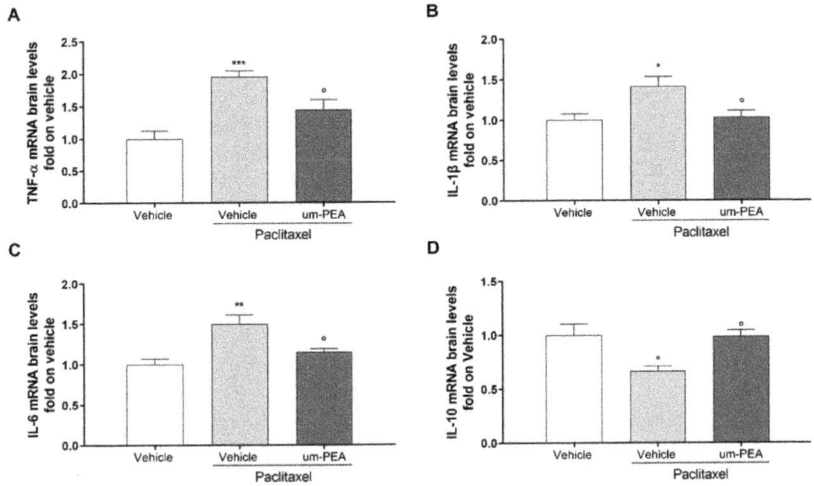

Figure 3. Pro-inflammatory cytokine levels in the hippocampi of vehicle- or um-PEA-treated mice injected with PTX. (**A**) Fold expression of mRNA for pro-inflammatory TNF-α; (**B**) fold expression of mRNA for pro-inflammatory IL-1β; (**C**) fold expression of mRNA for pro-inflammatory IL-6; (**D**) fold expression of mRNA for anti-inflammatory IL-10; * $p < 0.05$, ** $p < 0.01$, *** $p < 0.001$ versus vehicle group; ° $p < 0.05$ versus PTX group. Data are presented as mean ± SEM ($n = 8$). Differences were evaluated by ANOVA followed by Bonferroni's post hoc test for multiple comparisons.

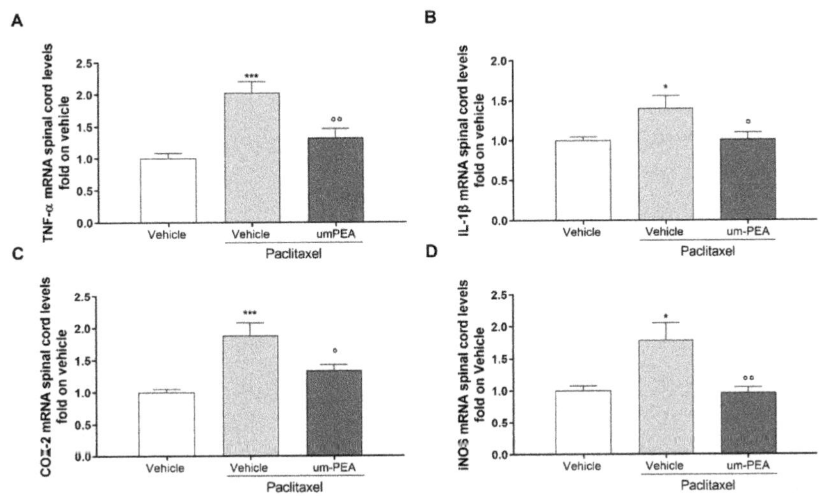

Figure 4. Pro-inflammatory cytokine levels in the spinal cords of vehicle- or um-PEA-treated mice injected with PTX. (**A**) Fold expression of mRNA for pro-inflammatory TNF-α; (**B**) fold expression of mRNA for pro-inflammatory IL-1β; (**C**) fold expression of mRNA for pro-inflammatory COX-2; (**D**) fold expression of mRNA for anti-inflammatory iNOS; * $p < 0.05$ and *** $p < 0.001$ versus vehicle group; ° $p < 0.05$ and °° $p < 0.01$ versus PTX group. Results are shown as mean ± SEM. Differences were analyzed using ANOVA followed by Bonferroni's post hoc test for multiple comparisons.

3.4. Effect of um-PEA on PTX-Induced Peripheral Neuropathy

PTX treatment induced neuropathic pain; indeed, PTX-treated mice showed a significant reduction in mechanical allodynia (** $p < 0.01$) and hyperalgesia (*** $p < 0.001$) compared to the vehicle group (Figure 5A–B). Um-PEA treatment significantly counteracted the effect of PTX in both tests (° $p < 0.05$; Figure 5A-B). PTX treatment also induced marked cold allodynia and thermal hyperalgesia, resulting in a significant increase in cold responses (*** $p < 0.001$) and significant decrease in thermal nociceptive thresholds (** $p < 0.01$) compared to vehicle group (Figure 5C,D). In contrast, um-PEA administration produced a significant reduction in the number of paw withdrawals in the acetone test, and enhanced the thermal withdrawal thresholds in the Hargreaves test (° $p < 0.05$) compared to PTX animals (Figure 5C,D).

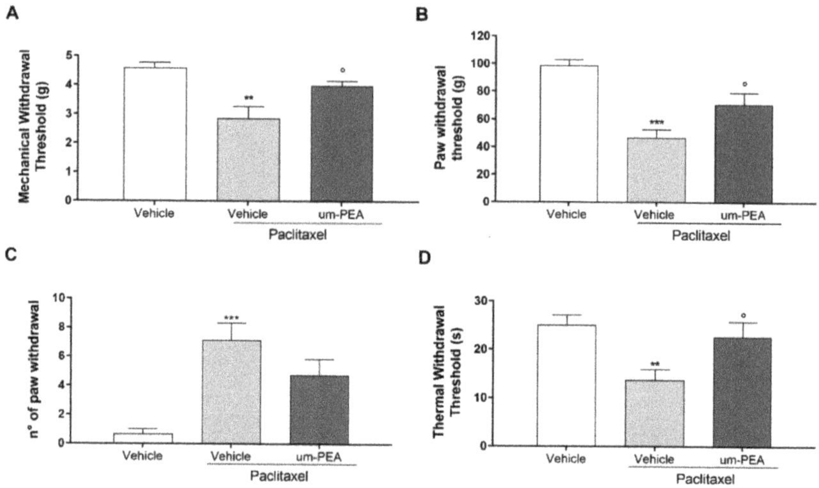

Figure 5. Effect of um-PEA on peripheral neuropathy. Results of the (**A**) von Frey test to evaluate mechanical allodynia; (**B**) Randall–Selitto test to assess mechanical hyperalgesia; (**C**) acetone evaporation test to evaluate thermal allodynia; (**D**) Hargreaves test to evaluate thermal hyperalgesia; ** $p < 0.01$; *** $p < 0.001$ versus vehicle group; ° $p < 0.05$ versus PTX group. Data are presented as mean ± SEM (n = 8). Differences were evaluated by ANOVA followed by Bonferroni's post hoc test for multiple comparisons.

3.5. Effect of um-PEA in PTX-Treated Mice Is PPAR-α and CB1 Mediated

The antinociceptive action of um-PEA was investigated in the presence of a selective CB1 receptor antagonist (AM281) and a selective PPAR-α antagonist (GW6471). As expected, the increase in paw withdrawal threshold of um-PEA-treated mice was reversed to a significant extent by AM281 in the Randall–Selitto, von Frey, and Hargreaves tests, compared to um-PEA-treated mice (## $p < 0.01$, Figure 6A–C). Additionally, we found that the analgesic effect of um-PEA was also significantly counteracted in mice treated with the selective PPAR-α antagonist (GW6471). Indeed, GW6471 administration in the um-PEA-treated group resulted in significant hyperalgesic and allodynic effects (## $p < 0.01$, Figure 6A–C).

Finally, in order to evaluate the roles of CB1 and PPAR-α receptors in PTX-induced neuropathy, ex vivo experiments were conducted. By Western blot analysis, we confirmed that PTX was involved in the maintenance of pain hypersensitivity by mechanical stimuli, since it was able to reduce CB1 receptor expression in the spinal cord compared to the vehicle group (Figure 7A–B, * $p < 0.05$). Um-PEA administration increased significantly CB1 receptor expression (Figure 7A, °° $p < 0.01$). No significant differences between vehicle and PTX+um-PEA groups were observed. Moreover, we also evaluated PPAR-α expression in the spinal cord, since this receptor has an important and well-known role in inflammation control. PTX reduced its expression (Figure 7B, * $p < 0.05$), while um-PEA was able to restore it by a significant degree ($^{\#}$ $p < 0.05$).

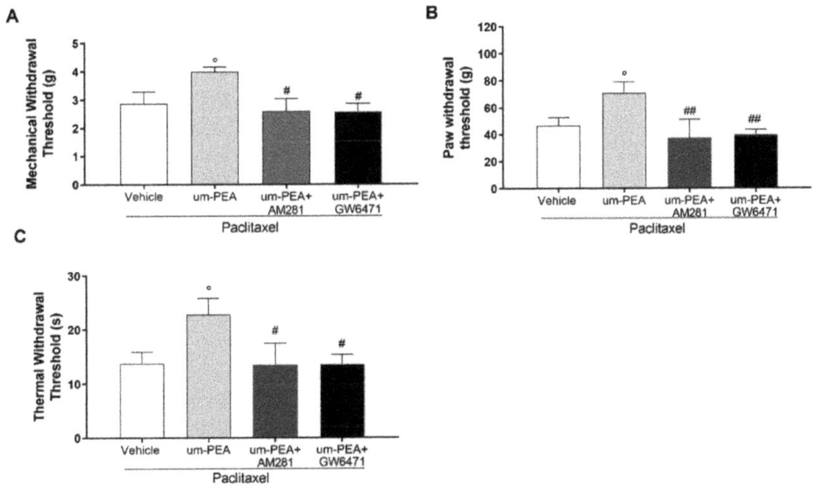

Figure 6. Effect of um-PEA on peripheral neuropathy in presence of CB1 or PPAR-α antagonists. Results of the (**A**) von Frey test to evaluate mechanical allodynia; (**B**) Randall–Selitto test to assess mechanical hyperalgesia; (**C**) Hargreaves test to evaluate thermal hyperalgesia; $^{\#}$ $p < 0.05$ and $^{\#\#}$ $p < 0.01$ versus PTX+um-PEA; ° $p < 0.05$ versus PTX-vehicle group. Data are presented as mean ± SEM ($n = 8$). Differences were evaluated by ANOVA followed by Bonferroni's post hoc test for multiple comparisons.

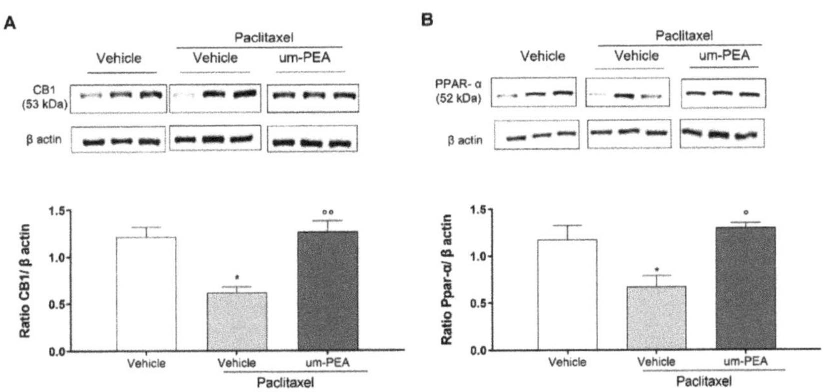

Figure 7. Western blotting analyses for CB1 (**A**) and PPAR-α (**B**) receptors in spinal cords. Immunoblots representative of all tissues analyzed are shown. Densitometric analysis of protein bands are reported: the levels are expressed as the density ratio of target to control protein bands (β-actin). * $p < 0.05$ versus vehicle group; ° $p < 0.05$ and °° $p < 0.005$ versus PTX group. Values are expressed as mean ± SEM ($n = 3$). Differences were evaluated by ANOVA followed by Bonferroni's post hoc test for multiple comparisons.

4. Discussion

The use of PTX as a chemotherapeutic agent has become a broadly accepted option in the treatment of patients with ovarian, breast, non-small-cell lung cancers, malignant brain tumors, and a variety of other solid tumors [1]. Significant toxicities in both the brain and periphery limit the effectiveness of PTX-based treatment regimens; this is a crucial limiting factor that can lead to a change or reduction in therapy or, in severe cases, to its total cessation [1]. In this study, we evaluated the efficacy of um-PEA in reducing PTX-related side effects, i.e., peripheral neuropathy and mood alteration, and we also investigated the possible mechanisms of action of um-PEA.

It is known that after the first cycle of PTX, it is already possible to observe the development of neuropathic pain and emotional disorders [55]. Accordingly, here we show that PTX-treated mice manifested anxiety-like behaviors in the EPM and OF tests, and depressive-like behaviors in the FST and TST. The possible mechanisms underlying these behavioral symptoms can be attributed to the development of neuroinflammation through glial cell activation [52] and/or the induction of central neurotoxicity. It is also possible that PTX sensitized the immune responses. Indeed, hypersensitivity to stimuli, not only in neuropathic pain, but also in inflammatory pain, can be explained by both peripheral and central sensitization of sensory nerve fibers [56].

It has been reported that during chronic inflammation, tissue levels of endogenous PEA are decreased, either due to reduced production or increased degradation [57] (or both). We thus hypothesized that treatment with a bioavailable form of PEA (i.e., um-PEA) could prevent or treat PTX-induced side effects. Our data show that 7 days of oral administration of um-PEA inhibited affective disorders in PTX mice. In detail, um-PEA reduced depressive- and anxiety-like behaviors, as shown in the TST, FST, and EPM test, respectively, while the OF test was not sensitive to PTX-induced changes. This result suggests that locomotion alterations did not influence the behavioral results. Several studies have indicated that PPAR signaling is involved in the regulation of anxiety responses. Indeed, Domi et al. [58] demonstrated that PPAR-γ antagonism induces an anxiogenic effect in mice, as detected both in the OF and EPM tests. Regarding the depressive-like behaviors associated with PTX administration, um-PEA treatment normalized the increase in the immobility time in the TST and FST. Several studies have already reported the potential antidepressant effects of PEA (either alone or in combination with antidepressants) [59,60], even in depressive-like behavior associated with neuropathic pain or traumatic injury [61]. Moreover, pharmacological inhibition of PEA degradation, as well as the upregulation of its biosynthesis, also resulted in antidepressant effects [62,63].

Although several molecular mechanisms have been suggested to explain PEA effects, its activity is mainly mediated by PPAR-α [59,64,65]. The activation of PPAR-α receptor initiates a cascade of events that causes the suppression of pain and inflammation, including decreases in pro-inflammatory cytokines such as IL-1β and IL-6, and TNF-α [57]. Moreover, low levels of PPAR-α are also responsible of several pathological conditions, neurodegenerative diseases, and stress-related disorders [66].

Here we found decreased levels of PPAR-α in the spinal cord of PTX mice, which was normalized by um-PEA. Although PPARs are not canonical endocannabinoid receptors, they are activated by several endocannabinoid mediators, and thus are considered as part of the enlarged endocannabinoid system, currently referred to as the endocannabinoidome [67]. Interestingly, evidence has shown that PEA, either by reducing anandamide (AEA) metabolism or binding to PPAR-α, upregulated the expression of CB receptors and increased TRPV1 activation, suggesting that PEA is able to interact both with the endocannabinoid and endovanilloid systems [68,69]. The CB1 receptor, which is highly expressed in the CNS, plays an important role in the regulation of stress and emotions [70,71]. Indeed, several studies have shown that CB1 agonists reduce neuroinflammation and have anxiolytic as well as antidepressant effects [72,73].

Therefore, one of the objectives of the study was to evaluate the protein expression of two main receptors involved in the beneficial effects of PEA, PPAR-α, and CB1 receptors. For instance, PEA exhibits analgesic effects via two different and distinct pathways, direct activation of PPAR-α or indirect activation of CB1 receptors, both of which relieve pain in different ways. Our results show that um-PEA counteracts the PTX-induced decrease in the expression not only of PPAR-α, but also CB1. Moreover, here we found that the effects of um-PEA were significantly inhibited by the administration of either AM281 (CB1 antagonist) or GW6471 (PPAR-α antagonist). As already mentioned, the reduction in or the absence of CB1 and PPAR-α could lead to neuroinflammation [66,74–77].

It is well documented that chronic administration of PEA is able to significantly reduce neuroinflammation [23,34,78], protect neurons from death [79,80], reduce oxygen radicals, and improve behavioral, motor, and cognitive deficits [81,82]. Neuroinflammation is a localized inflammation occurring in the PNS and CNS in response to trauma, bacterial/viral infection, autoimmunity, and/or toxins [83,84]. In particular, neuroinflammation is a common feature across different conditions, including neurodegenerative diseases, fibromyalgia, and chronic pain [83,85,86]. Different studies have reported the effect of PEA in the management of pain and inflammatory conditions [87–89]. The relationship between inflammation and pain is bidirectional, since the activation of pain circuits can also regulate neuroinflammation in the CNS [84]. Although acute neuroinflammation plays a protective role [90], its chronicization (i.e., non-resolving neuroinflammation) is detrimental, since the over-release of pro-inflammatory factors and cytokines can alter brain structure and function [91,92]. Non-resolving neuroinflammation is a key factor in the pathogenesis of CIPN, as shown by the significant increase in plasma levels of pro-inflammatory cytokines and chemokines involved in hypersensitivity and pain (e.g., IL-1β and TNF-α) in PTX-treated mice [93].

The decreased hippocampal gene expression of pro-inflammatory cytokines, which was here observed following 7-day oral administration of um-PEA, clearly shows the protective role of PEA-um against PTX-induced neuroinflammation. This central protective effect was also observed at the spinal cord level, since um-PEA treatment significantly decreased COX-2, iNOS, TNF-α, and IL-1β compared to the vehicle group.

Based on these findings, we also evaluated the efficacy um-PEA in PTX-induced peripheral neuropathy. In fact, Donvito et al. [41] had already found that PEA reversed PTX-induced neuropathy in a dose dependent manner. In their study, a single administration of PEA was able to reverse mechanical allodynia through a PPAR-α-mediated mechanism. In our study, mice receiving um-PEA treatment beginning at the last PTX injection show a reduction not only in allodynia signs but also hyperalgesia. In agreement with our data, Di Cesare Mannelli and coworkers reported analgesic proprieties of um-PEA in oxaliplatin-induced neuropathy, and showed that this acylethanolamine prevented the development of mechanical hypersensitivity, with a significant anti-inflammatory effect also being observed [40]. Recently, we have confirmed that um-PEA exerts its analgesic and anti-inflammatory effects primarily through direct activation of the transcription factor PPAR-α [50]. In particular, PEA has been found to switch off the nuclear factor kB signaling pathway, a crucial element in the transcription of genes, leading to the synthesis of pro-inflammatory and pro-analgesic mediators [64,94].

Moreover, in a chronic constriction injury model of neuropathic pain, PEA not only reduced edema and macrophage infiltration, but also counteracted the decrease in axon diameter and myelin thickness, the effects being lost in PPAR-α-null mice [95]. Recently, it has also been reported that the activation of PPARs may interfere with the production of pro-inflammatory cytokines in CIPN, potentially attenuating and preventing the symptoms of neuropathy [96]. In agreement with all these studies, we here demonstrate that um-PEA reduces pro-inflammatory cytokines by PPAR-α activation. In fact, repeated um-PEA administration increased the spinal cord expression of the nuclear receptor, and the PPAR-α antagonist GW6471 reversed um-PEA analgesic effects.

In addition, it is also interesting to note that the restoration of CB1 receptor levels in the spinal cord, following um-PEA treatment, seem to be important for pain control and inflammation reduction. We have also confirmed the correlation between PEA and the endocannabinoid system via both in vivo and ex vivo experiments, using AM281, a CB1 antagonist able to reverse um-PEA analgesic effects; moreover, repeated um-PEA administration preserves CB1 receptor expression in the spinal cord. Accordingly, a recent study reported that the administration of an analog of PEA, N-(4-methoxy-2-nitrophenyl)hexadecanamide (HD), produced a dose-dependent antinociceptive effect in rats, which was significantly counteracted by AM281 administration [97]. Taken together, these findings provide an overview of the crosstalk between PPARs and cannabinoids, and the importance of their reciprocal regulation in the control of major physiological and pathophysiological functions.

5. Conclusions

In summary, our results demonstrate that 7-day oral administration of um-PEA significantly reduced PTX side effects. Due to its anti-inflammatory activity and marked analgesic proprieties, as well as its ability to activate PPAR-α and influence the endocannabinoid system, um-PEA is a good candidate for the management of neuropathic pain and mood disorders produced by chemotherapy.

Author Contributions: Conceptualization, C.C., A.C. and R.R.; methodology, M.C., F.M.L. and C.A.; validation, C.C.; formal analysis, C.C.; investigation, C.C.; resources, R.R.; data curation, R.R.; writing—original draft preparation, C.C., C.A. and R.R.; writing—review and editing, R.R. and A.C.; supervision, A.C. and R.R.; project administration, R.R. All authors have read and agreed to the published version of the manuscript.

Funding: This research received no external funding.

Institutional Review Board Statement: The animal study protocol was approved by the Institutional Review Board (or Ethics Committee) of University of Naples Federico II (protocol n° 595/2019-PR, approved 30 July 2019).

Data Availability Statement: Not applicable.

Acknowledgments: The authors thank Giovanni Esposito and Angelo Russo for animal care and assistance.

Conflicts of Interest: The authors declare no conflict of interest.

References

1. Flatters, S.J.L.; Dougherty, P.M.; Colvin, L.A. Clinical and Preclinical Perspectives on Chemotherapy-Induced Peripheral Neuropathy (CIPN): A Narrative Review. *Br. J. Anaesth.* **2017**, *119*, 737–749. [CrossRef] [PubMed]
2. Dranitsaris, G.; King, J.; Kaura, S.; Yu, B.; Zhang, A. Nab-Paclitaxel, Docetaxel, or Solvent-Based Paclitaxel in Metastatic Breast Cancer: A Cost-Utility Analysis from a Chinese Health Care Perspective. *Clin. Outcomes Res.* **2015**, *7*, 249–256. [CrossRef] [PubMed]
3. Seretny, M.; Currie, G.L.; Sena, E.S.; Ramnarine, S.; Grant, R.; MacLeod, M.R.; Colvin, L.A.; Fallon, M. Incidence, Prevalence, and Predictors of Chemotherapy-Induced Peripheral Neuropathy: A Systematic Review and Meta-Analysis. *Pain* **2014**, *155*, 2461–2470. [CrossRef] [PubMed]
4. Banach, M.; Juranek, J.K.; Zygulska, A.L. Chemotherapy-Induced Neuropathies-a Growing Problem for Patients and Health Care Providers. *Brain Behav.* **2017**, *7*, e00558. [CrossRef] [PubMed]
5. Staff, N.P.; Grisold, A.; Grisold, W.; Windebank, A.J. Chemotherapy-Induced Peripheral Neuropathy: A Current Review: CIPN. *Ann. Neurol.* **2017**, *81*, 772–781. [CrossRef] [PubMed]
6. Bonhof, C.S.; Poll-Franse, L.V.; Vissers, P.A.J.; Wasowicz, D.K.; Wegdam, J.A.; Révész, D.; Vreugdenhil, G.; Mols, F. Anxiety and Depression Mediate the Association between Chemotherapy-induced Peripheral Neuropathy and Fatigue: Results from the Population-based PROFILES Registry. *Psycho-Oncology* **2019**, *28*, 1926–1933. [CrossRef]
7. Massie, M.J. Prevalence of Depression in Patients With Cancer. *J. Natl. Cancer Inst. Monogr.* **2004**, *2004*, 57–71. [CrossRef]
8. Mehnert, A.; Brähler, E.; Faller, H.; Härter, M.; Keller, M.; Schulz, H.; Wegscheider, K.; Weis, J.; Boehncke, A.; Hund, B.; et al. Four-Week Prevalence of Mental Disorders in Patients With Cancer Across Major Tumor Entities. *J. Clin. Oncol.* **2014**, *32*, 3540–3546. [CrossRef]
9. Furman, D.; Campisi, J.; Verdin, E.; Carrera-Bastos, P.; Targ, S.; Franceschi, C.; Ferrucci, L.; Gilroy, D.W.; Fasano, A.; Miller, G.W.; et al. Chronic Inflammation in the Etiology of Disease across the Life Span. *Nat. Med.* **2019**, *25*, 1822–1832. [CrossRef]
10. Duggett, N.A.; Griffiths, L.A.; McKenna, O.E.; de Santis, V.; Yongsanguanchai, N.; Mokori, E.B.; Flatters, S.J.L. Oxidative Stress in the Development, Maintenance and Resolution of Paclitaxel-Induced Painful Neuropathy. *Neuroscience* **2016**, *333*, 13–26. [CrossRef]
11. Kidd, J.F.; Pilkington, M.F.; Schell, M.J.; Fogarty, K.E.; Skepper, J.N.; Taylor, C.W.; Thorn, P. Paclitaxel Affects Cytosolic Calcium Signals by Opening the Mitochondrial Permeability Transition Pore. *J. Biol. Chem.* **2002**, *277*, 6504–6510. [CrossRef] [PubMed]

12. Carozzi, V.A.; Canta, A.; Chiorazzi, A. Chemotherapy-Induced Peripheral Neuropathy: What Do We Know about Mechanisms? *Neurosci. Lett.* **2015**, *596*, 90–107. [CrossRef] [PubMed]
13. Vendrell, I.; Macedo, D.; Alho, I.; Dionísio, M.R.; Costa, L. Treatment of Cancer Pain by Targeting Cytokines. *Mediat. Inflamm.* **2015**, *2015*, 984570. [CrossRef] [PubMed]
14. Bluthé, R.M.; Layé, S.; Michaud, B.; Combe, C.; Dantzer, R.; Parnet, P. Role of Interleukin-1beta and Tumour Necrosis Factor-Alpha in Lipopolysaccharide-Induced Sickness Behaviour: A Study with Interleukin-1 Type I Receptor-Deficient Mice. *Eur. J. Neurosci.* **2000**, *12*, 4447–4456. [CrossRef] [PubMed]
15. Bluthé, R. Role of IL-6 in Cytokine-Induced Sickness Behavior a Study with IL-6 Deficient Mice. *Physiol. Behav.* **2000**, *70*, 367–373. [CrossRef]
16. Szajnik, M.; Szczepanski, M.J.; Czystowska, M.; Elishaev, E.; Mandapathil, M.; Nowak-Markwitz, E.; Spaczynski, M.; Whiteside, T.L. TLR4 Signaling Induced by Lipopolysaccharide or Paclitaxel Regulates Tumor Survival and Chemoresistance in Ovarian Cancer. *Oncogene* **2009**, *28*, 4353–4363. [CrossRef]
17. Rajput, S.; Volk-Draper, L.D.; Ran, S. TLR4 is a Novel Determinant of the Response to Paclitaxel in Breast Cancer. *Mol. Cancer Ther.* **2013**, *12*, 1676–1687. [CrossRef]
18. Wang, A.-C.; Ma, Y.-B.; Wu, F.-X.; Ma, Z.-F.; Liu, N.-F.; Gao, R.; Gao, Y.-S.; Sheng, X.-G. TLR4 Induces Tumor Growth and Inhibits Paclitaxel Activity in MyD88-Positive Human Ovarian Carcinoma in Vitro. *Oncol. Lett.* **2014**, *7*, 871–877. [CrossRef]
19. Varrassi, G.; Alon, E.; Bagnasco, M.; Lanata, L.; Mayoral-Rojals, V.; Paladini, A.; Pergolizzi, J.V.; Perrot, S.; Scarpignato, C.; Tölle, T. Towards an Effective and Safe Treatment of Inflammatory Pain: A Delphi-Guided Expert Consensus. *Adv. Ther.* **2019**, *36*, 2618–2637. [CrossRef]
20. Ghazizadeh-Hashemi, M.; Ghajar, A.; Shalbafan, M.-R.; Ghazizadeh-Hashemi, F.; Afarideh, M.; Malekpour, F.; Ghaleiha, A.; Ardebili, M.E.; Akhondzadeh, S. Palmitoylethanolamide as Adjunctive Therapy in Major Depressive Disorder: A Double-Blind, Randomized and Placebo-Controlled Trial. *J. Affect. Disord.* **2018**, *232*, 127–133. [CrossRef]
21. Nau, R.; Ribes, S.; Djukic, M.; Eiffert, H. Strategies to Increase the Activity of Microglia as Efficient Protectors of the Brain against Infections. *Front. Cell. Neurosci.* **2014**, *8*, 138. [CrossRef] [PubMed]
22. Bronzuoli, M.R.; Facchinetti, R.; Steardo, L.; Romano, A.; Stecca, C.; Passarella, S.; Steardo, L.; Cassano, T.; Scuderi, C. Palmitoylethanolamide Dampens Reactive Astrogliosis and Improves Neuronal Trophic Support in a Triple Transgenic Model of Alzheimer's Disease: In Vitro and In Vivo Evidence. *Oxidative Med. Cell. Longev.* **2018**, *2018*, 4720532. [CrossRef] [PubMed]
23. Beggiato, S.; Tomasini, M.C.; Ferraro, L. Palmitoylethanolamide (PEA) as a Potential Therapeutic Agent in Alzheimer's Disease. *Front. Pharmacol.* **2019**, *10*, 821. [CrossRef] [PubMed]
24. Rankin, L.; Fowler, C.J. The Basal Pharmacology of Palmitoylethanolamide. *Int. J. Mol. Sci.* **2020**, *21*, 7942. [CrossRef] [PubMed]
25. Impellizzeri, D.; Bruschetta, G.; Cordaro, M.; Crupi, R.; Siracusa, R.; Esposito, E.; Cuzzocrea, S. Micronized/Ultramicronized Palmitoylethanolamide Displays Superior Oral Efficacy Compared to Nonmicronized Palmitoylethanolamide in a Rat Model of Inflammatory Pain. *J. Neuroinflammation* **2014**, *11*, 136. [CrossRef] [PubMed]
26. Costa, B.; Comelli, F.; Bettoni, I.; Colleoni, M.; Giagnoni, G. The Endogenous Fatty Acid Amide, Palmitoylethanolamide, Has Anti-Allodynic and Anti-Hyperalgesic Effects in a Murine Model of Neuropathic Pain: Involvement of CB1, TRPV1 and PPARγ Receptors and Neurotrophic Factors. *Pain* **2008**, *139*, 541–550. [CrossRef]
27. Bettoni, I.; Comelli, F.; Colombo, A.; Bonfanti, P.; Costa, B. Non-Neuronal Cell Modulation Relieves Neuropathic Pain: Efficacy of the Endogenous Lipid Palmitoylethanolamide. *CNS Neurol. Disord. -Drug Targets* **2013**, *12*, 34–44. [CrossRef]
28. D'Amico, R.; Impellizzeri, D.; Cuzzocrea, S.; Di Paola, R. ALIAmides Update: Palmitoylethanolamide and Its Formulations on Management of Peripheral Neuropathic Pain. *Int. J. Mol. Sci.* **2020**, *21*, 5330. [CrossRef]
29. Skaper, S.D.; Facci, L.; Fusco, M.; della Valle, M.F.; Zusso, M.; Costa, B.; Giusti, P. Palmitoylethanolamide, a Naturally Occurring Disease-Modifying Agent in Neuropathic Pain. *Inflammopharmacol* **2014**, *22*, 79–94. [CrossRef]
30. Peritore, A.F.; Siracusa, R.; Fusco, R.; Gugliandolo, E.; D'Amico, R.; Cordaro, M.; Crupi, R.; Genovese, T.; Impellizzeri, D.; Cuzzocrea, S.; et al. Ultramicronized Palmitoylethanolamide and Paracetamol, a New Association to Relieve Hyperalgesia and Pain in a Sciatic Nerve Injury Model in Rat. *Int. J. Mol. Sci.* **2020**, *21*, 3509. [CrossRef]
31. Impellizzeri, D.; Peritore, A.F.; Cordaro, M.; Gugliandolo, E.; Siracusa, R.; Crupi, R.; D'Amico, R.; Fusco, R.; Evangelista, M.; Cuzzocrea, S.; et al. The Neuroprotective Effects of Micronized PEA (PEA-m) Formulation on Diabetic Peripheral Neuropathy in Mice. *FASEB J.* **2019**, *33*, 11364–11380. [CrossRef] [PubMed]
32. Steels, E.; Venkatesh, R.; Steels, E.; Vitetta, G.; Vitetta, L. A Double-Blind Randomized Placebo Controlled Study Assessing Safety, Tolerability and Efficacy of Palmitoylethanolamide for Symptoms of Knee Osteoarthritis. *Inflammopharmacol* **2019**, *27*, 475–485. [CrossRef] [PubMed]
33. Marini, I.; Bartolucci, M.L.; Bortolotti, F.; Gatto, M.R.; Bonetti, G.A. Palmitoylethanolamide versus a Nonsteroidal Anti-Inflammatory Drug in the Treatment of Temporomandibular Joint Inflammatory Pain. *J. Orofac. Pain* **2012**, *26*, 99–104. [PubMed]
34. Keppel Hesselink, J.; Kopsky, D. Palmitoylethanolamide, a Neutraceutical, in Nerve Compression Syndromes: Efficacy and Safety in Sciatic Pain and Carpal Tunnel Syndrome. *J. Pain Res.* **2015**, *8*, 729–734. [CrossRef]
35. Schifilliti, C.; Cucinotta, L.; Fedele, V.; Ingegnosi, C.; Luca, S.; Leotta, C. Micronized Palmitoylethanolamide Reduces the Symptoms of Neuropathic Pain in Diabetic Patients. *Pain Res. Treat.* **2014**, *2014*, 849623. [CrossRef]
36. Bacci, C.; Cassetta, G.; Emanuele, B.; Berengo, M. Randomized Split-Mouth Study on Postoperative Effects of Palmitoylethanolamide for Impacted Lower Third Molar Surgery. *ISRN Surg.* **2011**, *2011*, 917350. [CrossRef]

37. Del Giorno, R.; Skaper, S.; Paladini, A.; Varrassi, G.; Coaccioli, S. Palmitoylethanolamide in Fibromyalgia: Results from Prospective and Retrospective Observational Studies. *Pain Ther.* **2015**, *4*, 169–178. [CrossRef]
38. Stochino-Loi, E.; Pontis, A.; Cofelice, V.; Pirarba, S.; Fais, M.F.; Daniilidis, A.; Melis, I.; Paoletti, A.M.; Angioni, S. Effect of Ultramicronized-Palmitoylethanolamide and Co-Micronized Palmitoylethanolamide/Polydatin on Chronic Pelvic Pain and Quality of Life in Endometriosis Patients: An Open-Label Pilot Study. *Int. J. Womens Health* **2019**, *11*, 443–449. [CrossRef]
39. Habib, A.M.; Okorokov, A.L.; Hill, M.N.; Bras, J.T.; Lee, M.-C.; Li, S.; Gossage, S.J.; van Drimmelen, M.; Morena, M.; Houlden, H.; et al. Microdeletion in a FAAH Pseudogene Identified in a Patient with High Anandamide Concentrations and Pain Insensitivity. *Br. J. Anaesth.* **2019**, *123*, e249–e253. [CrossRef]
40. Di Cesare Mannelli, L.; Pacini, A.; Corti, F.; Boccella, S.; Luongo, L.; Esposito, E.; Cuzzocrea, S.; Maione, S.; Calignano, A.; Ghelardini, C. Antineuropathic Profile of N-Palmitoylethanolamine in a Rat Model of Oxaliplatin-Induced Neurotoxicity. *PLoS ONE* **2015**, *10*, e0128080. [CrossRef]
41. Donvito, G.; Wilkerson, J.L.; Damaj, M.I.; Lichtman, A.H. Palmitoylethanolamide Reverses Paclitaxel-Induced Allodynia in Mice. *J. Pharmacol. Exp. Ther.* **2016**, *359*, 310–318. [CrossRef] [PubMed]
42. Caillaud, M.; Patel, N.H.; Toma, W.; White, A.; Thompson, D.; Mann, J.; Tran, T.H.; Roberts, J.L.; Poklis, J.L.; Bigbee, J.W.; et al. A Fenofibrate Diet Prevents Paclitaxel-Induced Peripheral Neuropathy in Mice. *Cancers* **2020**, *13*, 69. [CrossRef] [PubMed]
43. D'Agostino, G.; Russo, R.; Avagliano, C.; Cristiano, C.; Meli, R.; Calignano, A. Palmitoylethanolamide Protects Against the Amyloid-B25-35-Induced Learning and Memory Impairment in Mice, an Experimental Model of Alzheimer Disease. *Neuropsychopharmacol* **2012**, *37*, 1784–1792. [CrossRef]
44. Esposito, E.; Cuzzocrea, S. Palmitoylethanolamide in Homeostatic and Traumatic Central Nervous System Injuries. *CNS Neurol. Disord. -Drug Targets* **2013**, *12*, 55–61. [CrossRef]
45. Herrera, M.I.; Udovin, L.D.; Toro-Urrego, N.; Kusnier, C.F.; Luaces, J.P.; Capani, F. Palmitoylethanolamide Ameliorates Hippocampal Damage and Behavioral Dysfunction After Perinatal Asphyxia in the Immature Rat Brain. *Front. Neurosci.* **2018**, *12*, 145. [CrossRef]
46. Brotini, S.; Schievano, C.; Guidi, L. Ultra-Micronized Palmitoylethanolamide: An Efficacious Adjuvant Therapy for Parkinson's Disease. *CNS Neurol. Disord. -Drug Targets* **2017**, *16*, 1–9. [CrossRef] [PubMed]
47. Kruk-Slomka, M.; Dzik, A.; Budzynska, B.; Biala, G. Endocannabinoid System: The Direct and Indirect Involvement in the Memory and Learning Processes—A Short Review. *Mol. Neurobiol.* **2017**, *54*, 8332–8347. [CrossRef]
48. Crupi, R.; Paterniti, I.; Ahmad, A.; Campolo, M.; Esposito, E.; Cuzzocrea, S. Effects of Palmitoylethanolamide and Luteolin in an Animal Model of Anxiety/Depression. *CNS Neurol. Disord. -Drug Targets* **2013**, *12*, 989–1001. [CrossRef]
49. Karabatsiakis, A.; Hamuni, G.; Wilker, S.; Kolassa, S.; Renu, D.; Kadereit, S.; Schauer, M.; Hennessy, T.; Kolassa, I.-T. Metabolite Profiling in Posttraumatic Stress Disorder. *J. Mol. Psychiatry* **2015**, *3*, 2. [CrossRef]
50. Cristiano, C.; Pirozzi, C.; Coretti, L.; Cavaliere, G.; Lama, A.; Russo, R.; Lembo, F.; Mollica, M.P.; Meli, R.; Calignano, A.; et al. Palmitoylethanolamide Counteracts Autistic-like Behaviours in BTBR T+tf/J Mice: Contribution of Central and Peripheral Mechanisms. *Brain Behav. Immun.* **2018**, *74*, 166–175. [CrossRef]
51. Fusco, R.; Gugliandolo, E.; Campolo, M.; Evangelista, M.; Di Paola, R.; Cuzzocrea, S. Effect of a New Formulation of Micronized and Ultramicronized N-Palmitoylethanolamine in a Tibia Fracture Mouse Model of Complex Regional Pain Syndrome. *PLoS ONE* **2017**, *12*, e0178553. [CrossRef] [PubMed]
52. Toma, W.; Caillaud, M.; Patel, N.H.; Tran, T.H.; Donvito, G.; Roberts, J.; Bagdas, D.; Jackson, A.; Lichtman, A.; Gewirtz, D.A.; et al. N-acylethanolamine-hydrolysing Acid Amidase: A New Potential Target to Treat Paclitaxel-induced Neuropathy. *Eur. J. Pain* **2021**, *25*, 1367–1380. [CrossRef] [PubMed]
53. Warden, A.; Truitt, J.; Merriman, M.; Ponomareva, O.; Jameson, K.; Ferguson, L.B.; Mayfield, R.D.; Harris, R.A. Localization of PPAR Isotypes in the Adult Mouse and Human Brain. *Sci. Rep.* **2016**, *6*, 27618. [CrossRef] [PubMed]
54. Busquets-Garcia, A.; Bains, J.; Marsicano, G. CB1 Receptor Signaling in the Brain: Extracting Specificity from Ubiquity. *Neuropsychopharmacology* **2018**, *43*, 4–20. [CrossRef]
55. Roch, G.; Batallé, G.; Bai, X.; Pouso-Vázquez, E.; Rodríguez, L.; Pol, O. The Beneficial Effects of Heme Oxygenase 1 and Hydrogen Sulfide Activation in the Management of Neuropathic Pain, Anxiety- and Depressive-like Effects of Paclitaxel in Mice. *Antioxidants* **2022**, *11*, 122. [CrossRef]
56. Fornasari, D. Pain Mechanisms in Patients with Chronic Pain. *Clin. Drug Investig.* **2012**, *32*, 45–52. [CrossRef]
57. Clayton, P.; Subah, S.; Venkatesh, R.; Hill, M.; Bogoda, N. Palmitoylethanolamide: A Potential Alternative to Cannabidiol. *J. Diet. Suppl.* **2021**, 1–26. [CrossRef]
58. Domi, E.; Uhrig, S.; Soverchia, L.; Spanagel, R.; Hansson, A.C.; Barbier, E.; Heilig, M.; Ciccocioppo, R.; Ubaldi, M. Genetic Deletion of Neuronal PPARγ Enhances the Emotional Response to Acute Stress and Exacerbates Anxiety: An Effect Reversed by Rescue of Amygdala PPARγ Function. *J. Neurosci.* **2016**, *36*, 12611–12623. [CrossRef]
59. Locci, A.; Pinna, G. Stimulation of Peroxisome Proliferator-Activated Receptor-α by N-Palmitoylethanolamine Engages Allopregnanolone Biosynthesis to Modulate Emotional Behavior. *Biol. Psychiatry* **2019**, *85*, 1036–1045. [CrossRef]
60. Yu, H.-L.; Deng, X.-Q.; Li, Y.-J.; Li, Y.-C.; Quan, Z.-S.; Sun, X.-Y. Short Communication—N-Palmitoylethanolamide, an Endocannabinoid, Exhibits Antidepressant Effects in the Forced Swim Test and the Tail Suspension Test in Mice. *Pharmacol. Rep.* **2011**, *63*, 834–839. [CrossRef]

61. Guida, F.; Luongo, L.; Marmo, F.; Romano, R.; Iannotta, M.; Napolitano, F.; Belardo, C.; Marabese, I.; D'Aniello, A.; De Gregorio, D.; et al. Palmitoylethanolamide Reduces Pain-Related Behaviors and Restores Glutamatergic Synapses Homeostasis in the Medial Prefrontal Cortex of Neuropathic Mice. *Mol. Brain* **2015**, *8*, 47. [CrossRef] [PubMed]
62. Adamczyk, P.; Gołda, A.; McCreary, A.C.; Filip, M.; Przegaliński, E. Activation of Endocannabinoid Transmission Induces Antidepressant-like Effects in Rats. *J. Physiol. Pharmacol.* **2008**, *59*, 217–228. [PubMed]
63. Melis, M.; Carta, G.; Pistis, M.; Banni, S. Physiological Role of Peroxisome Proliferator-Activated Receptors Type Alpha on Dopamine Systems. *CNS Neurol. Disord. -Drug Targets* **2013**, *12*, 70–77. [CrossRef] [PubMed]
64. Lo Verme, J.; Fu, J.; Astarita, G.; La Rana, G.; Russo, R.; Calignano, A.; Piomelli, D. The Nuclear Receptor Peroxisome Proliferator-Activated Receptor-α Mediates the Anti-Inflammatory Actions of Palmitoylethanolamide. *Mol. Pharmacol.* **2005**, *67*, 15–19. [CrossRef]
65. Mattace Raso, G.; Russo, R.; Calignano, A.; Meli, R. Palmitoylethanolamide in CNS Health and Disease. *Pharmacol. Res.* **2014**, *86*, 32–41. [CrossRef]
66. Matrisciano, F.; Pinna, G. PPAR-α Hypermethylation in the Hippocampus of Mice Exposed to Social Isolation Stress Is Associated with Enhanced Neuroinflammation and Aggressive Behavior. *Int. J. Mol. Sci.* **2021**, *22*, 10678. [CrossRef]
67. Maione, S.; Costa, B.; Di Marzo, V. Endocannabinoids: A Unique Opportunity to Develop Multitarget Analgesics. *Pain* **2013**, *154*, S87–S93. [CrossRef]
68. Guida, F.; Luongo, L.; Boccella, S.; Giordano, M.E.; Romano, R.; Bellini, G.; Manzo, I.; Furiano, A.; Rizzo, A.; Imperatore, R.; et al. Palmitoylethanolamide Induces Microglia Changes Associated with Increased Migration and Phagocytic Activity: Involvement of the CB2 Receptor. *Sci. Rep.* **2017**, *7*, 375. [CrossRef]
69. Ambrosino, P.; Soldovieri, M.V.; De Maria, M.; Russo, C.; Taglialatela, M. Functional and Biochemical Interaction between PPARα Receptors and TRPV1 Channels: Potential Role in PPARα Agonists-Mediated Analgesia. *Pharmacol. Res.* **2014**, *87*, 113–122. [CrossRef]
70. Hillard, C.J. Stress Regulates Endocannabinoid-CB1 Receptor Signaling. *Semin. Immunol.* **2014**, *26*, 380–388. [CrossRef]
71. Pinna, G. Endocannabinoids and Precision Medicine for Mood Disorders and Suicide. *Front. Psychiatry* **2021**, *12*, 658433. [CrossRef] [PubMed]
72. Patel, S.; Hillard, C.J. Pharmacological Evaluation of Cannabinoid Receptor Ligands in a Mouse Model of Anxiety: Further Evidence for an Anxiolytic Role for Endogenous Cannabinoid Signaling. *J. Pharmacol. Exp. Ther.* **2006**, *318*, 304–311. [CrossRef] [PubMed]
73. Scherma, M.; Medalie, J.; Fratta, W.; Vadivel, S.K.; Makriyannis, A.; Piomelli, D.; Mikics, E.; Haller, J.; Yasar, S.; Tanda, G.; et al. The Endogenous Cannabinoid Anandamide Has Effects on Motivation and Anxiety That Are Revealed by Fatty Acid Amide Hydrolase (FAAH) Inhibition. *Neuropharmacology* **2008**, *54*, 129–140. [CrossRef]
74. Esmaeili, M.A.; Yadav, S.; Gupta, R.K.; Waggoner, G.R.; Deloach, A.; Calingasan, N.Y.; Beal, M.F.; Kiaei, M. Preferential PPAR-α Activation Reduces Neuroinflammation, and Blocks Neurodegeneration in Vivo. *Hum. Mol. Genet.* **2016**, *25*, 317–327. [CrossRef]
75. Rolland, B.; Deguil, J.; Jardri, R.; Cottencin, O.; Thomas, P.; Bordet, R. Therapeutic Prospects of PPARs in Psychiatric Disorders: A Comprehensive Review. *Curr. Drug Targets* **2013**, *14*, 724–732. [CrossRef] [PubMed]
76. Racke, M.K.; Drew, P.D. PPARs in Neuroinflammation. *PPAR Res.* **2008**, *2008*, 638356. [CrossRef]
77. O'Leary, A. Stress, Emotion, and Human Immune Function. *Psychol. Bull.* **1990**, *108*, 363–382. [CrossRef]
78. Scuderi, C.; Esposito, G.; Blasio, A.; Valenza, M.; Arietti, P.; Steardo Jr, L.; Carnuccio, R.; De Filippis, D.; Petrosino, S.; Iuvone, T.; et al. Palmitoylethanolamide Counteracts Reactive Astrogliosis Induced by β-Amyloid Peptide. *J. Cell. Mol. Med.* **2011**, *15*, 2664–2674. [CrossRef]
79. Bergandi, L.; Apprato, G.; Silvagno, F. Antioxidant and Anti-Inflammatory Activity of Combined Phycocyanin and Palmitoylethanolamide in Human Lung and Prostate Epithelial Cells. *Antioxidants* **2022**, *11*, 201. [CrossRef]
80. D'Aloia, A.; Molteni, L.; Gullo, F.; Bresciani, E.; Artusa, V.; Rizzi, L.; Ceriani, M.; Meanti, R.; Lecchi, M.; Coco, S.; et al. Palmitoylethanolamide Modulation of Microglia Activation: Characterization of Mechanisms of Action and Implication for Its Neuroprotective Effects. *Int. J. Mol. Sci.* **2021**, *22*, 3054. [CrossRef]
81. Antonucci, N.; Cirillo, A.; Siniscalco, D. Beneficial Effects of Palmitoylethanolamide on Expressive Language, Cognition, and Behaviors in Autism: A Report of Two Cases. *Case Rep. Psychiatry* **2015**, *2015*, 325061. [CrossRef]
82. Altamura, C.; Ventriglia, M.; Martini, M.G.; Montesano, D.; Errante, Y.; Piscitelli, F.; Scrascia, F.; Quattrocchi, C.; Palazzo, P.; Seccia, S.; et al. Elevation of Plasma 2-Arachidonoylglycerol Levels in Alzheimer's Disease Patients as a Potential Protective Mechanism against Neurodegenerative Decline. *JAD* **2015**, *46*, 497–506. [CrossRef] [PubMed]
83. Ji, R.-R.; Xu, Z.-Z.; Gao, Y.-J. Emerging Targets in Neuroinflammation-Driven Chronic Pain. *Nat. Rev. Drug Discov.* **2014**, *13*, 533–548. [CrossRef]
84. Xanthos, D.N.; Sandkühler, J. Neurogenic Neuroinflammation: Inflammatory CNS Reactions in Response to Neuronal Activity. *Nat. Rev. Neurosci.* **2014**, *15*, 43–53. [CrossRef] [PubMed]
85. Subhramanyam, C.S.; Wang, C.; Hu, Q.; Dheen, S.T. Microglia-Mediated Neuroinflammation in Neurodegenerative Diseases. *Semin. Cell Dev. Biol.* **2019**, *94*, 112–120. [CrossRef] [PubMed]
86. Landolfo, E.; Cutuli, D.; Petrosini, L.; Caltagirone, C. Effects of Palmitoylethanolamide on Neurodegenerative Diseases: A Review from Rodents to Humans. *Biomolecules* **2022**, *12*, 667. [CrossRef] [PubMed]

87. Gabrielsson, L.; Mattsson, S.; Fowler, C.J. Palmitoylethanolamide for the Treatment of Pain: Pharmacokinetics, Safety and Efficacy. *Br. J. Clin. Pharmacol.* **2016**, *82*, 932–942. [CrossRef]
88. Seol, T.-K.; Lee, W.; Park, S.; Kim, K.N.; Kim, T.Y.; Oh, Y.N.; Jun, J.H. Effect of Palmitoylethanolamide on Inflammatory and Neuropathic Pain in Rats. *Korean J. Anesthesiol.* **2017**, *70*, 561. [CrossRef]
89. Re, G.; Barbero, R.; Miolo, A.; Di Marzo, V. Palmitoylethanolamide, Endocannabinoids and Related Cannabimimetic Compounds in Protection against Tissue Inflammation and Pain: Potential Use in Companion Animals. *Vet. J.* **2007**, *173*, 21–30. [CrossRef]
90. Basbaum, A.I.; Bautista, D.M.; Scherrer, G.; Julius, D. Cellular and Molecular Mechanisms of Pain. *Cell* **2009**, *139*, 267–284. [CrossRef]
91. Block, M.L.; Zecca, L.; Hong, J.-S. Microglia-Mediated Neurotoxicity: Uncovering the Molecular Mechanisms. *Nat. Rev. Neurosci.* **2007**, *8*, 57–69. [CrossRef] [PubMed]
92. Ji, R.-R.; Chamessian, A.; Zhang, Y.-Q. Pain Regulation by Non-Neuronal Cells and Inflammation. *Science* **2016**, *354*, 572–577. [CrossRef] [PubMed]
93. Al-Mazidi, S.; Alotaibi, M.; Nedjadi, T.; Chaudhary, A.; Alzoghaibi, M.; Djouhri, L. Blocking of Cytokines Signalling Attenuates Evoked and Spontaneous Neuropathic Pain Behaviours in the Paclitaxel Rat Model of Chemotherapy-Induced Neuropathy. *Eur. J. Pain* **2018**, *22*, 810–821. [CrossRef] [PubMed]
94. D'Agostino, G.; La Rana, G.; Russo, R.; Sasso, O.; Iacono, A.; Esposito, E.; Mattace Raso, G.; Cuzzocrea, S.; LoVerme, J.; Piomelli, D.; et al. Central Administration of Palmitoylethanolamide Reduces Hyperalgesia in Mice via Inhibition of NF-KB Nuclear Signalling in Dorsal Root Ganglia. *Eur. J. Pharmacol.* **2009**, *613*, 54–59. [CrossRef]
95. Di Cesare Mannelli, L.; D'Agostino, G.; Pacini, A.; Russo, R.; Zanardelli, M.; Ghelardini, C.; Calignano, A. Palmitoylethanolamide Is a Disease-Modifying Agent in Peripheral Neuropathy: Pain Relief and Neuroprotection Share a PPAR-Alpha-Mediated Mechanism. *Mediat. Inflamm.* **2013**, *2013*, 328797. [CrossRef]
96. Khasabova, I.A.; Seybold, V.S.; Simone, D.A. The Role of PPARγ in Chemotherapy-Evoked Pain. *Neurosci. Lett.* **2021**, *753*, 135845. [CrossRef]
97. Roa-Coria, J.E.; Navarrete-Vázquez, G.; Fowler, C.J.; Flores-Murrieta, F.J.; Déciga-Campos, M.; Granados-Soto, V. N-(4-Methoxy-2-Nitrophenyl)Hexadecanamide, a Palmitoylethanolamide Analogue, Reduces Formalin-Induced Nociception. *Life Sci.* **2012**, *91*, 1288–1294. [CrossRef]

biomolecules

MDPI

Review

Targeting Neuroinflammation in Osteoarthritis with Intra-Articular Adelmidrol

Francesca Guida [1], Monica Rocco [2,3], Livio Luongo [1,4], Pietro Persiani [5], Maria Chiara Vulpiani [2,6], Sveva Maria Nusca [2,6], Sabatino Maione [1,4] and Flaminia Coluzzi [3,7,*]

1 Department of Experimental Medicine, University of Campania Luigi Vanvitelli, 80100 Naples, Italy
2 Department of Surgical and Medical Science and Translational Medicine, Sapienza University of Rome, 00189 Rome, Italy
3 Unit of Anesthesia, Intensive Care and Pain Medicine, Sant'Andrea University Hospital, 00189 Rome, Italy
4 I.R.C.C.S. Neuromed, 86077 Pozzilli, Italy
5 Department of Anatomical, Histological, Forensic Medicine and Orthopedic Science, Sapienza University of Rome, 00161 Rome, Italy
6 Physical Medicine and Rehabilitation Unit, Sant'Andrea University Hospital, 00189 Rome, Italy
7 Department of Medical and Surgical Sciences and Biotechnologies, Sapienza University of Rome, 04100 Latina, Italy
* Correspondence: flaminia.coluzzi@uniroma1.it

Abstract: Neuroinflammation is an emerging therapeutic target in chronic degenerative and autoimmune diseases, such as osteoarthritis (OA) and rheumatoid arthritis. Mast cells (MCs) play a key role in the homeostasis of joints and the activation of MCs induces the release of a huge number of mediators, which fuel the fire of neuroinflammation. Particularly, synovial MCs release substances which accelerate the degradation of the extra-cellular matrix causing morphological joint changes and cartilage damage and inducing the proliferation of synovial fibroblasts, angiogenesis, and the sprouting of sensory nerve fibers, which mediate chronic pain. Palmitoylethanolamide (PEA) is a well-known MCs modulator, but in osteoarthritic joints, its levels are significantly reduced. Adelmidrol, a synthetic derivate of azelaic acid belonging to the ALIAmides family, is a PEA enhancer. Preclinical and clinical investigations showed that the intra-articular administration of Adelmidrol significantly reduced MC infiltration, pro-inflammatory cytokine release, and cartilage degeneration. The combination of 1% high molecular weight hyaluronic acid and 2% Adelmidrol has been effectively used for knee osteoarthritis and, a significant improvement in analgesia and functionality has been recorded.

Keywords: palmitoylethanolamide; adelmidrol; hyaluronic acid; visco-induction; osteoarthritis; neuroinflammation; mast cells; joint degeneration and pain

Citation: Guida, F.; Rocco, M.; Luongo, L.; Persiani, P.; Vulpiani, M.C.; Nusca, S.M.; Maione, S.; Coluzzi, F. Targeting Neuroinflammation in Osteoarthritis with Intra-Articular Adelmidrol. *Biomolecules* **2022**, *12*, 1453. https://doi.org/10.3390/biom12101453

Academic Editor: Rosalia Crupi

Received: 27 July 2022
Accepted: 4 October 2022
Published: 11 October 2022

1. Introduction

Osteoarthritis (OA) is a leading cause of musculoskeletal chronic pain, affecting nearly 16% of the population, characterized by cartilage degeneration and stiffness. A mechanism-based approach to OA should include at least three targets: the peripheral mechanisms of inflammation, the central mechanisms of pain hypersensitivity, and the prevention of joint destruction [1]. In the last few years, mast cells (MCs) have been widely investigated as a possible target to be modulated for controlling peripheral and central nervous system inflammation [2,3].

Joints are a target of neuroinflammation in chronic degenerative and autoimmune diseases, such as OA and rheumatoid arthritis, and MC activation has been recognized as a prominent feature of the synovial tissue in patients with OA [4]. The aim of this review is to present preclinical and clinical evidence of a new intra-articular formulation containing Adelmidrol for targeting neuroinflammation in OA diseases.

2. Neuro-Immune Mechanisms Underlying OA

The pathogenesis of OA is correlated with different interrelated mechanisms including inflammation, neuroplasticity, cartilage disruption, and bone damage. Neuro-immune signaling may occur when innate immune cells produce algogenic factors that act on the pain pathway [5]. Indeed, during OA progression, the nociceptors innervating joints can be sensitized by locally generated mediators, such as inflammatory cytokines and chemokines, nerve growth factor (NGF) [6], and disease-associated molecular patterns (DAMPs) [7,8] (Figure 1).

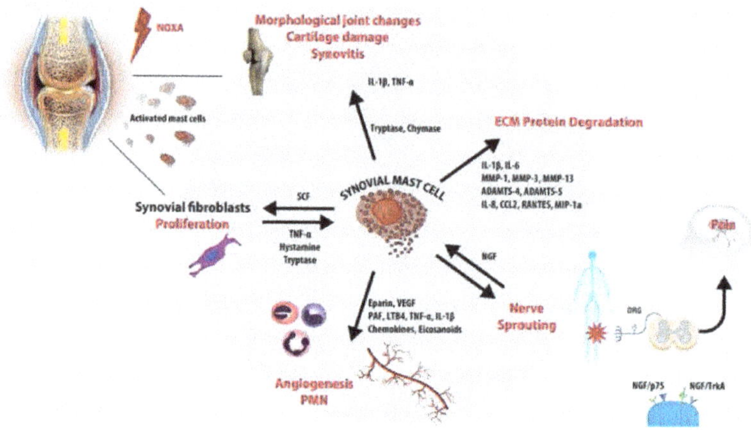

Figure 1. Schematic representation of the main contribution of MCs to degenerative joint diseases. Hyperactivated synovial MCs trigger the inflammatory process increasing vascular permeability and angiogenesis, recalling circulating PMN cells at the lesion site and activating synovial fibroblasts. Moreover, synovial MCs degranulation releases a massive amount of pro-inflammatory cytokines and chemokines, MMPs, and aggrecanases responsible for ECM protein degradation and, cartilage and bone remodeling.

The pathogenesis of OA is largely determined by the disrupted balance of pro- and anti-inflammatory mediators leading to a chronic low-grade inflammation which contributes to the associated pain condition.

The enhancement of proinflammatory cytokines causes the secretion of enzymes and other inflammatory factors involved in the pathogenesis of OA responsible for morphological joint changes, such as cartilage degeneration, osteophyte formation, and synovitis. One of the most important inflammatory mediators involved in this process is the interleukin (IL)-1β, a member of the IL-1 superfamily implicated in the pathogenesis of numerous neuroinflammatory diseases [9].

In OA, through the activation of the IL-1 receptor I (IL-1RI), IL-1β mediates cartilage destruction by inhibiting new formations of proteoglycans and degrading existing proteoglycans by stimulating degradative enzyme production. Widely distributed in different cells in the knee joint, IL-1RI has been shown to be upregulated in isolated human OA chondrocytes in vitro [10,11]. In OA synovium, a relative deficit in the production of natural antagonists of the IL-1 receptor (IL-1Ra) has been detected and it could possibly be related to the additional production of nitric oxide in OA tissues [12]. IL-1β activates several signaling pathways which overall contribute to the progression of OA. In particular, through mitogen-activated protein kinase (MAPK) signaling, IL-1β induces catabolic events including the cartilage extracellular matrix (ECM) degradation mediated by metalloproteinases (MMPs), such as MMP-1, MMP-3, and MMP-13, and aggrecanases such as ADAMTS-4 and ADAMTS-5. Additionally, the IL-1β-mediated NF-κB pathway activation leads to the synthesis and secretion of proinflammatory cytokines [IL-6 and tumor necrosis factor (TNF)-α], and chemokines [IL-8, monocyte chemoattractant protein-1 (MCP-1 or CCL2), CCL5 (RANTES), and macrophage inflammatory protein-1a (MIP-1a)] [13]. In OA, higher concentrations of IL-6 in synovial fluid have been correlated with the pain experience of OA patients [14] and in an experimental study, de Hooge et al. [15] showed that IL-6 gene knock out mice developed more advanced OA, compared to control animals.

High concentrations of TNF-α have been observed in structures such as synovial fluid, synovial membrane, cartilage, and subchondral bone layer [16]. As with IL-1β, TNF-α, via its conjugate receptors (TNFR-1 and TNFR-2), promotes the activation of the JNK kinase and the transcription factor NF-κB [16], by leading to the degeneration of cartilage and morphological joint changes.

Chemokines contribute to the inflammatory processes, stimulating the chemotaxis of inflammatory cells that, in turn, secrete proinflammatory cytokines, by representing thus the major challenge in treating and slowing the progression of osteoarthritis [11]. Increased levels of CCL2 have been detected in the synovial fluid of patients with knee OA [17]. Borzi et al. [18] demonstrated that CCL2 induces MMP-3 expression enhancement, contributing to proteoglycan loss and the degradation of cartilaginous tissue. In an experimental study, Miller et al., [19] showed that both CCL2 and CCR2 were upregulated in the innervating dorsal root ganglia (DRG) of the knee 8 weeks after the surgical model of OA while the absence of CCL2 or CCR2 suppresses selective inflammatory genes in joints 6 h post-de-stabilization of the medial meniscus (DMM) model, including arginase 1, prostaglandin synthase 2, nitric oxide synthase 2, IL-6, MMP-3, and the tissue inhibitor of metalloproteinase (TIMP)-1, by delaying pain development. CCL3 and CCL5 levels have been shown to be elevated in OA synovial fluid when compared with controls [17,20]. IL-1β-incubated human chondrocytes showed a significant upregulation of CCL3, CCL4, and CCL5 [21].

Chemokines from the CXC family also participate in the pathogenesis of OA. CXCL8 (IL-8) has been shown to be upregulated in the serum and in the synovial fluid of OA patients [22], and the study conducted by Guo et al. [23] reported that CXCL12 levels in the synovial fluid were closely related to the radiographic severity of OA. Interestingly, experimental evidence suggests a role for fractalkine (CX3CL1) and its receptor (CX3CR1) in the development of chronic pain in OA. Indeed, after DMM surgery, fractalkine release by DRG neurons is upregulated in the late phase of the model and this timing correlates with the development of microgliosis in the dorsal horn, where microglial cells express CX3CR1 [24]. The involvement of fractalkine and its receptor CX3CR1 in several preclinical models of chronic and neuropathic pain has been previously described [25,26]. In particular, this pathway has been highlighted as one of the most important cross-talk pathways involved in neuron-microglia communication [27].

As a member of the neurotrophin family, NGF is essential for the development of nociceptive primary neurons. NGF can bind the general neurotrophin receptor p75, as well its high-affinity cognate receptor, tropomyosin-related kinase (Trk)-A [28], which are expressed by joint cells, including chondrocytes, and increased in OA cartilage [29]. Triggered by a proinflammatory environment, the NGF synthesis is highly correlated with the degree of OA cartilage degradation in humans [29]. The NGF/TrkA signaling activation induces the release of a variety of inflammatory mediators such as serotonin, histamine, and NGF itself, which are known to cause sensitization of nociceptors. The injection of an anti-NGF monoclonal antibody shortly after model induction (day 2) was able to reverse the deficits in burrowing one day later compared to saline [30]. Moreover, the effects of a long-term NGF blockade in rat meniscal surgery models have been reported [31]. The authors showed that therapeutic treatment (day 14–day 21) reversed weight-bearing asymmetry and mechanical allodynia of the hind paw, while no short-term effects on histologic cartilage degeneration were observed. Similar findings were reported using Tanuzemab in a 28-day rat medial meniscal tear (MMT) model [32], where the authors described a significant increase in tibial cartilage degeneration, subchondral bone sclerosis, and tibial osteophytes induced by the treatment.

Damage-associated molecular patterns (DAMPs) are a group of molecules released from ECM to the joint cavity during cartilage degradation. Together with so-called pathogen-associated molecular patterns (PAMPs), DAMPs signal to the immune system a "dangerous state" requiring a protective response to a pathological state. Indeed, joint trauma results in the production of DAMPs and intracellular alarmins that signal through pattern recognition receptors (PRRs), which include Toll-like receptors (TLRs) on synovial macrophages, fibroblast-like synoviocytes (FLS), or chondrocytes to induce the local production of inflammatory mediators of various factors, like catabolic factors [matrix metalloproteinase, cytokines, chemokines cathepsins (B, K, and L)], and the complement cascade [7,33].

High levels of NGF, released by MCs, influence sensory nerve activity and its interaction with the high-affinity TrkA receptor (NGF/TrkA), and amplify inflammatory processes contributing to the development of a nerve sprouting of sensory pain fibers which mediate chronic pain maintenance.

3. Role of Mast Cells in OA

The important role of MCs in allergic reactions is well-known. Moreover, the involvement of MCs in physiological bone turnover and bone disorders has been described in detail [34,35]. MCs usually reside quietly within tissues throughout the body, whereas when they become activated, they secrete granules containing histamine and other inflammatory substances. The hyperactivation of MCs is often accompanied by pain syndromes [36]. Indeed, pain-like behaviors have been found to be MC-associated in several experimental models of pain [37].

In healthy joints, synovial tissue macrophages (STM) or MCs represent nearly 3% of the cellular population of the synovia and act as sentinels, monitoring possible pathogens and mediating the immune response [38]. In chronically inflamed joints, the synovial membrane becomes hypertrophic and the MC density significantly increases, as a result of the recruitment of circulating progenitors, the maturation of local MC precursors, the differentiation of local mesenchymal stem cells (MSCs) into MCs, and the release of neurotrophins as NGF that stimulate further MC proliferation. Farinelli et al. have recently shown, in hip and knee tissue from OA patients undergoing arthroplasty, a significant increase in the MCs and the vessel number, which significantly correlated with the synovitis score and disease severity [39]. The synovial MC infiltration has been associated with a variety of inflammatory changes which may play a role in the pathophysiology of OA [40,41]. The increased vascular permeability mediated by the histamine and prostaglandin D2 released by MCs, the recruitment of other circulating polymorphonuclear cells (PMN), such as leukocytes, and macrophage activation induced by cytokines and chemokines, contribute to the early phases of inflammatory arthritis [42]. Additionally, MCs may participate in joint destruction by the induction of MMPs from fibroblast-like synoviocytes, chondrocytes, and other cell types in the arthritic cartilage [42,43]. MCs have been detected even in osteophytes, where they play a role in accentuating the inflammatory pathology of OA [44]. Similarly, in the rheumatoid-arthritis (RA) joint, activated MCs release pro-inflammatory and immunomodulatory mediators (TNF-A, IL-6, IL-8, VEGF, histamine, heparin, tryptase) that induce angiogenesis, promote cartilage erosions, synoviocyte proliferation, and pain sensitization. Synovial MC concentration has been shown to correlate with the different stages of RA, particularly with the maintenance or lack thereof of the remission phase, suggesting a possible role as a biomarker to predict the severity and progression of the disease [45]. Cytokines and growth factors released during inflammation are the main mediators of the cross-talk between the subchondral bone, cartilage, and synovia. In vitro studies reveal that MCs synthesize and release NGF in response to antigen/IgE stimulation and it is involved in MCs maturation and degranulation [46]. High levels of NGF, released by MCs, induce the sprouting of sensory pain fibers and contribute to peripheral and central sensitization, which are the cornerstone of chronic pain maintenance [47]. The injection of complete Freund's adjuvant (CFA) induces inflammation of the joint and sprouting of CGRP+ nerve fibers in the synovium. This sprouting is significantly attenuated by anti-NGF therapy [48]. NGF binds TrkA receptors on inflammatory cells inducing the release of inflammatory mediators, such as serotonin, histamine, and even further NGF [49]. A more recent study suggested that NGF released from MCs may be implicated in osteoarthritis pain [50]. The authors have indicated the NGF responsible for MCs-PGD2-mediated increased nociceptive signaling, suggesting the TrkA blockade in MCs as a potential target for OA pain. Finally, MCs seem to play a role in the metabolism of hyaluronic acid (HA), which is the main component of the ECM and play a key role in joint lubrification during movement. MCs influence HA synthesis and degradation through two mechanisms of action. Firstly, MCs release tryptases, which may accelerate HA destruction. Secondly, granules of synovial MCs contain heparin, which represents the preliminary stage in the synthesis of HA. In vitro studies showed that MC chymase and tryptase accelerated the degradation of HA in the synovial fluid [51]. Moreover, MCs express the hyaluronic-acid-binding isoform of CD44 [52]. These observations could explain why the modulation of MCs could be a key factor for reducing intra-articular inflammation and ensuring the physiological metabolism of HA, filling the unmet needs of the currently available intra-articular therapies.

4. Intra-Articular Adelmidrol: A New Therapeutic Option for OA

Intra-articular treatments in OA management are warranted as their effects are limited to damaged joints, without significant systemic adverse events. The most commonly used intra-articular treatments are high-molecular-weight HA, corticosteroids, leukocyte-poor platelet-rich plasma (LP-PRP), leukocyte-rich platelet-rich plasma (LR-PRP), bone marrow mesenchymal stem cells (BM-MSCs), and adipose mesenchymal stem cells (AD-MSCs) [53]. Intra-articular steroids are widely used for their anti-inflammatory effect. However, they have a mild to moderate effect on pain severity [54] and their chronic use has been associated with a number of clinical concerns, the most relevant being the alteration of gene expression and immunomodulatory effects, which are associated with additional joint damage [55]. Furth more, corticosteroids may inhibit the anabolic activity of chondrocytes, decrease collagen expression, and, when chronically used, be chondrotoxic [56].

Recent evidence showed that in symptomatic patients with knee OA, a combination of PRP and HA induces a significantly greater improvement of pain and function compared with HA injections only [57]. PRP infiltration in OA allows for the modulation of inflammation and the induction of cartilage regeneration, while HA is a pure visco-supplementation therapy, and is thus prone to degradation by proinflammatory cytokines, free radicals, and proteinases [58]. Although commonly used, this treatment presents extreme variability during the preparations, often influenced by donor-specific factors or by processing methods. As conventional treatments provide limited therapeutic benefits, novel strategies for managing OA are required.

Adelmidrol has been recently investigated as a possible innovative intra-articular therapy for OA. Adelmidrol is a synthetic derivate of azelaic acid which works as a palmitoylethanolamide (PEA) enhancer increasing endogenous PEA levels [59]. Recently, it has been demonstrated that Adelmidrol is able to significantly increase the endogenous PEA level in the duodenum and colon of healthy mice in a dose/time-dependent and PPAR-γ-independent manner [60]. As with PEA, Adelmidrol belongs to the family of Autacoid Local Injury Antagonist Amides (ALIAmides) and its amphiphilic and amphipathic characteristics make Adelmidrol particularly soluble and suitable for topical [61,62] and intra-articular administration. In healthy synovia, an elevated amount of PEA is present, while its levels are significantly reduced in the synovial fluid of patients with OA and rheumatoid arthritis [63]. In OA joints, the reduction of endogenous PEA levels is responsible for MC hyperactivation, which, by triggering a cascade of neuroinflammatory events, alters the physiological functionality of nerve endings, chondrocytes, and joint synvioblasts. Hypertrophic synvioblasts become unable to produce basic components of cartilage, leading to the progression of joint damage [42,64,65]. Therefore, the restoration of normal endogenous PEA levels, necessary to ensure MC normo-reactivity, becomes essential for the maintenance of normal joint function. These findings suggest that Adelmidrol, due to its properties, could represent a valid therapeutic approach for all diseases characterized by a significant reduction in endogenous PEA levels.

5. Adelmidrol in OA: Preclinical Evidence

The anti-inflammatory and immunomodulatory properties of both PEA and Adelmidrol have been shown in numerous experimental studies [66,67]. Particularly, Adelmidrol has been shown to reduce acute inflammation in the carrageenan (CAR) model, when systemically injected [68]. In the same study, the beneficial effect of Adelmidrol was also proved in collagen-induced arthritis (CIA), an animal model of rheumatoid arthritis. Adelmidrol was shown to reduce clinical signs of OA, such as periarticular erythema and paw edema. X-ray examinations have shown that Adelmidrol reduced bone erosions in the femoral growth plate, as well as in the tibiotarsal joints with a consequent reduction of pain and proinflammatory cytokines (TNF-α, IL-6, and IL-1β). Interestingly, the significant presence of MCs, with an increased chymase and tryptase expression in the joint tissues after CIA induction, was also reduced [68]. Preclinical observations of the combination of Adelmidrol and HA are available in a rat model injected with monosodium-iodoacetate (MIA) to induce OA. Di Paola et al. demonstrated that the combination of Adelmidrol with HA in a dose/time-dependent manner reduced the plasma levels of pro-inflammatory cytokines (TNF-a and IL-1b), MMPs, and NGF production in the osteoarthritic knee of the rats. Moreover, the combination significantly reduced MC infiltration and histological alterations induced by intra-articular MIA injection [69]. Among animal models developed for the examination of OA pathophysiology [70], MIA represents the most used model to study chronic pain associated with OA [71]. Along with joint damage, MIA injections induce mechanical sensitivity in the ipsilateral hind paw and weight-bearing deficits. Di Paola et al. observed that mechanical allodynia and motor functioning along with the degeneration of articular cartilage were reduced by the combination of HA and Adelmidrol [69].

6. Intra-Articular Adelmidrol: Clinical Evidence

A new intra-articular medical device containing Adelmidrol has been recently introduced onto the market (Hyadrol®, 2 mL pre-loaded syringe for intra-articular injection, Epitech Group SpA, Milano, Italy). In this product, 2% Adelmidrol is administered together with 1% high-molecular-weight (from 1.3 to 2.0 × 106 dalton) HA, which physiologically binds the CD44+ receptor [72]. Intra-articular Adelmidrol is supposed to be a valid therapeutic option for normalizing MC function and ensuring the visco-induction of HA, through a dual mechanism of action. Firstly, Adelmidrol, by increasing PEA levels in the OA joint, might switch MCs from the hyperactivated to the physiological status; thus, it may reduce the HA degradation, by inhibiting the release by MCs of lytic enzymes, such as tryptases, hyaluronidases, and β-hexosaminidases. Secondly, Adelmidrol, by normalizing the MCs' release of heparin, which is a precursor of HA, induces HA production, leading to a visco-inductive effect in the OA joint (Figure 2).

A multi-center clinical trial has been published on the use of this combination of Adelmidrol and HA for intra-articular injection in patients suffering from Kellgren and Lawrence grade II-III osteoarthritis of the knee [73]. Eligible patients were $\geq$40-year-old subjects suffering from pain for more than 6 months, with an average pain score > 20 on the Western Ontario and McMaster Universities (WOMAC) Osteoarthritis Index pain sub-scale, and must have performed a washout of at least 2 weeks from other analgesics and/or non-steroidal anti-inflammatory drugs (NSAIDs). The principal exclusion criteria were: (a) the presence of concomitant severe systemic inflammatory joint diseases; (b) the use of corticosteroids in the previous 3 months; (c) the use of chondroprotective drugs, visco-supplementation injections, or arthroscopies within 6 months. The Italian WOMAC version was used in its VAS format: all 24 items were rated by the subject on a 10 cm VAS ranging from 0 (indicating no pain, stiffness, or functional limitation) to 10 (indicating extreme pain, stiffness, or functional limitation). Ranges of the WOMAC scores were: pain (0–50); stiffness (0–20), and functional limitation (0–170) [74]. The enrolled patients received 4 weekly intra-articular injections of 1% high-molecular-weight hyaluronic acid and 2% Adelmidrol in a 2 mL pre-loaded syringe. Patients were assessed using the WOMAC Osteoarthritis Index, the 12-Item Short Form Health Survey (SF-12) for the quality of life, and the Likert Patient Global Impression of Change (PGIC). In total, 102 patients (female n 60, 58.8%), with a mean age of 63.2 ± 0.6 years were recruited in this study, and follow-ups were conducted for 4 weeks after the last intra-articular injection. Only 7 patients withdrew prematurely from the trial: 3 for accidental trauma not related to the treatment, 3 due to knee pain and/or swelling after infiltration, and 1 was lost at the follow-up stage.

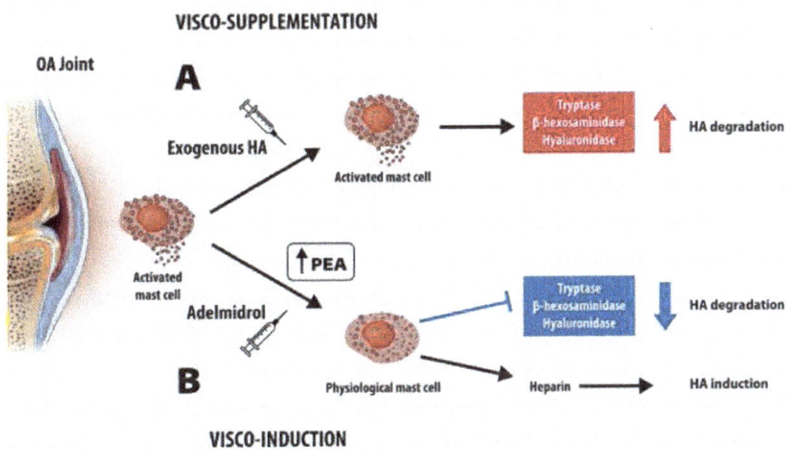

Figure 2. Visco-supplementation vs. Visco-induction. (**A**) Visco-supplementation: Exogenous HA is injected in the OA joints where activated MCs release tryptases, hyaluronidases, and β-hexosaminidases which in turn may accelerate HA degradation; (**B**) Visco-induction: Adelmidrol, by increasing endogenous PEA levels, downregulates MCs activation and shift them towards a physiological condition. Adelmidrol reduces the MCs degranulation, the release of lytic enzymes, and therefore the HA degradation, normalizing the physiological release of heparin, a HA precursor.

The 95 patients included in the final evaluation 4 weeks after treatment showed a significant improvement in all 3 elements of the WOMAC subscale (pain, stiffness, and functional limitation) (Table 1). Furthermore, a significant improvement in the quality of life was recorded in both the mental and physical components of the SF-12 (Figure 3).

Table 1. Changes in the WOMAC Osteoarthritis Index. Adapted from [73].

	T0	T1	T2	T3	T4	T5
	Basal (before 1st infiltration)	1st w	2nd ws	3rd ws	1st w after 4th infiltration	4th w after 4th infiltration
N patients	102	102	101	98	97	95
Pain	22.5 ± 0.76	16.8 ± 0.87	12.2 ± 0.81	9.4 ± 0.82	7.0 ± 0.76	5.2 ± 0.63
Stiffness	7.8 ± 0.49	6.3 ± 0.48	4.7 ± 0.43	3.6 ± 0.39	2.5 ± 0.33	1.9 ± 0.29
Articular function	67.3 ± 3.24	58.4 ± 3.10	46.3 ± 2.84	36.0 ± 2.83	28.1 ± 2.66	21.4 ± 2.13
Total WOMAC score	97.6 ± 4.11	81.5 ± 4.08	63.3 ± 3.78	49.1 ± 3.77	37.6 ± 3.53	28.5 ± 2.85

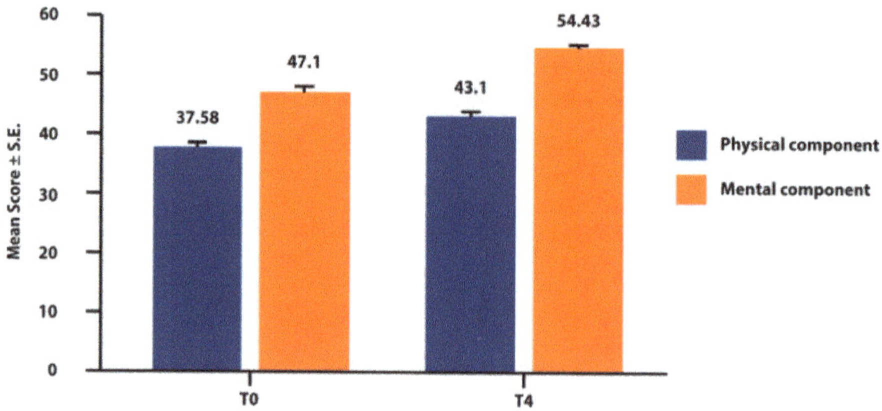

Figure 3. Changes in the SF12. Adapted from [73].

Over 90% of the treated subjects reported a clinical improvement according to the PGIC subjective evaluation, among which 76.4% felt "very much" or "much" improved (Figure 4). Intra-articular injections of 1% high-molecular-weight hyaluronic acid and 2% Adelmidrol were well tolerated. No severe adverse events were observed. Seven patients (7.4%) reported local pain at the site of injection, accompanied by swelling in 2 patients (2.1%), regarded as related to the infiltrative procedure. Other adverse events, such as diarrhea, headache, abdominal cramps, and trauma were observed in 10 patients judged by the investigators as not related to the treatment [73].

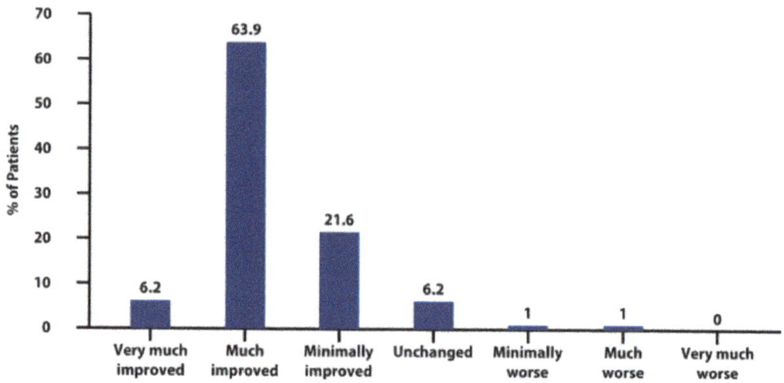

Figure 4. Patient Global Impression of Change. Adapted from [69].

Vulpiani et al. recently published the follow-up of the described clinical trial, limited to a group of patients enrolled in a single hospital (Sant'Andrea University Hospital, Rome, Italy), available for re-evaluation at 6 months, 1 year, and 2 years following the last intra-articular infiltration [75]. Forty-seven patients (77% of those enrolled in the first trial in a single center) were evaluated using both the WOMAC scale and the SF-12 questionnaire. The results obtained after 4 infiltrations, as measured with the WOMAC score, have been maintained for the whole follow-up period, including the effects on functional limitation. Similarly, no statistically significant modifications of both the physical and mental components of SF-12 have been observed in the 2-year follow-up period. These results suggest a long-term effect of Adelmidrol infiltration, which is not common for other intra-articular therapies.

7. Conclusions

OA is a complex degenerative disease, which requires a multimodal approach [1]. Systemic analgesia plays a role in targeting recurrent episodes of acute inflammation, which, by definition, lasts only a few days. NSAIDs may be useful for mitigating the dangerous effects of the destructive enzymes, cytokines, and prostaglandins released in the inflamed synovial membrane, after leukocyte infiltration [76,77]. Central analgesics, such as acetaminophen, opioids, and adjuvants, may be used for targeting maladaptive neuronal plasticity [78,79] and central mechanisms of pain [80]. Intra-articular therapies may support systemic analgesia by affecting the progression of joint destruction. In the real-world experience, however, there are still many unmet needs in the currently available intra-articular treatments for OA, in terms of analgesic efficacy, duration, tolerability, anti-inflammatory effect, and the modulation of cartilage degeneration.

Adelmidrol represents an innovative intra-articular treatment, which targets neuroinflammation and provides "visco-induction" to OA joints. It is time to "think outside the box": we can prevent endogenous HA degradation, rather than merely provide HA through exogenous administration. Traditional exogenous HA is intra-articularly administered in order to provide "visco-supplementation" and to improve joint lubrification. However, when considering the inflammatory environment where the HA is injected, it is not surprising that its efficacy can be invalidated by the huge number of substances locally released, including MC-released tryptases and chymase, hyaluronidases, and β-hexosaminidases. Furthermore, MCs have played a key role in joint homeostasis, and their degranulation induces neuroinflammation.

MC-released pro-inflammatory cytokines fuel the fire of inflammation, promoting cartilage damage, angiogenesis, nerve sprouting, and pain sensitization.

PEA is a well-known substance that physiologically down-regulates MC activation and in OA joints, its endogenous levels are significantly reduced. Therefore, one of the targets of OA treatment, alongside the prevention of the progression of cartilage damage, should be to extinguish the fire of joint neuroinflammation.

Adelmidrol, by increasing and normalizing PEA levels in the joints, could be the first innovative molecule for "visco-induction", through two mechanisms of action: (a) by reducing degranulation of MCs and therefore improving the efficacy of HA; (b) by normalizing MC function and restoring the physiological production of heparin, a HA precursor.

Preclinical and clinical data available on Adelmidrol administered together with HA support these hypotheses, by showing a significant reduction in cartilage degeneration, MC infiltration, and pro-inflammatory cytokine and chemokine plasma levels, and by improving analgesia and the functionality of OA patients.

Unfortunately, there are still many limitations to this emerging evidence. Firstly, further studies are needed to address the exact molecular mechanisms underlying the effects of Adelmidrol in the inflamed joint, despite the expected results demonstrated both in animal and clinical studies. Secondly, nowadays, clinical experience is still limited to the reported multicenter clinical trial by Vulpiani et al. [73] and the subsequent follow-up study [75], which confirmed the long-term results of this treatment. Therefore, further studies are warranted to confirm these interesting preliminary results.

In conclusion, in the context of a multimodal approach to OA, where systemic analgesia plays a role in targeting recurrent acute inflammation and central neuroplasticity leading to chronic pain [1], intra-articular treatment with Adelmidrol can be considered a promising treatment for preventing the progression of joint degeneration and neuro-inflammatory processes.

Author Contributions: All authors (F.G., M.R., L.L., P.P., M.C.V., S.M.N., S.M. and F.C.) designed the review, critically reviewed the contents, and contributed to manuscript preparation. All authors have read and agreed to the published version of the manuscript.

Funding: Fee for open access publication was unconditioned supported by Epitech Group S.pA. Milano (Italy).

Institutional Review Board Statement: Not applicable.

Informed Consent Statement: Not applicable.

Data Availability Statement: Not applicable.

Acknowledgments: The authors gratefully thank EPITECH Group SpA for the information support and Chiara Gomiero for the design and the realization of Figures 1 and 2.

Conflicts of Interest: FC served as a consultant for EPITECH Group SpA.

References

1. Cohen, E.; Lee, Y.C. A Mechanism-Based Approach to the Management of Osteoarthritis Pain. *Curr. Osteoporos. Rep.* **2015**, *13*, 399–406. [CrossRef] [PubMed]
2. Mittal, A.; Sagi, V.; Gupta, M.; Gupta, K. Mast Cell Neural Interactions in Health and Disease. *Front. Cell. Neurosci.* **2019**, *13*, 110. [CrossRef]
3. Miller, R.J.; Malfait, A.-M.; Miller, R.E. The innate immune response as a mediator of osteoarthritis pain. *Osteoarthr. Cartil.* **2020**, *28*, 562–571. [CrossRef] [PubMed]
4. Buckley, M.G.; Gallagher, P.J.; Walls, A.F. Mast cell subpopulations in the synovial tissue of patients with osteoarthritis: Selective increase in numbers of tryptase-positive, chymase-negative mast cells. *J. Pathol.* **1998**, *186*, 67–74. [CrossRef]
5. Yu, H.; Huang, T.; Lu, W.W.; Tong, L.; Chen, D. Osteoarthritis Pain. *Int. J. Mol. Sci.* **2022**, *23*, 4642. [CrossRef] [PubMed]
6. Seidel, M.F.; Hügle, T.; Morlion, B.; Koltzenburg, M.; Chapman, V.; MaassenVanDenBrink, A.; Lane, N.E.; Perrot, S.; Zieglgänsberger, W. Neurogenic inflammation as a novel treatment target for chronic pain syndromes. *Exp. Neurol.* **2022**, *356*, 114108. [CrossRef]
7. Lambert, C.; Zappia, J.; Sanchez, C.; Florin, A.; Dubuc, J.-E.; Henrotin, Y. The Damage-Associated Molecular Patterns (DAMPs) as Potential Targets to Treat Osteoarthritis: Perspectives from a Review of the Literature. *Front. Med.* **2020**, *7*, 607186. [CrossRef]
8. Terkawi, M.A.; Ebata, T.; Yokota, S.; Takahashi, D.; Endo, T.; Matsumae, G.; Shimizu, T.; Kadoya, K.; Iwasaki, N. Low-Grade Inflammation in the Pathogenesis of Osteoarthritis: Cellular and Molecular Mechanisms and Strategies for Future Therapeutic Intervention. *Biomedicines* **2022**, *10*, 1109. [CrossRef] [PubMed]
9. Dinarello, C.A. Overview of the IL-1 family in innate inflammation and acquired immunity. *Immunol. Rev.* **2018**, *281*, 8–27. [CrossRef]
10. Martel-Pelletier, J.; McCollum, R.; DiBattista, J.; Faure, M.P.; Chin, J.A.; Fournier, S.; Sarfati, M.; Pelletier, J.P. The interleukin-1 receptor in normal and osteoarthritic human articular chondrocytes. Identification as the type I receptor and analysis of binding kinetics and biologic function. *Arthritis Rheum.* **1992**, *35*, 530–540. [CrossRef] [PubMed]
11. Molnar, V.; Matišić, V.; Kodvanj, I.; Bjelica, R.; Jeleč, Ž.; Hudetz, D.; Rod, E.; Čukelj, F.; Vrdoljak, T.; Vidović, D.; et al. Cytokines and Chemokines Involved in Osteoarthritis Pathogenesis. *Int. J. Mol. Sci.* **2021**, *22*, 9208. [CrossRef] [PubMed]
12. Fernandes, J.C.; Martel-Pelletier, J.; Pelletier, J.-P. The role of cytokines in osteoarthritis pathophysiology. *Biorheology* **2002**, *39*, 237–246. [PubMed]
13. Chevalier, X.; Eymard, F.; Richette, P. Biologic agents in osteoarthritis: Hopes and disappointments. *Nat. Rev. Rheumatol.* **2013**, *9*, 400–410. [CrossRef]
14. Li, L.; Li, Z.; Li, Y.; Hu, X.; Zhang, Y.; Fan, P. Profiling of inflammatory mediators in the synovial fluid related to pain in knee osteoarthritis. *BMC Musculoskelet. Disord.* **2020**, *21*, 99. [CrossRef] [PubMed]
15. De Hooge, A.S.K.; Van de Loo, F.A.J.; Bennink, M.B.; Arntz, O.J.; De Hooge, P.; Van den Berg, W.B. Male IL-6 gene knock out mice developed more advanced osteoarthritis upon aging. *Osteoarthr. Cartil.* **2005**, *13*, 66–73. [CrossRef] [PubMed]
16. Wojdasiewicz, P.; Poniatowski, L.A.; Szukiewicz, D. The role of inflammatory and anti-inflammatory cytokines in the pathogenesis of osteoarthritis. *Mediat. Inflamm.* **2014**, *2014*, 561459. [CrossRef] [PubMed]
17. Monibi, F.; Roller, B.L.; Stoker, A.; Garner, B.; Bal, S.; Cook, J.L. Identification of Synovial Fluid Biomarkers for Knee Osteo-arthritis and Correlation with Radiographic Assessment. *J. Knee Surg.* **2016**, *29*, 242–247. [CrossRef] [PubMed]
18. Borzi, R.M.; Mazzetti, I.; Macor, S.; Silvestri, T.; Bassi, A.; Cattini, L.; Facchini, A. Flow cytometric analysis of intracellular chemokines in chondrocytes in vivo: Constitutive expression and enhancement in osteoarthritis and rheumatoid arthritis. *FEBS Lett.* **1999**, *455*, 238–242. [CrossRef]
19. Miller, R.E.; Tran, P.B.; Das, R.; Ghoreishi-Haack, N.; Ren, D.; Miller, R.J.; Malfait, A.-M. CCR2 chemokine receptor signaling mediates pain in experimental osteoarthritis. *Proc. Natl. Acad. Sci. USA* **2012**, *109*, 20602–20607. [CrossRef]
20. Beekhuizen, M.; Gierman, L.M.; Van Spil, W.E.; Van Osch, G.J.V.M.; Huizinga, T.W.J.; Saris, D.B.F.; Creemers, L.B.; Zuurmond, A.-M. An explorative study comparing levels of soluble mediators in control and osteoarthritic synovial fluid. *Osteoarthr. Cartil.* **2013**, *21*, 918–922. [CrossRef]
21. Sandell, L.J.; Xing, X.; Franz, C.; Davies, S.; Chang, L.-W.; Patra, D. Exuberant expression of chemokine genes by adult human articular chondrocytes in response to IL-1beta. *Osteoarthr. Cartil.* **2008**, *16*, 1560–1571. [CrossRef] [PubMed]
22. Koh, S.M.; Chan, C.K.; Teo, S.H.; Singh, S.; Merican, A.; Ng, W.M.; Abbas, A.; Kamarul, T. Elevated plasma and synovial fluid interleukin-8 and interleukin-18 may be associated with the pathogenesis of knee osteoarthritis. *Knee* **2020**, *27*, 26–35. [CrossRef]

23. Guo, X.; Xu, T.; Zheng, J.; Cui, X.; Li, M.; Wang, K.; Su, M.; Zhang, H.; Zheng, K.; Sun, C.; et al. Accumulation of synovial fluid CD19+CD24hiCD27+ B cells was associated with bone destruction in rheumatoid arthritis. *Sci. Rep.* **2020**, *10*, 14386. [CrossRef] [PubMed]
24. Tran, P.B.; Miller, R.E.; Ishihara, S.; Miller, R.J.; Malfait, A.M. Spinal microglial activation in a murine surgical model of knee osteoarthritis. *Osteoarthr. Cartil.* **2017**, *25*, 718–726. [CrossRef] [PubMed]
25. Milligan, E.D.; Sloane, E.M.; Watkins, L.R. Glia in pathological pain: A role for fractalkine. *J. Neuroimmunol.* **2008**, *198*, 113–120. [CrossRef]
26. Luongo, L.; Sajic, M.; Grist, J.; Clark, A.K.; Maione, S.; Malcangio, M. Spinal changes associated with mechanical hypersensi-tivity in a model of Guillain-Barré syndrome. *Neurosci. Lett.* **2008**, *437*, 98–102. [CrossRef]
27. Clark, A.K.; Grist, J.; Al-Kashi, A.; Perretti, M.; Malcangio, M. Spinal cathepsin S and fractalkine contribute to chronic pain in the collagen-induced arthritis model. *Arthritis Rheum.* **2012**, *64*, 2038–2047. [CrossRef]
28. Miller, R.E.; Malfait, A.-M.; Block, J.A. Current status of nerve growth factor antibodies for the treatment of osteoarthritis pain. *Clin. Exp. Rheumatol.* **2017**, *35* (Suppl. S1), 85–87.
29. Iannone, F.; De Bari, C.; Dell'Accio, F.; Covelli, M.; Patella, V.; Lo Bianco, G.; Lapadula, G. Increased expression of nerve growth factor (NGF) and high affinity NGF receptor (p140 TrkA) in human osteoarthritic chondrocytes. *Rheumatology* **2002**, *41*, 1413–1418. [CrossRef]
30. Bryden, L.A.; Nicholson, J.R.; Doods, H.; Pekcec, A. Deficits in spontaneous burrowing behavior in the rat bilateral mono-sodium iodoacetate model of osteoarthritis: An objective measure of pain-related behavior and analgesic efficacy. *Osteoarthr. Cartil.* **2015**, *23*, 1605–1612. [CrossRef]
31. Xu, L.; Nwosu, L.N.; Burston, J.J.; Millns, P.J.; Sagar, D.R.; Mapp, P.I.; Meesawatsom, P.; Li, L.; Bennett, A.J.; Walsh, D.A.; et al. The anti-NGF antibody muMab 911 both prevents and reverses pain behaviour and subchondral osteoclast numbers in a rat model of osteoarthritis pain. *Osteoarthr. Cartil.* **2016**, *24*, 1587–1595. [CrossRef] [PubMed]
32. LaBranche, T.P.; Bendele, A.M.; Omura, B.C.; Gropp, K.E.; Hurst, S.I.; Bagi, C.M.; Cummings, T.R.; Ii, L.E.G.; Shelton, D.L.; Zorbas, M.A. Nerve growth factor inhibition with tanezumab influences weight-bearing and subsequent cartilage damage in the rat medial meniscal tear model. *Ann. Rheum. Dis.* **2017**, *76*, 295–302. [CrossRef]
33. Sokolove, J.; Lepus, C.M. Role of inflammation in the pathogenesis of osteoarthritis: Latest findings and interpretations. *Ther. Adv. Musculoskelet. Dis.* **2013**, *5*, 77–94. [CrossRef] [PubMed]
34. Kroner, J.; Kovtun, A.; Kemmler, J.; Messmann, J.J.; Strauss, G.; Seitz, S.; Schinke, T.; Amling, M.; Kotrba, J.; Froebel, J.; et al. Mast Cells Are Critical Regulators of Bone Fracture-Induced Inflammation and Osteoclast Formation and Activity. *J. Bone Miner. Res.* **2017**, *32*, 2431–2444. [CrossRef] [PubMed]
35. Ragipoglu, D.; Dudeck, A.; Haffner-Luntzer, M.; Voss, M.; Kroner, J.; Ignatius, A.; Fischer, V. The Role of Mast Cells in Bone Metabolism and Bone Disorders. *Front. Immunol.* **2020**, *11*, 163. [CrossRef] [PubMed]
36. Giannetti, A.; Filice, E.; Caffarelli, C.; Ricci, G.; Pession, A. Mast Cell Activation Disorders. *Medicina* **2021**, *57*, 124. [CrossRef] [PubMed]
37. Mai, L.; Liu, Q.; Huang, F.; He, H.; Fan, W. Involvement of Mast Cells in the Pathophysiology of Pain. *Front. Cell. Neurosci.* **2021**, *15*, 665066. [CrossRef] [PubMed]
38. Sellam, J.; Berenbaum, F. Is osteoarthritis a metabolic disease? *Jt. Bone Spine* **2013**, *80*, 568–573. [CrossRef] [PubMed]
39. Farinelli, L.; Aquili, A.; Mattioli-Belmonte, M.; Manzotti, S.; D'Angelo, F.; Ciccullo, C.; Gigante, A. Synovial mast cells from knee and hip osteoarthritis: Histological study and clinical correlations. *J. Exp. Orthop.* **2022**, *9*, 13. [CrossRef]
40. De Lange-Brokaar, B.J.E.; Kloppenburg, M.; Andersen, S.N.; Dorjée, A.L.; Yusuf, E.; Herb-van Toorn, L.; Kroon, H.M.; Zuur-mond, A.-M.; Stojanovic-Susulic, V.; Bloem, J.L.; et al. Characterization of synovial mast cells in knee osteoarthritis: Association with clinical parameters. *Osteoarthr. Cartil.* **2016**, *24*, 664–671. [CrossRef] [PubMed]
41. Dan, J.; Izumi, M.; Habuchi, H.; Habuchi, O.; Takaya, S.; Kasai, Y.; Hayashi, R.; Aso, K.; Ushida, T.; Ikeuchi, M. A novel mice model of acute flares in osteoarthritis elicited by intra-articular injection of cultured mast cells. *J. Exp. Orthop.* **2021**, *8*, 75. [CrossRef]
42. Nigrovic, P.A.; Lee, D.M. Mast cells in inflammatory arthritis. *Arthritis Res. Ther.* **2005**, *7*, 1–11. [CrossRef] [PubMed]
43. Magarinos, N.J.; Bryant, K.J.; Fosang, A.J.; Adachi, R.; Stevens, R.L.; McNeil, H.P. Mast cell-restricted, tetramer-forming tryptases induce aggrecanolysis in articular cartilage by activating matrix metalloproteinase-3 and -13 zymogens. *J. Immunol.* **2013**, *191*, 1404–1412. [CrossRef] [PubMed]
44. Kulkarni, P.; Harsulkar, A.; Märtson, A.; Suutre, S.; Märtson, A.; Koks, S. Mast cells differentiated in synovial fluid and resident in osteophytes exalt the inflammatory pathology of osteoarthritis. *Int. J. Mol. Sci.* **2022**, *23*, 541. [CrossRef] [PubMed]
45. Rivellese, F.; Rossi, F.W.; Galdiero, M.R.; Pitzalis, C.; De Paulis, A. Mast cells in early rheumatoid arthritis. *Int. J. Mol. Sci.* **2019**, *28*, 2040. [CrossRef]
46. Skaper, S.D. Nerve growth factor: A neuroimmune crosstalk mediator for all seasons. *Immunology* **2017**, *151*, 1–15. [CrossRef] [PubMed]
47. Jimenez-Andrade, J.M.; Mantyh, P.W. Sensory and sympathetic nerve fibers undergo sprouting and neuroma formation in the painful arthritic joint of geriatric mice. *Arthritis Res. Ther.* **2012**, *14*, R101. [CrossRef] [PubMed]
48. Ghilardi, J.R.; Freeman, K.T.; Jimenez-Andrade, J.M.; Coughlin, K.A.; Kaczmarska, M.J.; Castaneda-Corral, G.; Bloom, A.P.; Kuskowski, M.A.; Mantyh, P.W. Neuroplasticity of sensory and sympathetic nerve fibers in a mouse model of a painful ar-thritic joint. *Arthritis Rheum.* **2012**, *64*, 2223–2232. [CrossRef] [PubMed]

49. Barker, P.A.; Mantyh, P.; Arendt-Nielsen, L.; Viktrup, L.; Tive, L. Nerve Growth Factor Signaling and Its Contribution to Pain. *J. Pain Res.* **2020**, *13*, 1223–1241. [CrossRef]
50. Sousa-Valente, J.; Calvo, L.; Vacca, V.; Simeoli, R.; Arévalo, J.C.; Malcangio, M. Role of TrkA signalling and mast cells in the initiation of osteoarthritis pain in the monoiodoacetate model. *Osteoarthr. Cartil.* **2018**, *26*, 84–94. [CrossRef] [PubMed]
51. Guo, Y.; Wei, T.; Hu, N.; Zhou, X. Disrupted homeostasis of synovial hyaluronic acid and its associations with synovial mast cell proteases of rheumatoid arthritis patients and collagen-induced arthritis rats. *Immunol. Res.* **2021**, *69*, 584–593. [CrossRef] [PubMed]
52. Fukui, M.; Whittlesey, K.; Metcalfe, D.D.; Dastych, J. Human mast cells express the hyaluronic-acid-binding isoform of CD44 and adhere to hyaluronic acid. *Clin. Immunol.* **2000**, *94*, 173–178. [CrossRef]
53. Zhao, D.; Pan, J.-K.; Yang, W.-Y.; Han, Y.-H.; Zeng, L.-F.; Liang, G.-H.; Liu, J. Intra-Articular Injections of Platelet-Rich Plasma, Adipose Mesenchymal Stem Cells, and Bone Marrow Mesenchymal Stem Cells Associated with Better Outcomes Than Hyaluronic Acid and Saline in Knee Osteoarthritis: A Systematic Review and Network Meta-analysis. *Arthrosc. J. Arthrosc. Relat. Surg.* **2021**, *37*, 2298–2314.e10. [CrossRef] [PubMed]
54. Saltychev, M.; Mattie, R.; McCormick, Z.; Laimi, K. The Magnitude and Duration of the Effect of Intra-articular Corticosteroid Injections on Pain Severity in Knee Osteoarthritis: A Systematic Review and Meta-Analysis. *Am. J. Phys. Med. Rehabil.* **2020**, *99*, 617–625. [CrossRef] [PubMed]
55. McAlindon, T.E.; LaValley, M.P.; Harvey, W.F.; Price, L.L.; Driban, J.B.; Zhang, M.; Ward, R.J. Effect of Intra-articular Tri-amcinolone vs. Saline on Knee Cartilage Volume and Pain in Patients with Knee Osteoarthritis: A Randomized Clinical Trial. *JAMA* **2017**, *317*, 1967–1975. [CrossRef] [PubMed]
56. Hartmann, K.; Koenen, M.; Schauer, S.; Wittig-Blaich, S.; Ahmad, M.; Baschant, U.; Tuckermann, J.P. Molecular Actions of Glucocorticoids in Cartilage and Bone During Health, Disease, and Steroid Therapy. *Physiol. Rev.* **2016**, *96*, 409–447. [CrossRef] [PubMed]
57. Karasavvidis, T.; Totlis, T.; Gilat, R.; Cole, B.J. Platelet-Rich Plasma Combined with Hyaluronic Acid Improves Pain and Function Compared with Hyaluronic Acid Alone in Knee Osteoarthritis: A Systematic Review and Meta-analysis. *Arthrosc. J. Arthrosc. Relat. Surg.* **2021**, *37*, 1277–1287.e1. [CrossRef] [PubMed]
58. Webb, D.; Naidoo, P. Viscosupplementation for knee osteoarthritis: A focus on Hylan G-F 20. *Orthop. Res. Rev.* **2018**, *10*, 73–81. [CrossRef]
59. Petrosino, S.; Puigdemont, A.; Della Valle, M.F.; Fusco, M.; Verde, R.; Allarà, M.; Aveta, T.; Orlando, P.; Di Marzo, V. Adelmidrol increases the endogenous concentrations of palmitoylethanolamide in canine keratinocytes and down-regulates an inflammatory reaction in an in vitro model of contact allergic dermatitis. *Vet. J.* **2016**, *207*, 85–91. [CrossRef] [PubMed]
60. Del Re, A.; Palenca, I.; Seguella, L.; Pesce, M.; Corpetti, C.; Steardo, L.; Rurgo, S.; Sarnelli, G.; Esposito, G. Oral Adelmidrol Administration Up-Regulates Palmitoylethanolamide Production in Mice Colon and Duodenum through a PPAR-γ Inde-pendent Action. *Metabolites* **2022**, *12*, 457. [CrossRef] [PubMed]
61. De Filippis, D.; D'Amico, A.; Cinelli, M.P.; Esposito, G.; Di Marzo, V.; Iuvone, T. Adelmidrol, a palmitoylethanolamide analogue, reduces chronic inflammation in a carrageenin-granuloma model in rats. *J. Cell. Mol. Med.* **2009**, *13*, 1086–1095. [CrossRef] [PubMed]
62. Cerrato, S.; Brazis, P.; Della Valle, M.F.; Miolo, A.; Puigdemont, A. Inhibitory effect of topical adelmidrol on antigen-induced skin wheal and mast cell behavior in a canine model of allergic dermatitis. *BMC Vet. Res.* **2012**, *8*, 230. [CrossRef] [PubMed]
63. Richardson, D.; Pearson, R.G.; Kurian, N.; Latif, M.L.; Garle, M.J.; Barrett, D.A.; Kendall, D.A.; Scammell, B.E.; Reeve, A.J.; Chapman, V. Characterisation of the cannabinoid receptor system in synovial tissue and fluid in patients with osteoarthritis and rheumatoid arthritis. *Arthritis Res. Ther.* **2008**, *10*, R43. [CrossRef] [PubMed]
64. Punzi, L.; Frigato, M.; Frallonardo, P.; Ramonda, R. Inflammatory osteoarthritis of the hand. *Best Pract. Res. Clin. Rheumatol.* **2010**, *24*, 301–312. [CrossRef] [PubMed]
65. Moon, P.M.; Beier, F. Novel Insights into Osteoarthritis Joint Pathology from Studies in Mice. *Curr. Rheumatol. Rep.* **2015**, *17*, 50. [CrossRef] [PubMed]
66. Luongo, L.; Guida, F.; Boccella, S.; Bellini, G.; Gatta, L.; Rossi, F.; De Novellis, V.; Maione, S. Palmitoylethanolamide reduces formalin-induced neuropathic-like behaviour through spinal glial/microglial phenotypical changes in mice. *CNS Neurol. Disord. Drug Targets* **2013**, *12*, 45–54. [CrossRef] [PubMed]
67. Guida, F.; Boccella, S.; Iannotta, M.; De Gregorio, D.D.; Giordano, C.; Belardo, C.; Romano, R.; Palazzo, E.; Scafuro, M.A.; Serra, N.; et al. Palmitoylethanolamide reduces neuropsychiatric behaviors by restoring cortical electrophysiological activity in a mouse model of mild traumatic brain injury. *Front. Pharmacol.* **2017**, *8*, 95. [CrossRef]
68. Impellizzeri, D.; Di Paola, R.; Cordaro, M.; Gugliandolo, E.; Casili, G.; Morittu, V.M.; Britti, D.; Esposito, E.; Cuzzocrea, S. Adelmidrol, a palmitoylethanolamide analogue, as a new pharmacological treatment for the management of acute and chronic inflammation. *Biochem. Pharmacol.* **2016**, *119*, 27–41. [CrossRef] [PubMed]
69. Di Paola, R.; Fusco, R.; Impellizzeri, D.; Cordaro, M.; Britti, D.; Morittu, V.M.; Evangelista, M.; Cuzzocrea, S. Adelmidrol, in combination with hyaluronic acid, displays increased anti-inflammatory and analgesic effects against monosodium iodo-acetate-induced osteoarthritis in rats. *Arthritis Res. Ther.* **2016**, *18*, 291. [CrossRef] [PubMed]
70. Zaki, S.; Blaker, C.L.; Little, C.B. OA foundations-experimental models of osteoarthritis. *Osteoarthr. Cartil.* **2022**, *30*, 357–380. [CrossRef]

71. Guzman, R.E.; Evans, M.G.; Bove, S.; Morenko, B.; Kilgore, K. Mono-iodoacetate-induced histologic changes in subchondral bone and articular cartilage of rat femorotibial joints: An animal model of osteoarthritis. *Toxicol. Pathol.* **2003**, *31*, 619–624. [CrossRef] [PubMed]
72. Banerji, S.; Wright, A.J.; Noble, M.; Mahoney, D.J.; Campbell, I.D.; Day, A.J.; Jackson, D.G. Structures of the Cd44-hyaluronan complex provide insight into a fundamental carbohydrate-protein interaction. *Nat. Struct. Mol. Biol.* **2007**, *14*, 234–239. [CrossRef] [PubMed]
73. Vulpiani, M.C.; Tucciarone, A.; Vetrano, M.; Trischitta, D.; Santoboni, F.; Nusca, S.M.; Godente, L.; Latini, E.; Ippolito, E. [Il trattamento della gonartrosi mediante infiltrazioni intra-articolari con un' associazione di acido ialuronico] Knee osteoarthritis treatment by intra-articular injections with a combination of hyaluronic acid and Adelmidrol (Hyadrol®). *Giot* **2021**, *47*, 169–179. [CrossRef]
74. Salaffi, F.; Leardini, G.; Canesi, B.; Mannoni, A.; Fioravanti, A.; Caporali, R.; Lapadula, G.; Punzi, L. Reliability and validity of the Western Ontario and McMaster Universities (WOMAC) Osteoarthritis Index in Italian patients with osteoarthritis of the knee. *Osteoarthr. Cartil.* **2003**, *11*, 551–560. [CrossRef]
75. Vulpiani, M.C.; Vetrano, M.; Trischitta, D.; Santoboni, F.; Nusca, S.M.; Latini, E.; Ippolito, E. [Risultati a lungo termine del trattamento della gonartrosi mediante infiltrazioni intra-articolari di Hyadrol ®] Long-term results of knee osteoarthritis treated by intra-articular injections of Hyadrol®. *Giot* **2022**, *48*, 100–106. [CrossRef]
76. Meissner, W.; Huygen, F.; Neugebauer, E.A.M.; Osterbrink, J.; Benhamou, D.; Betteridge, N.; Coluzzi, F.; De Andres, J.; Fawcett, W.; Fletcher, D.; et al. Management of acute pain in the postoperative setting: The importance of quality indicators. *Curr. Med. Res. Opin.* **2018**, *34*, 187–196. [CrossRef]
77. Ambrosio, F.; Finco, G.; Mattia, C.; Mediati, R.; Paoletti, F.; Coluzzi, F.; Piacevoli, Q.; Savoia, G.; Amantea, B.; Aurilio, C.; et al. SIAARTI recommendations for chronic noncancer pain. *Minerva Anestesiol.* **2006**, *72*, 859–880.
78. Pergolizzi, J.V., Jr.; Varrassi, G.; Magnusson, P.; Breve, F.; Raffa, R.B.; Christo, P.J.; Chopra, M.; Paladini, A.; LeQuang, J.A.; Mitchell, K.; et al. Pharmacologic agents directed at the treatment of pain associated with maladaptive neuronal plasticity. *Expert Opin. Pharmacother.* **2022**, *23*, 105–116. [CrossRef]
79. Morlion, B.; Coluzzi, F.; Aldington, D.; Kocot-Kepska, M.; Pergolizzi, J.; Mangas, A.C.; Ahlbeck, K.; Kalso, E. Pain chronification: What should a non-pain medicine specialist know? *Curr. Med. Res. Opin.* **2018**, *34*, 1169–1178. [CrossRef]
80. Coluzzi, F.; Fornasari, D.; Pergolizzi, J.; Romualdi, P. From acute to chronic pain: Tapentadol in the progressive stages of this disease entity. *Eur. Rev. Med. Pharmacol. Sci.* **2017**, *21*, 1672–1683.

biomolecules

Review

Palmitoylethanolamide and White Matter Lesions: Evidence for Therapeutic Implications

Marta Valenza [1], Roberta Facchinetti [1], Luca Steardo [1,2,*] and Caterina Scuderi [1,*]

1 Department of Physiology and Pharmacology "Vittorio Erspamer", SAPIENZA University of Rome—P.le A. Moro, 5, 00185 Rome, Italy
2 Università Giustino Fortunato, 82100 Benevento, Italy
* Correspondence: luca.steardo@uniroma1.it (L.S.); caterina.scuderi@uniroma1.it (C.S.)

Abstract: Palmitoylethanolamide (PEA), the naturally occurring amide of ethanolamine and palmitic acid, is an endogenous lipid compound endowed with a plethora of pharmacological functions, including analgesic, neuroprotective, immune-modulating, and anti-inflammatory effects. Although the properties of PEA were first characterized nearly 65 years ago, the identity of the receptor mediating these actions has long remained elusive, causing a period of research stasis. In the last two decades, a renewal of interest in PEA occurred, and a series of interesting studies have demonstrated the pharmacological properties of PEA and clarified its mechanisms of action. Recent findings showed the ability of formulations containing PEA in promoting oligodendrocyte differentiation, which represents the first step for the proper formation of myelin. This evidence opens new and promising research opportunities. White matter defects have been detected in a vast and heterogeneous group of diseases, including age-related neurodegenerative disorders. Here, we summarize the history and pharmacology of PEA and discuss its therapeutic potential in restoring white matter defects.

Keywords: palmitoylethanolamide; luteolin; oligodendrocyte; astrocyte; myelin; demyelinating diseases; multiple sclerosis; Alzheimer's disease

Citation: Valenza, M.; Facchinetti, R.; Steardo, L.; Scuderi, C. Palmitoylethanolamide and White Matter Lesions: Evidence for Therapeutic Implications. *Biomolecules* **2022**, *12*, 1191. https://doi.org/10.3390/biom12091191

Academic Editors: Salvatore Cuzzocrea and Rosalia Crupi

Received: 29 July 2022
Accepted: 23 August 2022
Published: 27 August 2022

1. How Nature Provides Therapeutic Molecules: The History of Palmitoylethanolamide

The discovery of palmitoylethanolamide (PEA) arises from an interesting clinical observation in the early 1940s, when some clinicians reported that adding dried chicken egg yolk to the diets of underprivileged children reduced recurrences of rheumatic fever, despite the presence of streptococcal infections [1]. Later on, anti-allergic effects of the lipid fractions purified from egg yolk [2,3], peanut oil, and soybean lecithin [4] were demonstrated in guinea pigs. The agent responsible for these properties was definitively recognized as PEA [3], which was also identified in mammalian tissues fifteen years later [5].

The following years witnessed a series of interesting clinical studies on this molecule with good results. During the late 1970s, PEA clinical use was approved in former Czechoslovakia for acute respiratory diseases. For unknown reasons, not related to its toxicity, PEA was withdrawn from clinical use after several years on the market. In fact, the molecule remained largely unnoticed by the rest of the world for more than 20 years. Indeed, the market withdrawal and the failure to identify PEA molecular targets caused a period of research stasis. In the 1990s, Nobel laureate Prof. Rita Levi-Montalcini demonstrated that PEA is an endogenously produced regulator of inflammation [6]. This evidence, together with the discovery of the endocannabinoid system, caused a renewal of interest in PEA. In particular, during the last two decades, formulations containing PEA have received increasing attention as drugs and dietary supplements to be used in chronic pain, inflammation, and certain brain diseases.

2. The Pharmacology of Palmitoylethanolamide

PEA exhibits a rather complex biological and pharmacological profile. Despite some structural and metabolic similarities with the endocannabinoid signaling molecules, PEA is not a classical cannabinoid. It is occasionally called a "promiscuous" molecule, as it binds several receptors and also interacts with non-receptor targets (Figure 1). This alleged lack of selectivity is, however, increasingly regarded as an advantage in certain diseases, where "multiple-target" molecules may exert more beneficial effects than classical "single-target" drugs [7].

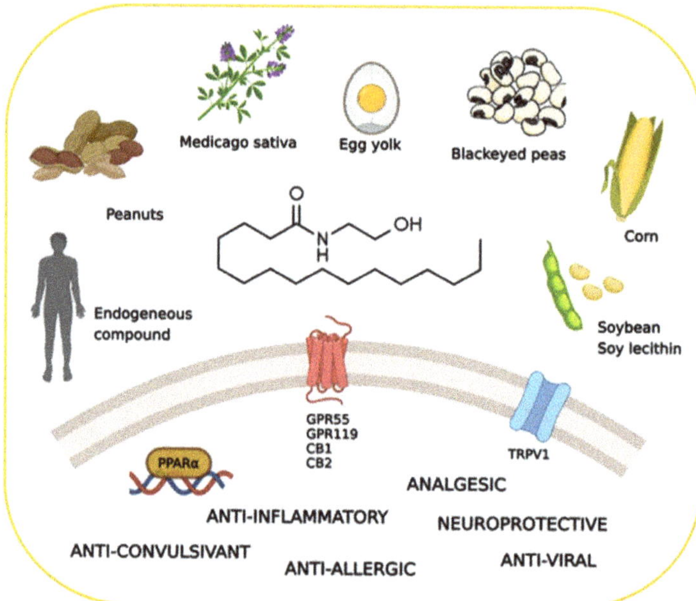

Figure 1. Key facts on palmitoylethanolamide (PEA), including its main sources, molecular targets, and effects. Created with BioRender.com (accessed on 22 July 2022).

2.1. *Biosynthesis, Degradation, and Pharmacokinetics of PEA*

PEA is an endogenous compound belonging to the family of N-acylethanolamines (NAEs) which include the endogenous cannabinoid receptor ligand anandamide (AEA, arachidonoylethanolamide) and the satiety agent oleoylethanolamide (OEA) [8,9].

Biosynthesis of PEA occurs on-demand within the lipid bilayer in two steps [10]. The first is the calcium- and cAMP-dependent transfer of palmitic acid from phosphatidylcholine to phosphatidylethanolamine to form N-acylphosphatidylethanolamine (NAPE); the second step is the cleavage of NAPE to release PEA, mediated by a NAPE-specific phospholipase D. The inactivation of PEA occurs via fatty acid amide amidohydrolase (FAAH) or PEA-preferring acid amidase (PAA) to form palmitic acid and ethanolamine.

All tissues, including the brain, have been found to contain PEA [11]. The physiological regulation of PEA levels in mammalian tissues is still under investigation. Different studies suggest that this compound accumulates after tissue injury, causing cellular stress; this supports the hypothesis that PEA is an endogenous mediator whose levels increase to exert a local reparative action [12].

Exogenous PEA administration presents some issues regarding its bioavailability. Given its lipid nature, PEA solubility in most aqueous solvents is indeed very low. It was shown that after intraperitoneal administration of radio-labeled-PEA to rats, the molecule was distributed mainly in some peripheral organs, whereas only low concentrations were detected in the brain and plasma [13]. Other groups demonstrated the ability of PEA to cross the blood–brain barrier (BBB) after oral administration, but only in very small amounts [14,15]. Of note, one clinical study reported that, in a dose-dependent manner, PEA administration leads to a two- to nine-fold increase in PEA plasma baseline concentrations [16].

To overcome the poor pharmacokinetics of PEA, different formulations containing PEA as micronized or ultra-micronized particles, or as ester derivatives (prepared by conjugating PEA with various amino acids), have been recently developed. These formulations improve PEA bioavailability in the central nervous system (CNS), without affecting its pharmacological efficacy [8].

2.2. *The Search for the Palmitoylethanolamide Receptor*

Between the 1950s and the 1980s, the mechanism(s) of action of PEA remained unidentified. This probably caused the lack of interest of the scientific community in PEA, despite its potential clinical significance. The renaissance of PEA originates from the work of Prof. Levi-Montalcini, who suggested that this endogenous compound exerts anti-inflammatory effects by serving as an autacoid local injury antagonist (ALIA) leading to a down-regulation of mast cell activation [6]. Moreover, the concurrent discovery of the endocannabinoid AEA and the cannabinoid receptors CB1 and CB2 shed new interest in PEA. Indeed, the similarity in chemical structure between AEA and PEA first suggested that these two endogenous mediators might share the same receptor. Several preclinical studies have definitely clarified that PEA does not bind the CB receptors and identified other mechanisms of action. The main molecular target of PEA seems to be the peroxisome proliferator-activated receptor-alpha (PPAR-α), through which PEA exerts its strong anti-inflammatory effects. Supporting this evidence, PEA effects vanished in mutant PPAR-α-null mice [10]. In particular, PPAR-α emerged as the key factor in mediating the ability of PEA to control neuroinflammation in different brain diseases [17–24].

Additionally, PEA can directly activate the G protein-coupled receptor 55 (GPR55) [25] and G protein-coupled receptor 119 (GPR119) [11]. Furthermore, it has been proposed that PEA indirectly potentiates the CB1 signal by inhibiting the degradation of AEA, a phenomenon known as the "entourage effect" [26,27]. Additionally, several studies have demonstrated the involvement of the transient receptor potential vanilloid 1 (TRPV1) in the actions of PEA [28–31]. The main PEA targets are schematized in Figure 1.

The pharmacological properties of PEA are numerous and widely acknowledged by the scientific community. However, among PEA properties, its therapeutic potential in pathologies characterized by myelin defects is scantly investigated. In this review, we summarize the most recent findings obtained with different formulations containing PEA in heterogeneous disease conditions, all of which are characterized by white matter defects.

3. Myelin Sheath Organization and Functions

The term myelin was coined by Rudolf Virchow in 1864 and derives from the Greek word "myeloid" (marrow). Myelin provides the structural basis for fast impulse propagation along neuronal axons, a process required for the proper performance of motor, sensory, and cognitive functions in the CNS [32]. The myelin sheath is an extension of the plasma membrane of mature oligodendrocytes, cells belonging to the family of glial cells [33]. Oligodendrocytes originate from oligodendrocyte precursor cells (OPCs). During development, OPCs migrate from the ventricular/subventricular zones of the CNS to the whole brain, where they proliferate [33]. Differentiation into oligodendrocytes was the first acknowledged function of OPCs [34]. However, a rather large percentage of OPCs never differentiate and live throughout adulthood, suggesting that OPCs may play additional important roles [35]. For instance, in adulthood, when a myelin lesion occurs (e.g., in cases of brain damage), local OPCs begin to proliferate and differentiate into myelinating oligodendrocytes [34,36].

The maturation of OPCs requires several stages, each characterized by distinctive morphological and functional aspects and by the expression of specific proteins (Figure 2).

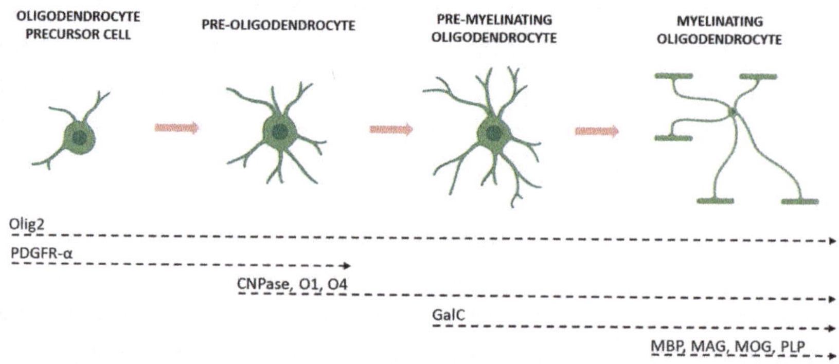

Figure 2. Schematic illustration of the various stages of OPC maturation and corresponding markers. Olig2 is expressed in all stages of the lineage; OPCs and pre-oligodendrocytes are characterized by PDGFR-α expression; CNPase, O1, and O4 are expressed during transition from pre-oligodendrocytes to differentiated oligodendrocytes (which also express GalC); axon-myelinating oligodendrocytes are characterized by myelin protein expression (MBP, MAG, MOG, and PLP). Created with BioRender.com (accessed on 19 August 2022).

Early OPCs are bipolar cells characterized mainly by the expression of the platelet-derived growth factor receptor-α (PDGFR-α). This receptor mediates the signal of the neuronal and astrocytic PDGF-α, which regulates the proliferation, migration, survival, and maturation of OPCs [37]. Once these cells keep contact with a target axon, they lose their bipolarity and start to develop filamentous myelin outgrowths, developing a complex shape. At this differentiation stage, pre-oligodendrocytes express 2′, 3′-cyclic-nucleotide 3′-phosphodiesterase (CNPase) and the cell surface markers O4 and O1. Finally, once fully mature, oligodendrocytes show a complex and ramified structure and form myelin sheaths. Myelinating oligodendrocytes can be identified by the expression of several markers including the myelin basic protein (MBP), the proteolipid protein (PLP), and the myelin-associated glycoprotein (MAG), as well as by the membrane marker galactocerebroside (GalC) and the surface marker myelin-oligodendrocyte glycoprotein (MOG) [38]. Olig2 is a transcription factor that drives oligodendrocyte-lineage cells to express myelin-related genes. It appears to be constantly expressed by oligodendrocytes regardless of the degree of differentiation, but progressively decreases at maturity [39].

The myelin structure consists of a multilayered battery of membranes organized in alternating electron-dense and electron-light layers [40]. The electron-dense layers represent the zone of adhesion between the tightly condensed cytoplasmic membranes, while the electron-light layers consist of extracellularly arranged myelin membranes [40]. The myelinated portions along axons are separated by the nodes of Ranvier, small myelin-free areas that promote rapid saltatory transmission of the nerve impulse by concentrating voltage-dependent sodium channels [41]. Myelin structure is stabilized by a multitude of adhesion mechanisms and proteins. The compaction of myelin is primarily performed by MBP, which condensates two adjacent cytoplasmic membrane surfaces into a major dense line of myelin [42,43]. The compaction begins in the very outer layers, close to the cell soma, and proceeds to the inner side in parallel with the growth of the membrane. In order to ensure the success of this process, MBP mRNA is trafficked from the nucleus to the myelin compartment where it is locally translated [44]. The integral membrane protein PLP allows the extracellular and cytoplasmic membranes to adhere to each other, enhancing the compaction of myelin [45]. The presence of the cytoplasm-rich compartments within myelin is ensured by CNPase which, by directly associating with cytoskeletal actin, antagonizes the adhesive forces exerted by polymerizing MBP [46]. Cytoplasmic channels allow the bidirectional movement of macromolecules under the myelin sheath maintaining vital such a compacted structure.

Through such channels, oligodendrocytes provide neurons with trophic and metabolic support. For example, oligodendrocytes use monocarboxylate transporters to deliver energy metabolites, like pyruvate and lactate, to neurons that use them to produce ATP [47].

4. The Role of the Astrocyte-Oligodendrocyte Cross-Talk in Myelination

The proliferation and differentiation of OPCs need the support of other glial cells, mainly microglia and astrocytes that release distinct patterns of secreted molecules to drive these processes. Microglia involvement in OPC differentiation, as well as in the context of myelin repair, is acknowledged, and the data in the literature highlight a dual role of these cells, providing evidence for both trophic and detrimental roles of microglia on oligodendrocytes and myelin (for a comprehensive review see [48]). Likewise, the role of astrocytes in the proliferation of OPCs and the formation and repair of myelin has long been recognized. It is believed that impairments of these star-shaped cells are implicated in the development of demyelinating diseases. This was first hypothesized by Müller in 1904, who was convinced that multiple sclerosis (MS), a demyelinating disease, was characterized by astrocytic dysfunction [49]. Research findings collected so far have confirmed this hypothesis, with various studies highlighting the key role of the cross-talk between astrocytes and oligodendrocytes in myelination, in both health and disease [50,51].

Astrocytes are the most abundant glial cell type of the CNS, found in both white and grey matter [52]. They were long considered secondary to neurons and were defined, from the Greek term glia, as "brain glue". However, research studies over the past two decades have demonstrated that astrocytes exert a plethora of different functions to maintain CNS homeostasis at molecular, cellular, organ, and system levels of organization [53]. Astrocytes provide both physical and metabolic support to neurons [50] and modulate synaptic transmission and information processing by neural circuits [54]. Astrocytes are key components of the BBB; in this way, they regulate cerebral blood flow and the communication between the CNS and the periphery [54]. Furthermore, they are involved in several processes, such as ion and water transport, pH buffering, neuroplasticity, and synapse pruning. They also release approximately 200 molecules, including neurotrophic factors and energy substrates, thus, providing trophic and metabolic support to all cells in the CNS [55].

Astrocytes and oligodendrocytes originate from a common lineage of neural progenitor cells within the neuroectoderm [56] and, after development, communicate in physical and functional ways. Physically, astrocytes are coupled to oligodendrocytes through gap junctions at the cell body level and the paranodes, directly connecting the outer layer of the myelin sheath with an astrocytic process [57,58]. Such intercellular channels are heterologous, since oligodendrocytes express connexin (Cx) 32, Cx47, and Cx29, whereas astrocytes have Cx26, Cx30, and Cx43 [59]. A large number of reports underscore the importance of gap junctions in myelination. For instance, mutations or genetic ablation of specific connexins could lead to myelin defects, demyelinating diseases, encephalopathies, and peripheral neuropathies [60].

Besides direct physical interaction, astrocytes can communicate with oligodendrocytes by releasing a variety of soluble factors, including PDGF, brain-derived neurotrophic factor (BDNF), ciliary neurotrophic factor (CNTF), transforming growth factor (TGF)-β, and basic fibroblast growth factor (FGF2). All of these are considered promoters of OPC proliferation and survival [48], despite some contrasting evidence suggesting that PDGF and FGF2 could also act as inhibitors [61,62]. Moreover, the tissue inhibitor of metalloproteinase-1 (TIMP-1), an endogenous regulator of matrix metalloproteinases, promotes oligodendrogenesis [63]. In addition, astrocytes maintain and support myelin sheath formation by synthesizing and delivering cholesterol to oligodendrocytes [64,65].

As astrocytes are critical for maintaining brain homeostasis [50], they promptly react to any CNS insult or damage by undergoing morphofunctional changes that are context-, time-, and disease-specific [66–69]. Astrocyte reactivity could have either beneficial or detrimental consequences for myelination, and the difference seems to be related to the severity of astrogliosis [70]. Mild astrogliosis leads to the release of CNTF, FGF2, and the proinflammatory interleukin (IL)-6, all associated with OPC survival, proliferation, and maturation [71,72]. During severe astrogliosis, astrocytes secrete tumor necrosis factor (TNF)-α, which correlates with the extent of demyelination in MS and with myelin and oligodendrocytes damage in vitro [73,74]. Additionally, in an animal model of experimental autoimmune encephalomyelitis (EAE) the release of interferon (IFN)-γ was shown to suppress remyelination and delay recovery [75].

The relevance of astrocyte contribution to oligodendrocyte function is also evident in several astrocytopathies, such as Alexander disease, vanishing white matter, megalencephalic leukoencephalopathy with subcortical cysts, Aicardi–Goutières syndrome, and oculodentodigital dysplasia [76]. All these diseases are characterized by genetic mutations that cause defective astrocyte function, with major consequences for oligodendrocyte physiology as low oligodendrocyte survival, impaired myelination, and absence of myelin with or without concurrent development of astrogliosis [77,78].

Based on this evidence, astrocytes appear as the major cells orchestrating cell-to-cell communication relevant for myelination. Thus, alterations in this cross-talk could lead to myelin defects (Figure 3).

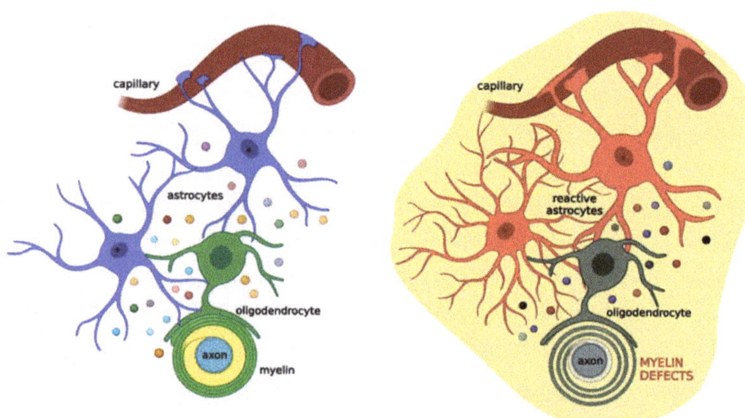

Figure 3. Schematic illustration of the coupling of oligodendrocytes, astrocytes, and neurons in physiological (left) and pathological (right) conditions. Astrocytes (blue) communicate with oligodendrocytes (green) through gap junctions and several released factors (colored circles), such as PDGF, BDNF, CNTF, TGF-β, FGF2, TIMP-1, ILs, TNF-α, and IFN-γ. This cross-talk allows the proper maturation of oligodendrocytes and their ability to form the myelin sheath. This, in turn, impacts neuronal activity. When brain damage occurs (see text for details), astrocytes (red) react promptly, affecting oligodendrocytes (grey) activity and ultimately the myelin sheath. Created with BioRender.com (accessed on 22 July 2022).

5. White Matter Defects in Demyelinating Diseases and the Therapeutic Potential of Palmitoylethanolamide

A vast and heterogeneous group of diseases are characterized by white matter abnormalities. Furthermore, any disorder that weakens or alters the myelin sheath, which surrounds the nerve fibers in the brain, optic nerves, and spinal cord, is referred to as a demyelinating disease. Several hereditary or acquired conditions can be distinguished, with the most common having inflammatory, infectious, toxic, and metabolic origins [36,79]. Myelination defects in humans typically result in substantial neurological symptoms, as would be predicted given its crucial involvement in the physiology of the mammalian nervous system. Indeed, when the myelin sheath is compromised, nerve impulses slow or even cease, leading to neurological issues. The most common symptoms of demyelinating disorders are vision loss, muscle weakness or stiffness, muscle spasms, and alteration in the bladder and/or bowel movements [80].

As mentioned above, a growing number of reports demonstrated the anti-inflammatory and neuroprotective effects of PEA in various models of brain diseases. This evidence opened up the possibility that PEA could be beneficial even in demyelinating diseases characterized by strong neuroinflammatory components. Among them, MS is the most common acquired demyelinating disease of the CNS [79,81]. Its etiology is not yet fully understood, but its inflammatory basis is known. The current therapeutic approach is based on the lifelong administration of immunosuppressive and immunomodulatory agents, in addition to corticosteroids to treat acute relapses [79,82]. Unfortunately, all these therapies have limited efficacy and many adverse effects. Moreover, they do not address all MS symptoms. There is, therefore, an urgent medical need for innovative therapies. In light of this, promising results were obtained in a small double-blind, randomized, placebo-controlled study in which relapsing–remitting MS patients took oral ultra-micronized PEA (um-PEA; NORMAST® 600 mg/day for up to one year), or placebo, in addition to subcutaneous IFNβ1a (Rebif® 132 μg/week). The um-PEA reduced pain sensation at the IFN-β1a injection site, improved results of a questionnaire assessing cognition, reduced IL-17 and TNF-α serum concentration, and increased AEA and OEA plasma levels. Overall, such chronic um-PEA supplementation did not show negative effects and the quality of life of patients seemed to improve [83].

Besides the neuroinflammatory component, oxidative stress also contributes to tissue injury in MS, and it promotes the inflammatory response [82]. For this reason, formulations containing PEA together with an antioxidant compound could be of advantage. Some in vitro studies indeed show that co-ultra-micronized palmitoylethanolamide/luteolin composite (co-ultra PEALut, 10:1 by mass) exerts beneficial effects that could be relevant for MS. For instance, pre-treatment of primary OPCs with co-ultra PEALut prevented the increase in gene expression of serum amyloid A protein (SSA) induced by challenging cells with the pro-inflammatory stimuli TNF-α [84]. This co-ultra PEALut effect is in line with the well-known anti-inflammatory effects of PEA and is relevant in both MS and Alzheimer's disease (AD) in which an elevation in SSA has been detected [85,86].

Of note, treatment of differentiating OPCs with co-ultra PEALut increases the transcription of major myelin proteins genes, as well as genes for enzymes involved in lipid biosynthesis, in a time-dependent manner. Neither PEA nor Lut was able to produce such effects on its own [87]. Co-ultra PEALut promotes MBP and PLP expression in primary OPCs cultured either in a medium favoring cell differentiation or in a medium favoring cell proliferation, and it does so in a rapamycin-dependent manner [88]. Co-ultra PEALut time-dependently increases gene expression of the myelin-associated proteins MBP, CNPase, and the tyrosine kinase TAM receptor Tyro3 in primary OPCs cultured in a pro-differentiating medium, while it induces a reduction in the gene expression level of other two TAM receptors (Axl and Mertk) compared to control cells [89]. All these effects were blunted by concurrent treatment with rapamycin, again supporting the hypothesis that an mTOR-dependent molecular pathway is involved in the co-ultra PEALut mechanism of action [87,89,90].

Two different research groups showed in vivo the potential benefit of administering formulations of PEA. In a model of EAE, commonly used to study features of MS in mice, daily systemic co-ultra PEALut administration, initiated at the first signs of sickness, progressively improved scores for neurological assessment compared to vehicle. Lower doses of co-ultra PEALut were less efficacious. Such behavioral improvement was accompanied by a co-ultra PEALut-mediated reduction in the disease-induced gene expression level of pro-inflammatory mediators and receptors [91]. These effects were likely driven by PEA since its administration alone showed similar beneficial effects in a separate and precedent study. Indeed, PEA ameliorated histological signs of the disease in spinal cord specimens from AEA mice compared to control animals. It also reduced the AEA-increased mRNA level of proinflammatory markers in spinal cord samples [92]. Chronic administration of PEA also showed a protective effect on lesioned peripheral nerves in a murine model of neuropathic pain, as the chronic constriction injury of the sciatic nerve (CCI). Furthermore, PEA, by engaging the PPAR-α, prevented the CCI-induced thinning of the myelin sheath and reduction in axonal diameter, as well as edema and macrophage infiltration [93]. Some authors have suggested that PEA exerts its antinociceptive activity at the spinal level by inhibiting responses of dorsal horn wide dynamic range neurons [94].

Of note, in a small group of patients with chemotherapy-induced painful neuropathy, a two-month treatment with PEA produced substantial pain relief and restored myelinated-fiber function in patients [95].

Taken together, these results highlight the beneficial effects of administering PEA in various conditions which share alteration or loss of myelin, acting at various levels and through different mechanisms.

6. White Matter Defects in Age-Related Neurodegenerative Diseases and the Therapeutic Potential of Palmitoylethanolamide

Aging is the main risk factor for most neurodegenerative diseases. Aging occurs at different rates in diverse species, while inter-individual variations exist within a species and in the different organs and tissues of a subject. The brain is primarily composed of postmitotic cells, so it is especially sensitive to the effects of aging. The cause of the cognitive decline exhibited by human and non-human primates during normal aging has long been considered the consequence of a loss of cortical neurons. Advances in brain imaging techniques have shown that a significant number of cortical neurons are not lost [96]. Therefore, other reasons for the cognitive decline have been sought. One of the putative contributing factors could be the age-related degenerative change in the morphology of myelinated nerve fibers [97]. It has been well established that white matter volume starts to decrease gradually from 50 years of age onwards [98]. Myelin sheaths exhibit degenerative changes with age. These include the formation of splits at the major dense line with electron-dense cytoplasm inclusions and balloons [99]. These balloons appear as holes, but electron microscopy reveals that they are spherical cavities causing the myelin sheaths to swell. Moreover, increased formation of sheaths with redundant myelin and circumferential splits in thick sheaths are considered as other age-related changes. Some of these multilamellar myelin fragments represent myelin outfolding, while others are engulfed by microglia, suggesting that microglia may actively strip off damaged myelin [100]. In the aged mouse brain, microglia display an expanded lysosomal compartment and accumulate autofluorescent material which may consist of remnants of indigestible myelin. Hence, the increased burden of myelin clearance in the aging brain may also contribute to age-associated microglial dysfunction.

Structural abnormalities in the white matter have been observed also in neurodegenerative diseases, including AD and Parkinson's disease (PD), which share symptoms, such as locomotor disorders and cognitive decline [101]. Evidence is accumulating that white matter lesions independently contribute to postural and gait disturbances typical of PD, as well as increase the risk of dementia in such patients [102–106]. Despite some data indicating that increased lesions of myelin are associated with worsening cognitive performance in PD, further studies are needed to examine the relationship between myelin lesions and PD-specific cognitive domains.

For a long time, AD has been considered a disease of grey matter. However, advancements in brain imaging techniques have now highlighted evident lesions to the white matter. Postmortem and in vivo imaging studies have demonstrated that AD brains show both reduced volume and microstructure alterations of the white matter, including defects in the physical organization of the myelin lipid bilayers, as reviewed in [107]. Studies have linked myelin damage and concurrent impairments in the maturation of oligodendrocytes with AD pathogenesis, appointing the cerebral deposition of beta-amyloid (Aβ) as a possible etiological factor [108–110]. Thus, myelin loss and the inability of oligodendrocytes to repair myelin damage have been suggested as additional central features of AD [111], besides the accumulation of toxic species of Aβ and the development of neurofibrillary tangles [112]. Biochemical analysis of total myelin fraction in AD patients revealed increased Aβ levels accompanied by a significant reduction in MBP, PLP, and CNPase, as well as alterations in myelin lipid content [113].

Despite this compelling evidence, very little is known about AD-related changes in oligodendrocytes maturation and myelin formation.

Counterintuitive effects of Aβ on oligodendrocyte differentiation have been reported. Some researchers showed that a direct treatment of oligodendrocytes with oligomeric Aβ1-42 promoted MBP upregulation in cultures of primary rat oligodendrocytes [114]. Furthermore, an accelerated differentiation of oligodendrocytes was found in different brain areas of transgenic models of AD (i.e., 3xTg-AD and APP/PS1 mice), compared with wild-type animals [115]. Very recently, we provided the very first data showing that, in a trans-well system (rat primary astrocytes and OPCs co-cultured separated by 0.4 μm pores), challenging astrocytes with Aβ1-42 increased the number of MBP^+ oligodendrocytes [116]. However, through an in-depth morphological analysis, we revealed that this push to the maturation operated by Aβ1-42 actually induces aberrant morphological changes in oligodendrocytes. Indeed, we noted a significant reduction in the cell surface area as well as in the number of intersections, suggesting that MBP^+ oligodendrocytes lose their proper dimension and complexity. These results agree with recent findings obtained in 3xTg-AD mice, showing that oligodendrocytes of six-month-old transgenic mice were atrophic compared with non-transgenic age-matched mice [117]. A conceivable explanation of the effects of Aβ on oligodendrocyte MBP expression could be related to the reaction of these cells to the toxic challenge [118] in an attempt to repair the damage and restore brain homeostasis [119]. To do so, oligodendrocytes must integrate a multitude of regulatory signals sent by neurons and other neuroglia cells, mainly astrocytes [120,121]. The communication between oligodendrocytes and neurons has been studied quite extensively, while their cross-talk with astrocytes has been scantly investigated in AD. As discussed in paragraph 4, growing evidence indicates that, during some pathological conditions, astrocytes modify their duties, hampering important physiological functions, including OPC differentiation and myelin sheath formation [122,123]. Thus, it is conceivable that exposure to Aβ alters the maturation of OPCs and the morphology of mature oligodendrocytes by impairing proper communication between oligodendrocytes and astrocytes. In support of such a hypothesis, we found a significant reactivity of astrocytes with the elevation of several pro-inflammatory mediators and the reduction in the production of astrocytic factors strictly implicated in OPC maturation, including FGF2 and TGF-β [116]. Our data support the hypothesis that in cases of brain damage, such as in AD, astrocytes react promptly by affecting oligodendrocytes maturation and, ultimately, myelin formation. Although the therapeutic potential of PEA in modulating different AD features has already been reported [124–129], no evidence is available so far on its effect on white matter defects.

Recent evidence, summarized in paragraph 5, shows the ability of co-ultra PEALut, but not of PEA alone, in promoting the morphological development of OPCs into mature oligodendrocytes [83,85]. In our model, we tested the effects of co-ultra PEALut. Obtained results indicate that co-ultra PEALut treatment almost completely restores the impairments caused by the exposure of cells to Aβ1-42 [116]. Moreover, we demonstrated that some of the effects of co-ultra PEAlut depend on the activation of the PPAR-α. This new and never explored ability of co-ultra PEALut in counteracting the Aβ-induced effect on oligodendrocyte homeostasis opens new and promising research opportunities.

7. Conclusions

Here, we discussed the potential benefits of administering PEA in white matter defects. Although still few and in some cases limited to preclinical evidence, the findings available thus far are promising. They broaden the therapeutic indications of PEA, a terrific endogenous molecule endowed with numerous pharmacological activities. Considering the safety profile of PEA, the opportunity to rapidly proceed to clinical use is reasonable.

It is the express wish of the authors to thank Doctor Francesco Della Valle, a visionary Italian entrepreneur in the pharmaceutical field who, with his enlightened competence, supported preclinical and clinical research on PEA. Dr. Della Valle has contributed significantly to scientific progress in this area of research. The world scientific community has benefited enormously from Francesco's intuitions. We address our dearest memory to you.

Author Contributions: Conceptualization, M.V. and C.S.; resources, L.S. and C.S.; writing—original draft preparation, M.V., R.F. and C.S.; writing—review and editing, M.V., L.S. and C.S. All authors have read and agreed to the published version of the manuscript.

Funding: This research received no external funding.

Institutional Review Board Statement: Not applicable.

Informed Consent Statement: Not applicable.

Conflicts of Interest: The authors declare no conflict of interest.

References

1. Coburn, A.; Moore, L. American Journal of Diseases of Children. *JAMA Pediatrics* **1943**, *65*, 744–756.
2. Coburn, A.F.; Graham, C.E.; Haninger, J. The effect of egg yolk in diets on anaphylactic arthritis (passive Arthus phenomenon) in the guinea pig. *J. Exp. Med.* **1954**, *100*, 425–435. [CrossRef] [PubMed]
3. Long, D.A.; Martin, A.J. Factor in arachis oil depressing sensitivity to tuberculin in B.C.G.-infected guineapigs. *Lancet* **1956**, *270*, 464–466. [CrossRef]
4. Long, D.; Miles, A. Opposite actions of thyroid and adrenal hormones in allergic hypersensitivity. *Lancet* **1950**, *255*, 492–495. [CrossRef]
5. Bachur, N.R.; Masek, K.; Melmon, K.L.; Udenfriend, S. Fatty Acid Amides of Ethanolamine in Mammalian Tissues. *J. Biol. Chem.* **1965**, *240*, 1019–1024. [CrossRef]
6. Aloe, L.; Leon, A.; Levi-Montalcini, R. A proposed autacoid mechanism controlling mastocyte behaviour. *Agents Actions* **1993**, *39*, C145–C147. [CrossRef]
7. Valenza, M.; Facchinetti, R.; Menegoni, G.; Steardo, L.; Scuderi, C. Alternative Targets to Fight Alzheimer's Disease: Focus on Astrocytes. *Biomolecules* **2021**, *11*, 600. [CrossRef]
8. Beggiato, S.; Tomasini, M.C.; Ferraro, L. Palmitoylethanolamide (PEA) as a Potential Therapeutic Agent in Alzheimer's Disease. *Front. Pharm.* **2019**, *10*, 821. [CrossRef]
9. Romano, A.; Di Bonaventura, M.V.M.; Gallelli, C.A.; Koczwara, J.B.; Smeets, D.; Giusepponi, M.E.; De Ceglia, M.; Friuli, M.; Di Bonaventura, E.M.; Scuderi, C.; et al. Oleoylethanolamide decreases frustration stress-induced binge-like eating in female rats: A novel potential treatment for binge eating disorder. *Neuropsychopharmacology* **2020**, *45*, 1931–1941. [CrossRef]
10. Lo Verme, J.; Fu, J.; Astarita, G.; La Rana, G.; Russo, R.; Calignano, A.; Piomelli, D. The nuclear receptor peroxisome proliferator-activated receptor-alpha mediates the anti-inflammatory actions of palmitoylethanolamide. *Mol. Pharm.* **2005**, *67*, 15–19. [CrossRef]
11. Rankin, L.; Fowler, C.J. The Basal Pharmacology of Palmitoylethanolamide. *Int. J. Mol. Sci.* **2020**, *21*, 7942. [CrossRef] [PubMed]
12. Darmani, N.A.; Izzo, A.A.; Degenhardt, B.; Valenti, M.; Scaglione, G.; Capasso, R.; Sorrentini, I.; Di Marzo, V. Involvement of the cannabimimetic compound, N-palmitoyl-ethanolamine, in inflammatory and neuropathic conditions: Review of the available pre-clinical data, and first human studies. *Neuropharmacology* **2005**, *48*, 1154–1163. [CrossRef] [PubMed]
13. Zhukov, O.D. [Distribution of N-([1-14C]-palmitoyl)ethanolamine in rat tissues]. *Ukr. Biokhimichnyi Zhurnal* **1999**, *71*, 124–125.
14. Artamonov, M.; Zhukov, O.; Shuba, I.; Storozhuk, L.; Khmel, T.; Klimashevsky, V.; Mikosha, A.; Gula, N. Incorporation of labelled N-acylethanolamine (NAE) into rat brain regions in vivo and adaptive properties of saturated NAE under X-ray irradiation. *Ukr. Biokhimichnyi Zhurnal* **2005**, *77*, 51–62.
15. Gabrielsson, L.; Mattsson, S.; Fowler, C.J. Palmitoylethanolamide for the treatment of pain: Pharmacokinetics, safety and efficacy. *Br. J. Clin. Pharm.* **2016**, *82*, 932–942. [CrossRef]
16. Balvers, M.G.; Verhoeckx, K.C.; Meijerink, J.; Wortelboer, H.M.; Witkamp, R.F. Measurement of palmitoylethanolamide and other N-acylethanolamines during physiological and pathological conditions. *CNS Neurol. Disord. Drug Targets* **2013**, *12*, 23–33. [CrossRef]
17. Straus, D.S.; Glass, C.K. Anti-inflammatory actions of PPAR ligands: New insights on cellular and molecular mechanisms. *Trends Immunol.* **2007**, *28*, 551–558. [CrossRef]
18. Scuderi, C.; Stecca, C.; Valenza, M.; Ratano, P.; Bronzuoli, M.R.; Bartoli, S.; Steardo, L.; Pompili, E.; Fumagalli, L.; Campolongo, P.; et al. Palmitoylethanolamide controls reactive gliosis and exerts neuroprotective functions in a rat model of Alzheimer's disease. *Cell Death Dis.* **2014**, *5*, e1419. [CrossRef]

19. Scuderi, C.; Valenza, M.; Stecca, C.; Esposito, G.; Carratù, M.R.; Steardo, L. Palmitoylethanolamide exerts neuroprotective effects in mixed neuroglial cultures and organotypic hippocampal slices via peroxisome proliferator-activated receptor-α. *J. Neuroinflamm.* **2012**, *9*, 49. [CrossRef]
20. Paterniti, I.; Impellizzeri, D.; Crupi, R.; Morabito, R.; Campolo, M.; Esposito, E.; Cuzzocrea, S. Molecular evidence for the involvement of PPAR-δ and PPAR-γ in anti-inflammatory and neuroprotective activities of palmitoylethanolamide after spinal cord trauma. *J. Neuroinflamm.* **2013**, *10*, 20. [CrossRef]
21. Cristiano, C.; Pirozzi, C.; Coretti, L.; Cavaliere, G.; Lama, A.; Russo, R.; Lembo, F.; Mollica, M.P.; Meli, R.; Calignano, A.; et al. Palmitoylethanolamide counteracts autistic-like behaviours in BTBR T+tf/J mice: Contribution of central and peripheral mechanisms. *Brain Behav. Immun.* **2018**, *74*, 166–175. [CrossRef] [PubMed]
22. Scuderi, C.; Esposito, G.; Blasio, A.; Valenza, M.; Arietti, P.; Steardo, L., Jr.; Carnuccio, R.; De Filippis, D.; Petrosino, S.; Iuvone, T.; et al. Palmitoylethanolamide counteracts reactive astrogliosis induced by β-amyloid peptide. *J. Cell Mol. Med.* **2011**, *15*, 2664–2674. [CrossRef] [PubMed]
23. Scuderi, C.; Steardo, L. Neuroglial roots of neurodegenerative diseases: Therapeutic potential of palmitoylethanolamide in models of Alzheimer's disease. *CNS Neurol. Disord. Drug Targets* **2013**, *12*, 62–69. [CrossRef] [PubMed]
24. D'Agostino, G.; Russo, R.; Avagliano, C.; Cristiano, C.; Meli, R.; Calignano, A. Palmitoylethanolamide protects against the amyloid-β25-35-induced learning and memory impairment in mice, an experimental model of Alzheimer disease. *Neuropsychopharmacology* **2012**, *37*, 1784–1792. [CrossRef] [PubMed]
25. Pertwee, R.G. GPR55: A new member of the cannabinoid receptor clan? *Br. J. Pharm.* **2007**, *152*, 984–986. [CrossRef] [PubMed]
26. Mattace Raso, G.; Russo, R.; Calignano, A.; Meli, R. Palmitoylethanolamide in CNS health and disease. *Pharm. Res.* **2014**, *86*, 32–41. [CrossRef] [PubMed]
27. Petrosino, S.; Di Marzo, V. The pharmacology of palmitoylethanolamide and first data on the therapeutic efficacy of some of its new formulations. *Br. J. Pharm.* **2017**, *174*, 1349–1365. [CrossRef]
28. De Petrocellis, L.; Davis, J.B.; Di Marzo, V. Palmitoylethanolamide enhances anandamide stimulation of human vanilloid VR1 receptors. *FEBS Lett.* **2001**, *506*, 253–256. [CrossRef]
29. Smart, D.; Jonsson, K.O.; Vandevoorde, S.; Lambert, D.M.; Fowler, C.J. 'Entourage' effects of N-acyl ethanolamines at human vanilloid receptors. Comparison of effects upon anandamide-induced vanilloid receptor activation and upon anandamide metabolism. *Br. J. Pharm.* **2002**, *136*, 452–458. [CrossRef]
30. Ambrosino, P.; Soldovieri, M.V.; Russo, C.; Taglialatela, M. Activation and desensitization of TRPV1 channels in sensory neurons by the PPARα agonist palmitoylethanolamide. *Br. J. Pharm.* **2013**, *168*, 1430–1444. [CrossRef]
31. Capasso, R.; Orlando, P.; Pagano, E.; Aveta, T.; Buono, L.; Borrelli, F.; Di Marzo, V.; Izzo, A.A. Palmitoylethanolamide normalizes intestinal motility in a model of post-inflammatory accelerated transit: Involvement of CB_1 receptors and TRPV1 channels. *Br. J. Pharm.* **2014**, *171*, 4026–4037. [CrossRef] [PubMed]
32. Snaidero, N.; Simons, M. Myelination at a glance. *J. Cell Sci.* **2014**, *127*, 2999–3004. [CrossRef] [PubMed]
33. Simons, M.; Nave, K.A. Oligodendrocytes: Myelination and Axonal Support. *Cold Spring Harb. Perspect. Biol.* **2015**, *8*, a020479. [CrossRef] [PubMed]
34. Gensert, J.M.; Goldman, J.E. Endogenous progenitors remyelinate demyelinated axons in the adult CNS. *Neuron* **1997**, *19*, 197–203. [CrossRef]
35. Fernandez-Castaneda, A.; Gaultier, A. Adult oligodendrocyte progenitor cells—Multifaceted regulators of the CNS in health and disease. *Brain Behav. Immun.* **2016**, *57*, 1–7. [CrossRef]
36. Stadelmann, C.; Timmler, S.; Barrantes-Freer, A.; Simons, M. Myelin in the Central Nervous System: Structure, Function, and Pathology. *Physiol. Rev.* **2019**, *99*, 1381–1431. [CrossRef]
37. Murtie, J.C.; Zhou, Y.X.; Le, T.Q.; Vana, A.C.; Armstrong, R.C. PDGF and FGF2 pathways regulate distinct oligodendrocyte lineage responses in experimental demyelination with spontaneous remyelination. *Neurobiol. Dis.* **2005**, *19*, 171–182. [CrossRef]
38. Kuhn, S.; Gritti, L.; Crooks, D.; Dombrowski, Y. Oligodendrocytes in Development, Myelin Generation and Beyond. *Cells* **2019**, *8*, 1424. [CrossRef]
39. Kitada, M.; Rowitch, D.H. Transcription factor co-expression patterns indicate heterogeneity of oligodendroglial subpopulations in adult spinal cord. *Glia* **2006**, *54*, 35–46. [CrossRef]
40. Hartline, D.K. What is myelin? *Neuron Glia Biol.* **2008**, *4*, 153–163. [CrossRef]
41. Fields, R.D.; Burnstock, G. Purinergic signalling in neuron-glia interactions. *Nat. Rev. NeuroSci.* **2006**, *7*, 423–436. [CrossRef]
42. Readhead, C.; Popko, B.; Takahashi, N.; Shine, H.D.; Saavedra, R.A.; Sidman, R.L.; Hood, L. Expression of a myelin basic protein gene in transgenic shiverer mice: Correction of the dysmyelinating phenotype. *Cell* **1987**, *48*, 703–712. [CrossRef]
43. Roach, A.; Takahashi, N.; Pravtcheva, D.; Ruddle, F.; Hood, L. Chromosomal mapping of mouse myelin basic protein gene and structure and transcription of the partially deleted gene in shiverer mutant mice. *Cell* **1985**, *42*, 149–155. [CrossRef]
44. Snaidero, N.; Möbius, W.; Czopka, T.; Hekking, L.H.; Mathisen, C.; Verkleij, D.; Goebbels, S.; Edgar, J.; Merkler, D.; Lyons, D.A.; et al. Myelin membrane wrapping of CNS axons by PI(3,4,5)P3-dependent polarized growth at the inner tongue. *Cell* **2014**, *156*, 277–290. [CrossRef] [PubMed]
45. Simons, M.; Krämer, E.M.; Thiele, C.; Stoffel, W.; Trotter, J. Assembly of myelin by association of proteolipid protein with cholesterol- and galactosylceramide-rich membrane domains. *J. Cell Biol.* **2000**, *151*, 143–154. [CrossRef] [PubMed]

46. Snaidero, N.; Velte, C.; Myllykoski, M.; Raasakka, A.; Ignatev, A.; Werner, H.B.; Erwig, M.S.; Möbius, W.; Kursula, P.; Nave, K.A.; et al. Antagonistic Functions of MBP and CNP Establish Cytosolic Channels in CNS Myelin. *Cell Rep.* **2017**, *18*, 314–323. [CrossRef]
47. Philips, T.; Rothstein, J.D. Oligodendroglia: Metabolic supporters of neurons. *J. Clin. Investig.* **2017**, *127*, 3271–3280. [CrossRef]
48. Traiffort, E.; Kassoussi, A.; Zahaf, A.; Laouarem, Y. Astrocytes and Microglia as Major Players of Myelin Production in Normal and Pathological Conditions. *Front. Cell NeuroSci.* **2020**, *14*, 79. [CrossRef]
49. Müller, E. Die multiple Sklerose des Gehirns und Rückenmarks: Ihre Pathologie und Behandlung. *Pathol. Anat. Und. Pathog.* **1904**, 300–344.
50. Sofroniew, M.V.; Vinters, H.V. Astrocytes: Biology and pathology. *Acta Neuropathol.* **2010**, *119*, 7–35. [CrossRef]
51. Barnett, S.C.; Linington, C. Myelination: Do astrocytes play a role? *Neuroscientist* **2013**, *19*, 442–450. [CrossRef] [PubMed]
52. Lundgaard, I.; Osório, M.J.; Kress, B.T.; Sanggaard, S.; Nedergaard, M. White matter astrocytes in health and disease. *Neuroscience* **2014**, *276*, 161–173. [CrossRef] [PubMed]
53. Verkhratsky, A.; Nedergaard, M. Physiology of Astroglia. *Physiol. Rev.* **2018**, *98*, 239–389. [CrossRef] [PubMed]
54. Verkhratsky, A.; Nedergaard, M.; Hertz, L. Why are astrocytes important? *NeuroChem. Res.* **2015**, *40*, 389–401. [CrossRef]
55. Verkhratsky, A.; Parpura, V.; Li, B.; Scuderi, C. Astrocytes: The Housekeepers and Guardians of the CNS. *Adv. Neurobiol.* **2021**, *26*, 21–53. [CrossRef] [PubMed]
56. Domingues, H.S.; Portugal, C.C.; Socodato, R.; Relvas, J.B. Oligodendrocyte, Astrocyte, and Microglia Crosstalk in Myelin Development, Damage, and Repair. *Front. Cell Dev. Biol.* **2016**, *4*, 71. [CrossRef]
57. Orthmann-Murphy, J.L.; Abrams, C.K.; Scherer, S.S. Gap junctions couple astrocytes and oligodendrocytes. *J. Mol. NeuroSci.* **2008**, *35*, 101–116. [CrossRef]
58. Nutma, E.; van Gent, D.; Amor, S.; Peferoen, L.A.N. Astrocyte and Oligodendrocyte Cross-Talk in the Central Nervous System. *Cells* **2020**, *9*, 600. [CrossRef]
59. Nagy, J.I.; Ionescu, A.V.; Lynn, B.D.; Rash, J.E. Connexin29 and connexin32 at oligodendrocyte and astrocyte gap junctions and in myelin of the mouse central nervous system. *J. Comp. Neurol.* **2003**, *464*, 356–370. [CrossRef]
60. Markoullis, K.; Sargiannidou, I.; Gardner, C.; Hadjisavvas, A.; Reynolds, R.; Kleopa, K.A. Disruption of oligodendrocyte gap junctions in experimental autoimmune encephalomyelitis. *Glia* **2012**, *60*, 1053–1066. [CrossRef]
61. Richardson, W.D.; Pringle, N.; Mosley, M.J.; Westermark, B.; Dubois-Dalcq, M. A role for platelet-derived growth factor in normal gliogenesis in the central nervous system. *Cell* **1988**, *53*, 309–319. [CrossRef]
62. McKinnon, R.D.; Waldron, S.; Kiel, M.E. PDGF alpha-receptor signal strength controls an RTK rheostat that integrates phosphoinositol 3′-kinase and phospholipase Cgamma pathways during oligodendrocyte maturation. *J. NeuroSci.* **2005**, *25*, 3499–3508. [CrossRef] [PubMed]
63. Moore, C.S.; Milner, R.; Nishiyama, A.; Frausto, R.F.; Serwanski, D.R.; Pagarigan, R.R.; Whitton, J.L.; Miller, R.H.; Crocker, S.J. Astrocytic tissue inhibitor of metalloproteinase-1 (TIMP-1) promotes oligodendrocyte differentiation and enhances CNS myelination. *J. NeuroSci.* **2011**, *31*, 6247–6254. [CrossRef] [PubMed]
64. Siegel, G.J.; Agranoff, B.W.; Albers, R.W.; Fisher, S.K.; Uhler, M.D. *Basic Neurochemistry: Molecular, Cellular and Medical Aspects*, 6th ed.; Lippincott-Raven: New York, NY, USA, 1999.
65. Pfrieger, F.W.; Ungerer, N. Cholesterol metabolism in neurons and astrocytes. *Prog. Lipid Res.* **2011**, *50*, 357–371. [CrossRef] [PubMed]
66. Sofroniew, M.V. Astrocyte Reactivity: Subtypes, States, and Functions in CNS Innate Immunity. *Trends Immunol.* **2020**, *41*, 758–770. [CrossRef] [PubMed]
67. Bronzuoli, M.R.; Facchinetti, R.; Valenza, M.; Cassano, T.; Steardo, L.; Scuderi, C. Astrocyte Function Is Affected by Aging and Not Alzheimer's Disease: A Preliminary Investigation in Hippocampi of 3xTg-AD Mice. *Front. Pharm.* **2019**, *10*, 644. [CrossRef] [PubMed]
68. Valenza, M.; Steardo, L., Jr.; Steardo, L.; Verkhratsky, A.; Scuderi, C. Systemic Inflammation and Astrocyte Reactivity in the Neuropsychiatric Sequelae of COVID-19: Focus on Autism Spectrum Disorders. *Front. Cell NeuroSci.* **2021**, *15*, 748136. [CrossRef]
69. Steardo, L., Jr.; Steardo, L.; Verkhratsky, A.; Scuderi, C. Post-COVID-19 neuropsychiatric syndrome: Is maladaptive glial recovery to blame? *Acta Physiol.* **2021**, *233*, e13717. [CrossRef]
70. Scuderi, C.; Verkhratsky, A.; Parpura, V.; Li, B. Neuroglia in Psychiatric Disorders. *Adv. Neurobiol.* **2021**, *26*, 3–19. [CrossRef]
71. Messersmith, D.J.; Murtie, J.C.; Le, T.Q.; Frost, E.E.; Armstrong, R.C. Fibroblast growth factor 2 (FGF2) and FGF receptor expression in an experimental demyelinating disease with extensive remyelination. *J. NeuroSci. Res.* **2000**, *62*, 241–256. [CrossRef]
72. Albrecht, P.J.; Dahl, J.P.; Stoltzfus, O.K.; Levenson, R.; Levison, S.W. Ciliary neurotrophic factor activates spinal cord astrocytes, stimulating their production and release of fibroblast growth factor-2, to increase motor neuron survival. *Exp. Neurol.* **2002**, *173*, 46–62. [CrossRef] [PubMed]
73. Selmaj, K.; Raine, C.S.; Cannella, B.; Brosnan, C.F. Identification of lymphotoxin and tumor necrosis factor in multiple sclerosis lesions. *J. Clin. Investig.* **1991**, *87*, 949–954. [CrossRef] [PubMed]
74. Cammer, W.; Zhang, H. Maturation of oligodendrocytes is more sensitive to TNF alpha than is survival of precursors and immature oligodendrocytes. *J. Neuroimmunol.* **1999**, *97*, 37–42. [CrossRef]

75. Ding, X.; Yan, Y.; Li, X.; Li, K.; Ciric, B.; Yang, J.; Zhang, Y.; Wu, S.; Xu, H.; Chen, W.; et al. Silencing IFN-γ binding/signaling in astrocytes versus microglia leads to opposite effects on central nervous system autoimmunity. *J. Immunol.* **2015**, *194*, 4251–4264. [CrossRef]
76. De Waard, D.M.; Bugiani, M. Astrocyte-Oligodendrocyte-Microglia Crosstalk in Astrocytopathies. *Front. Cell NeuroSci.* **2020**, *14*, 608073. [CrossRef]
77. Dooves, S.; Bugiani, M.; Postma, N.L.; Polder, E.; Land, N.; Horan, S.T.; van Deijk, A.L.; van de Kreeke, A.; Jacobs, G.; Vuong, C.; et al. Astrocytes are central in the pathomechanisms of vanishing white matter. *J. Clin. Investig.* **2016**, *126*, 1512–1524. [CrossRef]
78. Li, L.; Tian, E.; Chen, X.; Chao, J.; Klein, J.; Qu, Q.; Sun, G.; Sun, G.; Huang, Y.; Warden, C.D.; et al. GFAP Mutations in Astrocytes Impair Oligodendrocyte Progenitor Proliferation and Myelination in an hiPSC Model of Alexander Disease. *Cell Stem Cell* **2018**, *23*, 239–251.e6. [CrossRef]
79. Love, S. Demyelinating diseases. *J. Clin. Pathol.* **2006**, *59*, 1151–1159. [CrossRef]
80. Touma, L.; Muccilli, A. Diagnosis and Management of Central Nervous System Demyelinating Disorders. *Neurol. Clin.* **2022**, *40*, 113–131. [CrossRef]
81. Henríquez, K.; Molt, F.; Gajardo, J.; Cortés, B.; Ramirez-Santana, M. Sociodemographic and clinical characteristics of people with multiple sclerosis and neuro-myelitis optica spectrum disorder in a central northern region of Chile: A prevalence study. *Mult. Scler. Relat. Disord.* **2022**, *61*, 103750. [CrossRef]
82. Pegoretti, V.; Swanson, K.A.; Bethea, J.R.; Probert, L.; Eisel, U.L.M.; Fischer, R. Inflammation and Oxidative Stress in Multiple Sclerosis: Consequences for Therapy Development. *Oxid. Med. Cell Longev.* **2020**, *2020*, 7191080. [CrossRef] [PubMed]
83. Orefice, N.S.; Alhouayek, M.; Carotenuto, A.; Montella, S.; Barbato, F.; Comelli, A.; Calignano, A.; Muccioli, G.G.; Orefice, G. Oral Palmitoylethanolamide Treatment Is Associated with Reduced Cutaneous Adverse Effects of Interferon-β1a and Circulating Proinflammatory Cytokines in Relapsing-Remitting Multiple Sclerosis. *Neurotherapeutics* **2016**, *13*, 428–438. [CrossRef] [PubMed]
84. Barbierato, M.; Borri, M.; Facci, L.; Zusso, M.; Skaper, S.D.; Giusti, P. Expression and Differential Responsiveness of Central Nervous System Glial Cell Populations to the Acute Phase Protein Serum Amyloid A. *Sci. Rep.* **2017**, *7*, 12158. [CrossRef] [PubMed]
85. Ristori, G.; Laurenti, F.; Stacchini, P.; Gasperini, C.; Buttinelli, C.; Pozzilli, C.; Salvetti, M. Serum amyloid A protein is elevated in relapsing-remitting multiple sclerosis. *J. Neuroimmunol.* **1998**, *88*, 9–12. [CrossRef]
86. Chung, T.F.; Sipe, J.D.; McKee, A.; Fine, R.E.; Schreiber, B.M.; Liang, J.S.; Johnson, R.J. Serum amyloid A in Alzheimer's disease brain is predominantly localized to myelin sheaths and axonal membrane. *Amyloid* **2000**, *7*, 105–110. [CrossRef]
87. Barbierato, M.; Facci, L.; Marinelli, C.; Zusso, M.; Argentini, C.; Skaper, S.D.; Giusti, P. Co-ultramicronized Palmitoylethanolamide/Luteolin Promotes the Maturation of Oligodendrocyte Precursor Cells. *Sci. Rep.* **2015**, *5*, 16676. [CrossRef]
88. Skaper, S.D.; Barbierato, M.; Facci, L.; Borri, M.; Contarini, G.; Zusso, M.; Giusti, P. Co-Ultramicronized Palmitoylethanolamide/Luteolin Facilitates the Development of Differentiating and Undifferentiated Rat Oligodendrocyte Progenitor Cells. *Mol. Neurobiol.* **2018**, *55*, 103–114. [CrossRef]
89. Facci, L.; Barbierato, M.; Fusco, M.; Giusti, P.; Zusso, M. Co-Ultramicronized Palmitoylethanolamide/Luteolin-Induced Oligodendrocyte Precursor Cell Differentiation is Associated With Tyro3 Receptor Upregulation. *Front. Pharm.* **2021**, *12*, 698133. [CrossRef]
90. Cassano, T.; Magini, A.; Giovagnoli, S.; Polchi, A.; Calcagnini, S.; Pace, L.; Lavecchia, M.A.; Scuderi, C.; Bronzuoli, M.R.; Ruggeri, L.; et al. Early intrathecal infusion of everolimus restores cognitive function and mood in a murine model of Alzheimer's disease. *Exp. Neurol.* **2019**, *311*, 88–105. [CrossRef]
91. Contarini, G.; Franceschini, D.; Facci, L.; Barbierato, M.; Giusti, P.; Zusso, M. A co-ultramicronized palmitoylethanolamide/luteolin composite mitigates clinical score and disease-relevant molecular markers in a mouse model of experimental autoimmune encephalomyelitis. *J. Neuroinflamm.* **2019**, *16*, 126. [CrossRef]
92. Rahimi, A.; Faizi, M.; Talebi, F.; Noorbakhsh, F.; Kahrizi, F.; Naderi, N. Interaction between the protective effects of cannabidiol and palmitoylethanolamide in experimental model of multiple sclerosis in C57BL/6 mice. *Neuroscience* **2015**, *290*, 279–287. [CrossRef] [PubMed]
93. Di Cesare Mannelli, L.; D'Agostino, G.; Pacini, A.; Russo, R.; Zanardelli, M.; Ghelardini, C.; Calignano, A. Palmitoylethanolamide is a disease-modifying agent in peripheral neuropathy: Pain relief and neuroprotection share a PPAR-alpha-mediated mechanism. *Mediat. Inflamm.* **2013**, *2013*, 328797. [CrossRef] [PubMed]
94. González-Hernández, A.; Martínez-Lorenzana, G.; Rodríguez-Jiménez, J.; Rojas-Piloni, G.; Condés-Lara, M. Intracisternal injection of palmitoylethanolamide inhibits the peripheral nociceptive evoked responses of dorsal horn wide dynamic range neurons. *J. Neural. Transm.* **2015**, *122*, 369–374. [CrossRef]
95. Truini, A.; Biasiotta, A.; Di Stefano, G.; La Cesa, S.; Leone, C.; Cartoni, C.; Federico, V.; Petrucci, M.T.; Cruccu, G. Palmitoylethanolamide restores myelinated-fibre function in patients with chemotherapy-induced painful neuropathy. *CNS Neurol. Disord. Drug Targets* **2011**, *10*, 916–920. [CrossRef] [PubMed]
96. Peters, A. Structural changes that occur during normal aging of primate cerebral hemispheres. *NeuroSci. Biobehav. Rev.* **2002**, *26*, 733–741. [CrossRef]
97. Peters, A. The structure of myelin sheaths in the central nervous system of Xenopus laevis (Daudin). *J. Biophys. BioChem. Cytol.* **1960**, *7*, 121–126. [CrossRef]

98. Sowell, E.R.; Peterson, B.S.; Thompson, P.M.; Welcome, S.E.; Henkenius, A.L.; Toga, A.W. Mapping cortical change across the human life span. *Nat. NeuroSci.* **2003**, *6*, 309–315. [CrossRef]
99. Feldman, M.L.; Peters, A. Ballooning of myelin sheaths in normally aged macaques. *J. Neurocytol.* **1998**, *27*, 605–614. [CrossRef]
100. Deber, C.M.; Reynolds, S.J. Central nervous system myelin: Structure, function, and pathology. *Clin. BioChem.* **1991**, *24*, 113–134. [CrossRef]
101. Kormas, P.; Moutzouri, A. Current Psychological Approaches in Neurodegenerative Diseases. In *Handbook of Computational Neurodegeneration*; Vlamos, P., Kotsireas, I.S., Tarnanas, I., Eds.; Springer International Publishing: Cham, Germany, 2020; pp. 1–29.
102. Sohn, Y.H.; Kim, J.S. The influence of white matter hyperintensities on the clinical features of Parkinson's disease. *Yonsei. Med. J.* **1998**, *39*, 50–55. [CrossRef]
103. Piccini, P.; Pavese, N.; Canapicchi, R.; Paoli, C.; Del Dotto, P.; Puglioli, M.; Rossi, G.; Bonuccelli, U. White matter hyperintensities in Parkinson's disease. Clinical correlations. *Arch. Neurol.* **1995**, *52*, 191–194. [CrossRef] [PubMed]
104. Lee, S.J.; Kim, J.S.; Lee, K.S.; An, J.Y.; Kim, W.; Kim, Y.I.; Kim, B.S.; Jung, S.L. The severity of leukoaraiosis correlates with the clinical phenotype of Parkinson's disease. *Arch. Gerontol. Geriatr.* **2009**, *49*, 255–259. [CrossRef] [PubMed]
105. Choi, S.A.; Evidente, V.G.; Caviness, J.N.; Shill, H.A.; Sabbagh, M.N.; Connor, D.J.; Hentz, J.G.; Adler, C.H.; Beach, T.G. Are there differences in cerebral white matter lesion burdens between Parkinson's disease patients with or without dementia? *Acta Neuropathol.* **2010**, *119*, 147–149. [CrossRef] [PubMed]
106. Beyer, M.K.; Aarsland, D.; Greve, O.J.; Larsen, J.P. Visual rating of white matter hyperintensities in Parkinson's disease. *Mov. Disord.* **2006**, *21*, 223–229. [CrossRef] [PubMed]
107. Bartzokis, G. Alzheimer's disease as homeostatic responses to age-related myelin breakdown. *Neurobiol. Aging* **2011**, *32*, 1341–1371. [CrossRef]
108. Dean, D.C., 3rd; Hurley, S.A.; Kecskemeti, S.R.; O'Grady, J.P.; Canda, C.; Davenport-Sis, N.J.; Carlsson, C.M.; Zetterberg, H.; Blennow, K.; Asthana, S.; et al. Association of Amyloid Pathology With Myelin Alteration in Preclinical Alzheimer Disease. *JAMA Neurol.* **2017**, *74*, 41–49. [CrossRef]
109. Benitez, A.; Fieremans, E.; Jensen, J.H.; Falangola, M.F.; Tabesh, A.; Ferris, S.H.; Helpern, J.A. White matter tract integrity metrics reflect the vulnerability of late-myelinating tracts in Alzheimer's disease. *Neuroimage Clin.* **2014**, *4*, 64–71. [CrossRef]
110. Zhan, X.; Jickling, G.C.; Ander, B.P.; Stamova, B.; Liu, D.; Kao, P.F.; Zelin, M.A.; Jin, L.W.; DeCarli, C.; Sharp, F.R. Myelin basic protein associates with AβPP, Aβ1-42, and amyloid plaques in cortex of Alzheimer's disease brain. *J. Alzheimers Dis.* **2015**, *44*, 1213–1229. [CrossRef]
111. Wang, S.S.; Zhang, Z.; Zhu, T.B.; Chu, S.F.; He, W.B.; Chen, N.H. Myelin injury in the central nervous system and Alzheimer's disease. *Brain Res. Bull.* **2018**, *140*, 162–168. [CrossRef]
112. Valenza, M.; Scuderi, C. How useful are biomarkers for the diagnosis of Alzheimer's disease and especially for its therapy? *Neural Regen. Res.* **2022**, *17*, 2205–2207. [CrossRef]
113. Roher, A.E.; Weiss, N.; Kokjohn, T.A.; Kuo, Y.M.; Kalback, W.; Anthony, J.; Watson, D.; Luehrs, D.C.; Sue, L.; Walker, D.; et al. Increased A beta peptides and reduced cholesterol and myelin proteins characterize white matter degeneration in Alzheimer's disease. *Biochemistry* **2002**, *41*, 11080–11090. [CrossRef] [PubMed]
114. Quintela-López, T.; Ortiz-Sanz, C.; Serrano-Regal, M.P.; Gaminde-Blasco, A.; Valero, J.; Baleriola, J.; Sánchez-Gómez, M.V.; Matute, C.; Alberdi, E. Aβ oligomers promote oligodendrocyte differentiation and maturation via integrin β1 and Fyn kinase signaling. *Cell Death Dis.* **2019**, *10*, 445. [CrossRef] [PubMed]
115. Wu, Y.; Ma, Y.; Liu, Z.; Geng, Q.; Chen, Z.; Zhang, Y. Alterations of myelin morphology and oligodendrocyte development in early stage of Alzheimer's disease mouse model. *NeuroSci. Lett.* **2017**, *642*, 102–106. [CrossRef] [PubMed]
116. Facchinetti, R.; Valenza, M.; Gomiero, C.; Mancini, G.F.; Steardo, L.; Campolongo, P.; Scuderi, C. Co-Ultramicronized Palmitoylethanolamide/Luteolin Restores Oligodendrocyte Homeostasis via Peroxisome Proliferator-Activated Receptor-α in an In Vitro Model of Alzheimer's Disease. *Biomedicines* **2022**, *10*, 1236. [CrossRef] [PubMed]
117. Vanzulli, I.; Papanikolaou, M.; De-La-Rocha, I.C.; Pieropan, F.; Rivera, A.D.; Gomez-Nicola, D.; Verkhratsky, A.; Rodríguez, J.J.; Butt, A.M. Disruption of oligodendrocyte progenitor cells is an early sign of pathology in the triple transgenic mouse model of Alzheimer's disease. *Neurobiol. Aging* **2020**, *94*, 130–139. [CrossRef]
118. Bergles, D.E.; Richardson, W.D. Oligodendrocyte Development and Plasticity. *Cold Spring Harb. Perspect. Biol.* **2015**, *8*, a020453. [CrossRef]
119. Franklin, R.J.M.; Ffrench-Constant, C. Regenerating CNS myelin—from mechanisms to experimental medicines. *Nat. Rev. NeuroSci.* **2017**, *18*, 753–769. [CrossRef]
120. Back, S.A.; Tuohy, T.M.; Chen, H.; Wallingford, N.; Craig, A.; Struve, J.; Luo, N.L.; Banine, F.; Liu, Y.; Chang, A.; et al. Hyaluronan accumulates in demyelinated lesions and inhibits oligodendrocyte progenitor maturation. *Nat. Med.* **2005**, *11*, 966–972. [CrossRef]
121. Hammond, T.R.; Gadea, A.; Dupree, J.; Kerninon, C.; Nait-Oumesmar, B.; Aguirre, A.; Gallo, V. Astrocyte-Derived Endothelin-1 Inhibits Remyelination through Notch Activation. *Neuron* **2014**, *81*, 1442. [CrossRef]
122. Hamaguchi, M.; Muramatsu, R.; Fujimura, H.; Mochizuki, H.; Kataoka, H.; Yamashita, T. Circulating transforming growth factor-β1 facilitates remyelination in the adult central nervous system. *Elife* **2019**, *8*, 41869. [CrossRef]
123. Linnerbauer, M.; Rothhammer, V. Protective Functions of Reactive Astrocytes Following Central Nervous System Insult. *Front. Immunol.* **2020**, *11*, 573256. [CrossRef] [PubMed]

124. Facchinetti, R.; Valenza, M.; Bronzuoli, M.R.; Menegoni, G.; Ratano, P.; Steardo, L.; Campolongo, P.; Scuderi, C. Looking for a Treatment for the Early Stage of Alzheimer's Disease: Preclinical Evidence with Co-Ultramicronized Palmitoylethanolamide and Luteolin. *Int. J. Mol. Sci.* **2020**, *21*, 3802. [CrossRef] [PubMed]
125. Cipriano, M.; Esposito, G.; Negro, L.; Capoccia, E.; Sarnelli, G.; Scuderi, C.; De Filippis, D.; Steardo, L.; Iuvone, T. Palmitoylethanolamide Regulates Production of Pro-Angiogenic Mediators in a Model of β Amyloid-Induced Astrogliosis In Vitro. *CNS Neurol. Disord. Drug Targets* **2015**, *14*, 828–837. [CrossRef] [PubMed]
126. Scuderi, C.; Bronzuoli, M.R.; Facchinetti, R.; Pace, L.; Ferraro, L.; Broad, K.D.; Serviddio, G.; Bellanti, F.; Palombelli, G.; Carpinelli, G.; et al. Ultramicronized palmitoylethanolamide rescues learning and memory impairments in a triple transgenic mouse model of Alzheimer's disease by exerting anti-inflammatory and neuroprotective effects. *Transl. Psychiatry* **2018**, *8*, 32. [CrossRef]
127. Bronzuoli, M.R.; Facchinetti, R.; Steardo, L., Jr.; Romano, A.; Stecca, C.; Passarella, S.; Steardo, L.; Cassano, T.; Scuderi, C. Palmitoylethanolamide Dampens Reactive Astrogliosis and Improves Neuronal Trophic Support in a Triple Transgenic Model of Alzheimer's Disease: In Vitro and In Vivo Evidence. *Oxidative Med. Cell. Longev.* **2018**, *2018*, 4720532. [CrossRef]
128. Manni, B.; Federzoni, L.; Zucchi, P.; Fabbo, A. Co-ultraPEALut Effect on Mild Cognitive Impairment: A Retrospective Observational Study. *Acta Sci. Neurol.* **2021**, *4*, 8–14. [CrossRef]
129. Calabrò, R.S.; Naro, A.; De Luca, R.; Leonardi, S.; Russo, M.; Marra, A.; Bramanti, P. PEALut efficacy in mild cognitive impairment: Evidence from a SPECT case study! *Aging Clin. Exp. Res.* **2016**, *28*, 1279–1282. [CrossRef]

biomolecules

MDPI

Review

Palmitoylethanolamide and Related ALIAmides for Small Animal Health: State of the Art

Giorgia della Rocca [1,*] and Giovanni Re [2]

1 Department of Veterinary Medicine, Centro di Ricerca sul Dolore Animale (CeRiDA), University of Perugia, 06123 Perugia, Italy
2 Department of Veterinary Sciences, Division of Pharmacology & Toxicology, University of Turin, 10095 Grugliasco, Torino, Italy
* Correspondence: giorgia.dellarocca@unipg.it

Abstract: ALIAmides are a family of fatty acid amides whose name comes from their mechanism of action, i.e., the Autacoid Local Injury Antagonism (ALIA). Actually, the ALIAmide parent molecule, palmitoylethanolamide (PEA), is locally produced on demand from a cell membrane precursor in order to control immune inflammatory cell responses, avert chronic non-resolving inflammation, and limit the resulting clinical signs. ALIAmide sister compounds, such as Adelmidrol and palmitoylglucosamine, share mechanisms of action with PEA and may also increase endogenous levels of PEA. Provided that their respective bioavailability is properly addressed (e.g., through decreasing the particle size through micronization), exogenously administered ALIAmides thus mimic or sustain the prohomeostatic functions of endogenous PEA. The aim of the present paper is to review the main findings on the use of ALIAmides in small animals as a tribute to the man of vision who first believed in this "according-to-nature" approach, namely Francesco della Valle. After briefly presenting some key issues on the molecular targets, metabolism, and pharmacokinetics of PEA and related ALIAmides, here we will focus on the preclinical and clinical studies performed in dogs and cats. Although more data are still needed, ALIAmides may represent a novel and promising approach to small animal health.

Keywords: ALIAmides; dogs; cats; atopic dermatitis; osteoarthritis; mast cells; palmitoylethanolamide; Adelmidrol; palmitoylglucosamine

Citation: della Rocca, G.; Re, G. Palmitoylethanolamide and Related ALIAmides for Small Animal Health: State of the Art. *Biomolecules* **2022**, *12*, 1186. https://doi.org/10.3390/biom12091186

Academic Editors: Salvatore Cuzzocrea and Rosalia Crupi

Received: 12 July 2022
Accepted: 23 August 2022
Published: 26 August 2022

1. Introduction

ALIAmides are a family of fatty acid amides sharing a common mechanism of action, i.e., the autacoid local injury antagonism (ALIA), originally proposed in the mid-1990s by the late Nobel prize winner Rita Levi Montalcini [1]. The term "autacoid" comes from the Greek "autos" (self) and "acos" (healing or remedy) and refers to cell-produced factors that act locally near their site of synthesis [2]. In particular, the autacoid mechanism of ALIAmides serves auto-protective purposes through the down-modulation of cell hyperactivity (mainly immune cells), thus controlling inflammatory responses and limiting tissue damage [3]. It was originally observed that the ALIAmide parent molecule, palmitoylethanolamide (PEA), down-modulates rat mast cell behavior after challenge [1,4], as later confirmed in companion animals [5–7]. Different cell populations were also shown to be targets of PEA, with macrophages, keratinocytes, T and B cells, and glial cells being negatively controlled by PEA once overactivated [8–18].

Palmitoylethanolamide is a body's own (endogenous) N-acylethanolamine, produced "on demand" by several cell types, including mast cells, astrocytes, and microglia [19–21]. Interestingly, the autoprotective function of PEA was first suggested in dogs. It was indeed found that (i) the canine myocardium produces PEA in response to ischemic injury [22,23], and (ii) the canine brain possesses the biosynthetic and degradative machinery for PEA [24]. Since the 1980s, knowledge has advanced considerably in the field of ALIAmides, mainly due to the renewed interest in these molecules driven by the discovery of the PEA congener and the endocannabinoid mediator anandamide arachidonoylethanolamide (AEA) [25].

In those days, an enlightened man, Francesco della Valle (to whom the present special issue is dedicated), was launching his own science-driven entrepreneurial activity in the field of human and animal health, focused on innovation and networking [26]. During his previous experience in managing a pharmaceutical firm, he had been actively cooperating with two eminent scientists, Rita Levi Montalcini [27] and Erminio Costa [28,29] (Figure 1).

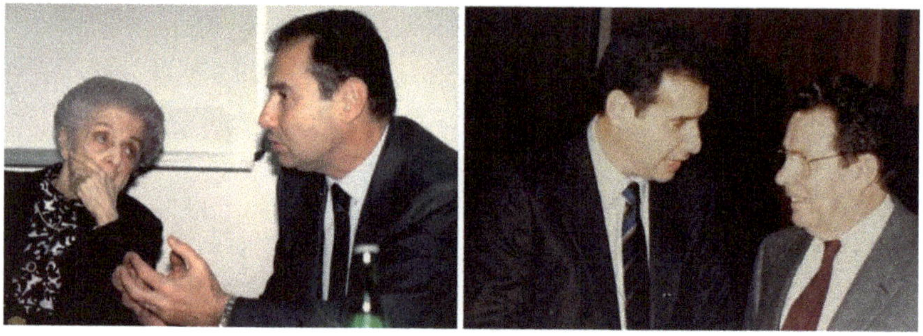

Figure 1. Francesco della Valle in the 1990s during brainstorming with his main scientific mentors, namely Rita Levi Montalcini (**left**) and Erminio Costa (**right**).

Both of them repeatedly invited della Valle to orientate the focus and efforts toward biological modulation mechanisms while learning from nature how to design a strategy of modulation [30,31]. Accordingly, della Valle based his strategic business plan on a "hypothetical-deductive" approach to inflammation and pain, according to regulatory pathways laid down by nature and intended to maintain a homeostatic balance in the body when challenged by stress or injury. This was the ALIAmide project. Although the historical view of ALIAmides is beyond the scope of the present review, it must be acknowledged that the ALIAmide story began in this particular framework, and most of the research data that will be reviewed here were born within it.

Besides PEA, ALIAmides currently comprise several lipid compounds, ranging from Adelmidrol (the diethanolamide derivative of azelaic acid) to palmitoylglucosamine (PGA), oleoylethanolamide, and many others (Figure 2).

Their respective mechanisms of action have been (and still are being) investigated and appear to be profoundly interconnected to the parent compound PEA, which is by far the most studied ALIAmide [3,32,33]. A brief overview of their molecular mechanisms will be given in the following paragraphs.

A large body of evidence has been accumulating on the prohomeostatic functions of ALIAmides in several diseases sustained by non-resolving inflammatory and neuroinflammatory responses. The findings have been reviewed by several excellent papers, to which the reader is encouraged to refer [3,32–37]. After addressing a few general key points on ALIAmides, here we will focus exclusively on the main studies performed on small animals.

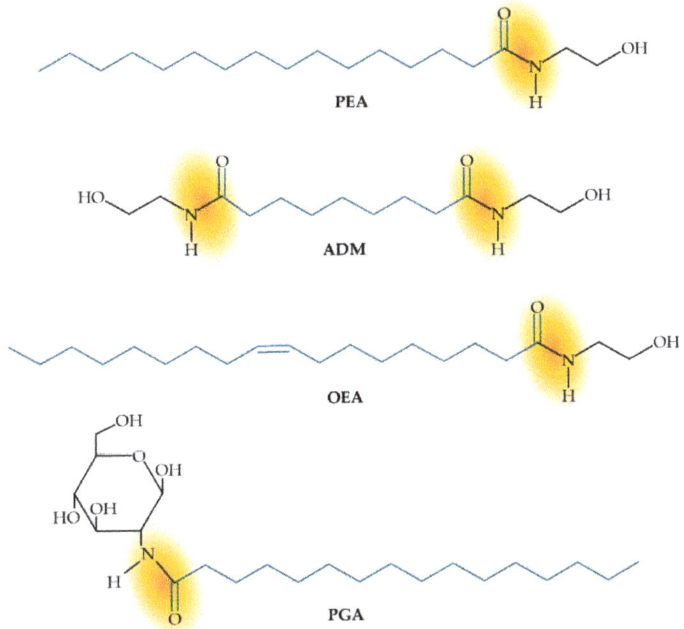

Figure 2. Chemical structure of the main ALIAmides. The amide bond (yellow shadow) and the fatty acid (blue color) are highlighted. ADM = Adelmidrol, OEA = oleoylethanolamide, PEA = palmitoylethanolamide, PGA = palmitoylglucosamine.

2. Mimicking and Supporting the Healing Power of Nature

Palmitoylethanolamide is produced starting from a glycerophospholipid precursor in the cell membrane and degraded by two amidases located in the cell membrane and lysosome, respectively, i.e., the fatty acid amide amidase (FAAH) and N-acylethanolamine acid amidase (NAAA) [38–41]. Although the endogenous levels of PEA are strictly regulated by these biosynthetic and degradative metabolic pathways [38], great deals of evidence suggest that PEA metabolism may be disturbed under certain conditions, such as chronic inflammatory disorders [42]. Indeed, the local levels of PEA change during disease states, and decreased levels are considered to contribute to disease development [8,43,44]. For example, a significant decrease in the local level of PEA has been found in different chronic pain models [45–47] as well as in human patients affected by visceral and somatic pain [48–50]. Interestingly, it was also shown that normalizing PEA levels through the inhibition of PEA degradative pathways resulted in reduced inflammation and pain relief in a rat model of osteoarthritis pain [47].

On the other side, PEA levels may increase in response to cell damage, as shown in epidermal cells subjected to UV irradiation [51] and the lesional skin of privately-owned dogs affected with atopic dermatitis [52] as well as the colons of dogs with chronic enteropathy [53].

It is currently accepted that changes in PEA levels are either suggestive of a loss of protection against inflammation/pain (i.e., decreased levels) or a compensatory synthesis in the attempt to limit disease severity (i.e., increased levels). Accordingly, the exogenous administration of PEA to effectively 'top up' the body's own supply is regarded as a promising approach [54]. Interestingly, other ALIAmides, such as Adelmidrol and PGA, have recently been found to increase the endogenous levels of PEA [55–57].

3. A Brief Insight into PEA Metabolism and Molecular Targets

As mentioned above, the biosynthesis of PEA occurs "on demand" in the cell membrane through the enzymatic hydrolysis of its glycerophospholipid precursor N-acyl-phosphatidylethanolamine [39,40]. Although early studies suggested the existence of a facilitated membrane transport [19,58], PEA can flip between the inner and outer leaflets of the plasma membrane thanks to its lipophilic nature [59]. Indeed intracellular binding proteins (i.e., fatty acid binding proteins and heat-shock proteins) are required for PEA trafficking within the cytosol [60]. Binding proteins transport PEA to catabolic enzymes (e.g., FAAH and NAAA) [41] and effector proteins [61–63].

Among the latter, the nuclear peroxisome proliferator-activated receptor alpha (PPARα) is of particular interest because it negatively interferes with inflammatory gene expression by regulating the IκBα/NF-κB pathway [64]. PPARα is not the only molecular target responsible for the prohomeostatic properties of PEA [65–69], as many other receptors are being increasingly recognized as mediating PEA functions, such as the GPR55 (G-protein-coupled receptor 55) [70,71], cannabinoid receptors type 1 and 2 (CB1 and CB2) [33,57,72,73] as well as the so-called "pain receptor" [74], i.e., the transient receptor potential vanilloid 1 (TRPV1) [75–78].

Interestingly, this heterogeneous family of PEA molecular targets is being extensively studied in companion animals, with their distribution being confirmed in several canine and feline cell types [79–92], as recently reviewed [3,32,93].

Notably, while PEA is a direct agonist of PPARα [66], its action on CB1, CB2, and even TRPV1 is indirect [73,76–78]. In particular, PEA can activate these latter three receptors thanks to its ability to (i) elevate their levels, (ii) reduce their degradation, or (iii) increase the receptor affinity of endocannabinoids, like AEA and 2-arachidonoylglycerol (2-AG) [35,57,72,73,76,78]. The mechanism has been termed the "entourage effect" [73,76,78] (Figure 3) and has been specifically shown in dogs [72]. In Beagle dogs, orally administered bioavailable micro-PEA (i.e., ultra-micronized, see below) resulted in a significant and up to ~20-fold increase in the plasma levels of 2-AG [72] (Figure 3B).

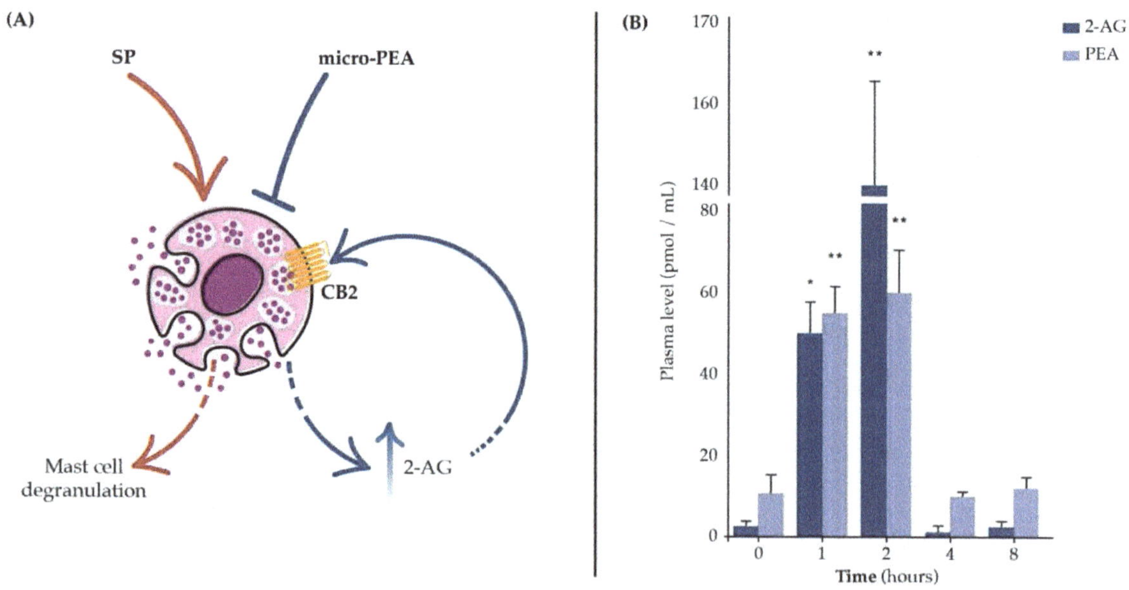

Figure 3. PEA may act on its molecular targets either directly or indirectly by increasing the agonism toward endocannabinoid receptors for which it has a low affinity. The latter mechanism is referred to as the "entourage effect". The figure illustrates the in vitro (**A**) and in vivo (**B**) demonstrations of the

entourage effect of bioavailable formulations of PEA (i.e., micro-PEA, please see next paragraph for further details) through increasing the levels of the endocannabinoid 2-AG. (**A**) Indirect agonism of micro-PEA on CB2 underlies the inhibitory effects on SP-induced mast cell degranulation, mediated by the stimulation of 2-AG biosynthesis [57]. (**B**) Following a single dietary supplementation with micro-PEA to hypersensitive Beagle dogs, not only plasma levels of PEA but also plasma levels of 2-AG significantly increase (* $p < 0.05$ and ** $p < 0.001$ versus the basal levels, time 0) [3]. (**B**) is slightly modified from [3]. 2-AG = 2-arachidonoylglycerol, CB2 = cannabinoid receptor type 2, micro-PEA = micronized or ultramicronized palmitoylethanolamide, SP = substance P.

To date, the molecular mechanisms of other ALIAmides are much less investigated than PEA's. Besides increasing PEA levels, as previously mentioned, these fatty acid amides are suggested to interact with different receptors. PGA, for example, is considered to exert its protective function through a toll-like receptor 4 antagonism [94], while the precise molecular targets of Adelmidrol are still debated [55,95].

4. Key Pharmacokinetic Issues

A key aspect that has to be taken into account when dealing with the use of ALIAmides for health purposes is their respective physicochemical features. Some ALIAmides are more appropriate for oral use, while others are particularly suitable for topical applications thanks to their amphipathic nature (e.g., Adelmidrol) [95,96].

PEA and PGA are both highly lipophilic compounds (log $p > 5$) [97,98], with their oral use being limited by their intrinsic low dissolution rate, absorption, and bioavailability [98,99]. Particle size reduction is one of the most compelling and practical strategies for improving pharmacokinetics and boosting functional properties following oral administration [100,101]. Provided the route of administration is oral, most of the studies presented below investigated "micro-PEA" and "micro-PGA" accordingly. Micro-ALIAmides result from micro-grinding a particular ALIAmide—either alone or together with adjuvants (typically antioxidants)—in order to downsize the particles to diameters in the range of 0.6–10 μm. Indeed, after the administration of micro-PEA, the plasma concentration of PEA was significantly higher compared to unprocessed (naïve) PEA [98]. Accordingly, superior effects have been shown for micro-PEA and micro -PGA compared to naïve PEA and PGA, respectively, in different inflammatory disease models [97,102,103].

Specifically, in dogs, a single oral administration of micro-PEA resulted in a five-fold increase in PEA plasma levels, with a peak between 1 and 2 h [72,104]. Interestingly, plasma levels correlated well with the clinical effects at different timepoints, although the latter lasted longer than the plasma elevation of PEA [104]. This was considered to depend on the ability of PEA to up-regulate the levels or enhance the action of other related bioactive endocannabinoids [104], according to the so-called "entourage hypothesis" briefly outlined in Figure 3.

5. Preclinical and Clinical Results in Small Animals

5.1. Dermatological Field

So far, most of the veterinary research on ALIAmides has been focused on the dermatological field [105]. Ex vivo and in vitro studies, performed on feline and canine skin mast cells, respectively, have confirmed that micro-PEA down-modulates allergic hyperactivity, prominently decreasing mediator release (i.e., degranulation) [5,7]. The ability of micro-PEA to down-modulate mast cell degranulation was also recently shown in canine skin organ cultures challenged with different concentrations of compound 48/80 (a well-known secretagogue which triggers mast cell degranulation) [6]. Not only did micro-PEA significantly counteract the increase of degranulating mast cells, but it also lowered the histamine content within the culture medium and the diameter of epidermal blood capillaries [6].

Moreover, down-modulation of skin mast cell releasability was observed in canine skin wounds (punch biopsies) topically treated with the ALIAmide Adelmidrol (2%) [106], with a parallel improvement in wound healing being detected [107].

Moving to in vivo studies, a growing body of evidence confirms that ALIAmides can efficiently benefit veterinary patients with hypersensitive skin disorders. In a double-blinded placebo-controlled cross-over study performed on dogs with experimental allergic dermatitis, the dietetic supplementation with micro-PEA at 15 mg/kg/day for 7 days delayed the development of clinical signs (i.e., pruritus and skin lesions) compared to the placebo-treated group [108]. Moreover, in a canine model of skin allergy, a single oral administration of micro-PEA (3, 10, and 30 mg/kg) significantly reduced the antigen-induced wheal area, with a maximum inhibitory effect at a 10 mg/kg dose [104]. Interestingly, topical application of Adelmidrol (2%) for 3 and 6 consecutive days gave similar results in terms of allergic wheal inhibition [96].

On the clinical side, two studies were performed on allergic cats. The first one investigated feline patients with eosinophilic plaques and eosinophilic granuloma, orally given micro-PEA (10 mg/kg daily) for 1 month as the sole intervention. Clinical improvement of pruritus, erythema, alopecia, and eosinophilic lesions was observed in 67% of them, with no side effects or adverse reactions being reported [7]. The second was conducted in 60 allergic cats with the aim of evaluating whether micro-PEA (15 mg/kg) could delay the relapse of clinical signs after steroid withdrawal [109]. A significant difference in the mean time-to-flare between the treated and placebo group was observed (40.5 days in the micro-PEA group vs. 22.2 days in the placebo group), suggesting that the ALIAmide exerts an excellent proactive function in preventing feline allergic flares after steroid withdrawal [109].

Some interesting clinical trials were also performed on allergic dogs. A double-blinded randomized placebo-controlled cross-over study in privately-owned dogs with either food-induced or non-food-induced atopic dermatitis showed that dietary integration with micro-PEA (15 mg/kg daily for 45 days) significantly decreased the severity of clinical signs (as assessed by the Canine Atopic Dermatitis Extension and Severity Index) [110].

An open multicentric study performed in 160 client-owned dogs with non-seasonal atopic dermatitis orally administered micro-PEA (10 mg/kg daily for 56 days) confirmed the ability of the ALIAmide to benefit allergic patients [111]. Pruritus (as measured on a Visual Analogue Scale) and clinically assessed skin lesions (Canine Atopic Dermatitis Lesion Index) were significantly reduced by the study end. Moreover, 45% of dogs reached the quality of life values described for healthy animals [111].

Finally, an open-label observational study was performed in privately-owned dogs with atopic dermatitis and pruritus lasting longer than 4 weeks, topically treated with Adelmidrol (2%) twice daily for 30 days. Not only a significant decrease in pruritus and erythema (both on owner and veterinarian assessment) was observed, but body odor and quality of life markedly improved by the study's end [112].

5.2. Other Health Needs

Although studies in small animals are still scarce, there is growing evidence that endocannabinoid-like ALIAmides play key roles in the health of different body organs, such as the gastrointestinal tract [113,114] and the nervous system [32,34,37], as well as the upper and lower urinary tract [115–118] and the musculoskeletal system [97,119,120]. In addition, the deep involvement of ALIAmides in obesity-induced metainflammation is becoming increasingly evident [69,113,121–123].

Actually, a preliminary study in dogs affected with chronic diarrhea demonstrated that dietetic supplementation with micro-PEA (10 mg/kg for 30 days) reduced the Canine Inflammatory Bowel Disease Activity Index (CIBDAI) score [53], in line with recent findings from animals with experimentally-induced colitis [124]. According to the experimental studies, the enteroprotective effect of PEA may depend upon the direct and indirect activation of PPAR-α and CB2 receptors [124–130], whose expression has been recently confirmed in the canine and feline gastrointestinal tract [86,87].

Interestingly, a dietetic supplement containing micro-PEA was also described to benefit a Syrian hamster with urolithiasis and diminish the disease recurrence after surgical treatment [131]. Moreover, micro-PGA has recently been shown to decrease inflammation and pain in a murine model of feline interstitial cystitis [132].

In the musculoskeletal field, an open-field trial on client-owned adult dogs with chronic osteoarthritis and persistent lameness has recently been performed. Dogs were supplemented for 4 weeks with a complementary feed containing PEA co-ultramicronized with the natural antioxidant quercetin (i.e., PEA-q, 24 mg/kg body weight). The severity of chronic pain and its interference with the dog's normal functioning significantly decreased as assessed with the Canine Brief Pain Inventory (CBPI) questionnaire. Moreover, lameness (either assessed on a 0–4 clinical scale or through a dynamic gait analysis) significantly improved [133].

Dogs with osteoarthritis also benefited from a long-term dietary integration with the ALIAmide PGA co-micronized with curcumin, administered as an add-on to conservative measures. One trial has been performed [134], where micro-PGA was added for 2 months to the individual management plan of 181 dogs with osteoarthritis. A significant decrease in lameness and pain as assessed by the veterinarian was observed. Moreover, owner-evaluated mobility impairment and pain behaviors also improved [134].

It is finally noteworthy that the topical administration of an Adelmidrol (2%) mucoadhesive gel in combination with dental prophylaxis resulted in less gingival inflammation and longer duration of dental scaling benefits in treated dogs compared to match untreated group [135].

Taken together, the data from preclinical and clinical trials point towards the promising role of ALIAmides in small animal health (Figure 4). Moreover, the presence of PEA and OEA, as well as other ALIAmides in food sources [136], in addition to their robust safety profile [36,97,137], are the foundation for their dietary use. Accordingly, several complementary feeds for dogs and cats have been developed and are being marketed in Europe and North America.

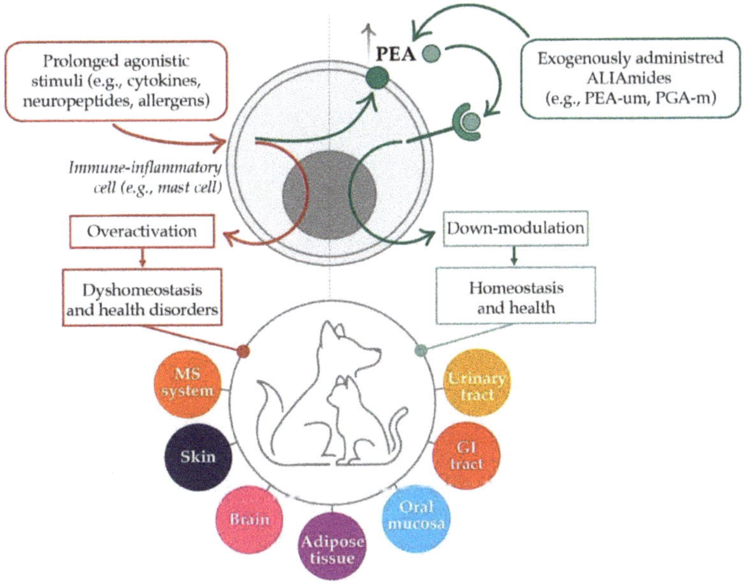

Figure 4. ALIAmides for small animal health—a global view. Upon prolonged stimulation, immune-inflammatory cells may become overactivated. If uncontrolled, their beneficial protective responses may instead turn harmful, leading to local dyshomeostasis and health disorders. In order to control the risk, autoprotective mechanisms are activated. The local production of PEA starting from a glycerophospholipid precursor (dark green circle) represents one of them. Once produced, PEA (light green circle) serves as a signaling molecule through its direct and indirect interactions with

multiple receptor targets resulting in cell down-modulation. Local homeostasis and body health are maintained accordingly. Exogenously administered ALIAmides mimic or sustain the autoprotective mechanism described above, mainly through restoring endogenous PEA levels. The main organs and body tissues purportedly benefiting from the aforementioned mechanism are listed in the colored circles on the bottom. GI = gastrointestinal, MS = musculoskeletal, PEA-um = ultramicronized palmitoylethanolamide, PGA-m = micronized palmitoylglucosamine.

6. Conclusions

Although the field is still in its infancy, the studies presented in this review highlight the promise that ALIAmides might play a broad role in small animal health. Their physiological prohomeostatic functions represent a key rationale for their use in promoting animals' health through an "according-to-nature" approach, i.e., mimicking or supporting the physiological mechanisms to maintain homeostasis.

Although further clinical studies are needed, ALIAmide-based products—either used as a sole intervention or associated with standard drugs—are emerging as a new and promising approach to veterinary patients.

Funding: This research received no external funding.

Institutional Review Board Statement: Not applicable.

Informed Consent Statement: Not applicable.

Data Availability Statement: Not applicable.

Acknowledgments: The authors wish to thank Sofia Pavan for graphical support and are indebted to Francesco della Valle for opening new horizons in the nature-driven approach to animal health and wellbeing.

Conflicts of Interest: The authors declare no conflict of interest.

References

1. Aloe, L.; Leon, A.; Levi-Montalcini, R. A Proposed Autacoid Mechanism Controlling Mastocyte Behaviour. *Agents Actions* **1993**, *39*, C145–C147. [CrossRef] [PubMed]
2. Melmon, K.L.; Rocklin, R.E.; Rosenkranz, R.P. Autacoids as Modulators of the Inflammatory and Immune Response. *Am. J. Med.* **1981**, *71*, 100–106. [CrossRef]
3. Gugliandolo, E.; Peritore, A.F.; Piras, C.; Cuzzocrea, S.; Crupi, R. Palmitoylethanolamide and Related ALIAmides: Prohomeostatic Lipid Compounds for Animal Health and Wellbeing. *Vet. Sci.* **2020**, *7*, 78. [CrossRef] [PubMed]
4. Mazzari, S.; Canella, R.; Petrelli, L.; Marcolongo, G.; Leon, A. N-(2-Hydroxyethyl)Hexadecanamide Is Orally Active in Reducing Edema Formation and Inflammatory Hyperalgesia by down-Modulating Mast Cell Activation. *Eur. J. Pharmacol.* **1996**, *300*, 227–236. [CrossRef]
5. Cerrato, S.; Brazis, P.; della Valle, M.F.; Miolo, A.; Puigdemont, A. Effects of Palmitoylethanolamide on Immunologically Induced Histamine, PGD2 and TNFalpha Release from Canine Skin Mast Cells. *Vet. Immunol. Immunopathol.* **2010**, *133*, 9–15. [CrossRef]
6. Abramo, F.; Lazzarini, G.; Pirone, A.; Lenzi, C.; Albertini, S.; Della Valle, M.F.; Schievano, C.; Vannozzi, I.; Miragliotta, V. Ultramicronized Palmitoylethanolamide Counteracts the Effects of Compound 48/80 in a Canine Skin Organ Culture Model. *Vet. Dermatol.* **2017**, *28*, 456-e104. [CrossRef]
7. Scarampella, F.; Abramo, F.; Noli, C. Clinical and Histological Evaluation of an Analogue of Palmitoylethanolamide, PLR 120 (Comicronized Palmidrol INN) in Cats with Eosinophilic Granuloma and Eosinophilic Plaque: A Pilot Study. *Vet. Dermatol.* **2001**, *12*, 29–39. [CrossRef]
8. Rinne, P.; Guillamat-Prats, R.; Rami, M.; Bindila, L.; Ring, L.; Lyytikäinen, L.-P.; Raitoharju, E.; Oksala, N.; Lehtimäki, T.; Weber, C.; et al. Palmitoylethanolamide Promotes a Proresolving Macrophage Phenotype and Attenuates Atherosclerotic Plaque Formation. *Arterioscler. Thromb. Vasc. Biol.* **2018**, *38*, 2562–2575. [CrossRef]
9. Petrosino, S.; Cristino, L.; Karsak, M.; Gaffal, E.; Ueda, N.; Tüting, T.; Bisogno, T.; De Filippis, D.; D'Amico, A.; Saturnino, C.; et al. Protective Role of Palmitoylethanolamide in Contact Allergic Dermatitis. *Allergy* **2010**, *65*, 698–711. [CrossRef]
10. Bettoni, I.; Comelli, F.; Colombo, A.; Bonfanti, P.; Costa, B. Non-Neuronal Cell Modulation Relieves Neuropathic Pain: Efficacy of the Endogenous Lipid Palmitoylethanolamide. *CNS Neurol. Disord. Drug Targets* **2013**, *12*, 34–44. [CrossRef]
11. Luongo, L.; Guida, F.; Boccella, S.; Bellini, G.; Gatta, L.; Rossi, F.; de Novellis, V.; Maione, S. Palmitoylethanolamide Reduces Formalin-Induced Neuropathic-like Behaviour through Spinal Glial/Microglial Phenotypical Changes in Mice. *CNS Neurol. Disord. Drug Targets* **2013**, *12*, 45–54. [CrossRef] [PubMed]

12. Guida, F.; Luongo, L.; Marmo, F.; Romano, R.; Iannotta, M.; Napolitano, F.; Belardo, C.; Marabese, I.; D'Aniello, A.; De Gregorio, D.; et al. Palmitoylethanolamide Reduces Pain-Related Behaviors and Restores Glutamatergic Synapses Homeostasis in the Medial Prefrontal Cortex of Neuropathic Mice. *Mol. Brain* **2015**, *8*, 47. [CrossRef] [PubMed]
13. Gabrielsson, L.; Gouveia-Figueira, S.; Häggström, J.; Alhouayek, M.; Fowler, C.J. The Anti-Inflammatory Compound Palmitoylethanolamide Inhibits Prostaglandin and Hydroxyeicosatetraenoic Acid Production by a Macrophage Cell Line. *Pharmacol. Res. Perspect.* **2017**, *5*, e00300. [CrossRef] [PubMed]
14. Chiurchiù, V.; Leuti, A.; Smoum, R.; Mechoulam, R.; Maccarrone, M. Bioactive Lipids ALIAmides Differentially Modulate Inflammatory Responses of Distinct Subsets of Primary Human T Lymphocytes. *FASEB J.* **2018**, *32*, 5716–5723. [CrossRef]
15. Bronzuoli, M.R.; Facchinetti, R.; Steardo, L.; Romano, A.; Stecca, C.; Passarella, S.; Steardo, L.; Cassano, T.; Scuderi, C. Palmitoylethanolamide Dampens Reactive Astrogliosis and Improves Neuronal Trophic Support in a Triple Transgenic Model of Alzheimer's Disease: In Vitro and In Vivo Evidence. *Oxidative Med. Cell. Longev.* **2018**, *2018*, 1–14. [CrossRef] [PubMed]
16. Scuderi, C.; Esposito, G.; Blasio, A.; Valenza, M.; Arietti, P.; Steardo, L.; Carnuccio, R.; De Filippis, D.; Petrosino, S.; Iuvone, T.; et al. Palmitoylethanolamide Counteracts Reactive Astrogliosis Induced by β-Amyloid Peptide. *J. Cell Mol. Med.* **2011**, *15*, 2664–2674. [CrossRef] [PubMed]
17. Ozaki, T.; Kamiyama, N.; Saechue, B.; Soga, Y.; Gotoh, R.; Nakayama, T.; Fukuda, C.; Dewayani, A.; Chalalai, T.; Ariki, S.; et al. Comprehensive Lipidomics of Lupus-Prone Mice Using LC-MS/MS Identifies the Reduction of Palmitoylethanolamide That Suppresses TLR9-Mediated Inflammation. *Genes Cells* **2022**, *27*, 493–504. [CrossRef] [PubMed]
18. Facchinetti, R.; Valenza, M.; Gomiero, C.; Mancini, G.F.; Steardo, L.; Campolongo, P.; Scuderi, C. Co-Ultramicronized Palmitoylethanolamide/Luteolin Restores Oligodendrocyte Homeostasis via Peroxisome Proliferator-Activated Receptor-α in an In Vitro Model of Alzheimer's Disease. *Biomedicines* **2022**, *10*, 1236. [CrossRef]
19. Bisogno, T.; Maurelli, S.; Melck, D.; De Petrocellis, L.; Di Marzo, V. Biosynthesis, Uptake, and Degradation of Anandamide and Palmitoylethanolamide in Leukocytes. *J. Biol. Chem.* **1997**, *272*, 3315–3323. [CrossRef]
20. Muccioli, G.G.; Stella, N. Microglia Produce and Hydrolyze Palmitoylethanolamide. *Neuropharmacology* **2008**, *54*, 16–22. [CrossRef]
21. Walter, L.; Franklin, A.; Witting, A.; Moller, T.; Stella, N. Astrocytes in Culture Produce Anandamide and Other Acylethanolamides. *J. Biol. Chem.* **2002**, *277*, 20869–20876. [CrossRef] [PubMed]
22. Natarajan, V.; Reddy, P.V.; Schmid, P.C.; Schmid, H.H. On the Biosynthesis and Metabolism of N-Acylethanolamine Phospholipids in Infarcted Dog Heart. *Biochim. Biophys. Acta* **1981**, *664*, 445–448. [CrossRef]
23. Epps, D.E.; Schmid, P.C.; Natarajan, V.; Schmid, H.H.O. N-Acylethanolamine Accumulation in Infarcted Myocardium. *Biochem. Biophys. Res. Commun.* **1979**, *90*, 628–633. [CrossRef]
24. Natarajan, V.; Schmid, P.C.; Reddy, P.V.; Schmid, H.H. Catabolism of N-Acylethanolamine Phospholipids by Dog Brain Preparations. *J. Neurochem.* **1984**, *42*, 1613–1619. [CrossRef] [PubMed]
25. Devane, W.A.; Hanus, L.; Breuer, A.; Pertwee, R.G.; Stevenson, L.A.; Griffin, G.; Gibson, D.; Mandelbaum, A.; Etinger, A.; Mechoulam, R. Isolation and Structure of a Brain Constituent That Binds to the Cannabinoid Receptor. *Science* **1992**, *258*, 1946–1949. [CrossRef]
26. della Valle, F.; Gambardella, A. 'Biological' Revolution and Strategies for Innovation in Pharmaceutical Companies. *R&D Manag.* **1993**, *23*, 287–302. [CrossRef]
27. Aloe, L. Rita Levi-Montalcini: The Discovery of Nerve Growth Factor and Modern Neurobiology. *Trends Cell. Biol.* **2004**, *14*, 395–399. [CrossRef]
28. Grayson, D.R.; Guidotti, A. Erminio Costa. *Neuropsychopharmacology* **2010**, *35*, 2646. [CrossRef]
29. Costa, E. *An Early Attempt to Foster Neuroscience Globalization: An Autobiography*; Good Life Press, Incorporated: Singapore, 2003; ISBN 978-0-9726121-1-1.
30. Levi-Montalcini, R.; Skaper, S.D.; Dal Toso, R.; Petrelli, L.; Leon, A. Nerve Growth Factor: From Neurotrophin to Neurokine. *Trends Neurosci.* **1996**, *19*, 514–520. [CrossRef]
31. Costa, E. Rrespecting Nature as a Strategy for the Development of New Drugs of the Nervous System. In Proceedings of the Drugs for Human Life Conference, Rome, Italy, 23–25 October 1986.
32. Della Rocca, G.; Gamba, D. Chronic Pain in Dogs and Cats: Is There Place for Dietary Intervention with Micro-Palmitoylethanolamide? *Animals* **2021**, *11*, 952. [CrossRef]
33. Re, G.; Barbero, R.; Miolo, A.; Di Marzo, V. Palmitoylethanolamide, Endocannabinoids and Related Cannabimimetic Compounds in Protection against Tissue Inflammation and Pain: Potential Use in Companion Animals. *Vet. J.* **2007**, *173*, 21–30. [CrossRef] [PubMed]
34. Petrosino, S.; Schiano Moriello, A. Palmitoylethanolamide: A Nutritional Approach to Keep Neuroinflammation within Physiological Boundaries-A Systematic Review. *Int. J. Mol. Sci.* **2020**, *21*, 9526. [CrossRef] [PubMed]
35. Petrosino, S.; Di Marzo, V. The Pharmacology of Palmitoylethanolamide and First Data on the Therapeutic Efficacy of Some of Its New Formulations. *Br. J. Pharmacol.* **2017**, *174*, 1349–1365. [CrossRef] [PubMed]
36. Esposito, E.; Cuzzocrea, S. Palmitoylethanolamide Is a New Possible Pharmacological Treatment for the Inflammation Associated with Trauma. *Mini Rev. Med. Chem.* **2013**, *13*, 237–255. [PubMed]
37. Scuderi, C.; Golini, L. Successful and Unsuccessful Brain Aging in Pets: Pathophysiological Mechanisms behind Clinical Signs and Potential Benefits from Palmitoylethanolamide Nutritional Intervention. *Animals* **2021**, *11*, 2584. [CrossRef]

38. Hussain, Z.; Uyama, T.; Tsuboi, K.; Ueda, N. Mammalian Enzymes Responsible for the Biosynthesis of N-Acylethanolamines. *Biochim. Biophys. Acta Mol. Cell Biol. Lipids* **2017**, *1862*, 1546–1561. [CrossRef]
39. Tsuboi, K.; Uyama, T.; Okamoto, Y.; Ueda, N. Endocannabinoids and Related N-Acylethanolamines: Biological Activities and Metabolism. *Inflamm. Regen.* **2018**, *38*, 28. [CrossRef]
40. Sun, Y.-X.; Tsuboi, K.; Okamoto, Y.; Tonai, T.; Murakami, M.; Kudo, I.; Ueda, N. Biosynthesis of Anandamide and N-Palmitoylethanolamine by Sequential Actions of Phospholipase A2 and Lysophospholipase D. *Biochem. J.* **2004**, *380*, 749–756. [CrossRef]
41. Ueda, N.; Tsuboi, K.; Uyama, T. N-Acylethanolamine Metabolism with Special Reference to N-Acylethanolamine-Hydrolyzing Acid Amidase (NAAA). *Prog. Lipid Res.* **2010**, *49*, 299–315. [CrossRef]
42. Alhouayek, M.; Muccioli, G.G. Harnessing the Anti-Inflammatory Potential of Palmitoylethanolamide. *Drug Discov. Today* **2014**, *19*, 1632–1639. [CrossRef]
43. Roviezzo, F.; Rossi, A.; Caiazzo, E.; Orlando, P.; Riemma, M.A.; Iacono, V.M.; Guarino, A.; Ialenti, A.; Cicala, C.; Peritore, A.; et al. Palmitoylethanolamide Supplementation during Sensitization Prevents Airway Allergic Symptoms in the Mouse. *Front. Pharmacol.* **2017**, *8*, 857. [CrossRef] [PubMed]
44. Skaper, S.D.; Facci, L.; Barbierato, M.; Zusso, M.; Bruschetta, G.; Impellizzeri, D.; Cuzzocrea, S.; Giusti, P. N-Palmitoylethanolamine and Neuroinflammation: A Novel Therapeutic Strategy of Resolution. *Mol. Neurobiol.* **2015**, *52*, 1034–1042. [CrossRef] [PubMed]
45. Petrosino, S.; Palazzo, E.; de Novellis, V.; Bisogno, T.; Rossi, F.; Maione, S.; Di Marzo, V. Changes in Spinal and Supraspinal Endocannabinoid Levels in Neuropathic Rats. *Neuropharmacology* **2007**, *52*, 415–422. [CrossRef] [PubMed]
46. Charrua, A.; Matos, R.; Oliveira, R.; Marczylo, T.; Nagy, I.; Cruz, F. Fatty Acid Amide Hydrolase Inhibition Normalises Bladder Function and Reduces Pain through Normalising the Anandamide/Palmitoylethanolamine Ratio in the Inflamed Bladder of Rats. *Naunyn. Schmiedebergs Arch. Pharmacol.* **2020**, *393*, 263–272. [CrossRef]
47. Zhou, P.; Xiang, L.; Yang, Y.; Wu, Y.; Hu, T.; Liu, X.; Lin, F.; Xiu, Y.; Wu, K.; Lu, C.; et al. N-Acylethanolamine Acid Amidase (NAAA) Inhibitor F215 as a Novel Therapeutic Agent for Osteoarthritis. *Pharmacol. Res.* **2019**, *145*, 104264. [CrossRef]
48. Fichna, J.; Wood, J.T.; Papanastasiou, M.; Vadivel, S.K.; Oprocha, P.; Sałaga, M.; Sobczak, M.; Mokrowiecka, A.; Cygankiewicz, A.I.; Zakrzewski, P.K.; et al. Endocannabinoid and Cannabinoid-like Fatty Acid Amide Levels Correlate with Pain-Related Symptoms in Patients with IBS-D and IBS-C: A Pilot Study. *PLoS ONE* **2013**, *8*, e85073. [CrossRef]
49. Sarnelli, G.; Pesce, M.; Seguella, L.; Lu, J.; Efficie, E.; Tack, J.; Elisa De Palma, F.D.; D'Alessandro, A.; Esposito, G. Impaired Duodenal Palmitoylethanolamide Release Underlies Acid-Induced Mast Cell Activation in Functional Dyspepsia. *Cell Mol. Gastroenterol. Hepatol.* **2021**, *11*, 841–855. [CrossRef]
50. Richardson, D.; Pearson, R.G.; Kurian, N.; Latif, M.L.; Garle, M.J.; Barrett, D.A.; Kendall, D.A.; Scammell, B.E.; Reeve, A.J.; Chapman, V. Characterisation of the Cannabinoid Receptor System in Synovial Tissue and Fluid in Patients with Osteoarthritis and Rheumatoid Arthritis. *Arthritis Res. Ther.* **2008**, *10*, R43. [CrossRef]
51. Berdyshev, E.V.; Schmid, P.C.; Dong, Z.; Schmid, H.H. Stress-Induced Generation of N-Acylethanolamines in Mouse Epidermal JB6 P+ Cells. *Biochem. J.* **2000**, *346 Pt 2*, 369–374. [CrossRef]
52. Abramo, F.; Campora, L.; Albanese, F.; della Valle, M.F.; Cristino, L.; Petrosino, S.; Di Marzo, V.; Miragliotta, V. Increased Levels of Palmitoylethanolamide and Other Bioactive Lipid Mediators and Enhanced Local Mast Cell Proliferation in Canine Atopic Dermatitis. *BMC Vet. Res.* **2014**, *10*, 21. [CrossRef]
53. Pengo, G.; Miolo, A. Utilizzo Di Palmitoiletanolamide Micronizzata Nell'infiammazione Gastrointestinale Idiopatica (IBD) Del Cane: Descrizione Di 7 Casi Clinici. In Proceedings of the 72 International SCIVAC Congress, Milan, Italy, 23–25 March 2012; pp. 299–301.
54. Skaper, S.D.; Facci, L.; Giusti, P. Mast Cells, Glia and Neuroinflammation: Partners in Crime? *Immunology* **2014**, *141*, 314–327. [CrossRef] [PubMed]
55. Del Re, A.; Palenca, I.; Seguella, L.; Pesce, M.; Corpetti, C.; Steardo, L.; Rurgo, S.; Sarnelli, G.; Esposito, G. Oral Adelmidrol Administration Up-Regulates Palmitoylethanolamide Production in Mice Colon and Duodenum through a PPAR-γ Independent Action. *Metabolites* **2022**, *12*, 457. [CrossRef] [PubMed]
56. Petrosino, S.; Puigdemont, A.; Della Valle, M.F.; Fusco, M.; Verde, R.; Allarà, M.; Aveta, T.; Orlando, P.; Di Marzo, V. Adelmidrol Increases the Endogenous Concentrations of Palmitoylethanolamide in Canine Keratinocytes and Down-Regulates an Inflammatory Reaction in an in Vitro Model of Contact Allergic Dermatitis. *Vet. J.* **2016**, *207*, 85–91. [CrossRef] [PubMed]
57. Petrosino, S.; Schiano Moriello, A.; Verde, R.; Allarà, M.; Imperatore, R.; Ligresti, A.; Mahmoud, A.M.; Peritore, A.F.; Iannotti, F.A.; Di Marzo, V. Palmitoylethanolamide Counteracts Substance P-Induced Mast Cell Activation in Vitro by Stimulating Diacylglycerol Lipase Activity. *J. Neuroinflammation* **2019**, *16*, 274. [CrossRef] [PubMed]
58. Jacobsson, S.O.; Fowler, C.J. Characterization of Palmitoylethanolamide Transport in Mouse Neuro-2a Neuroblastoma and Rat RBL-2H3 Basophilic Leukaemia Cells: Comparison with Anandamide. *Br. J. Pharmacol.* **2001**, *132*, 1743–1754. [CrossRef] [PubMed]
59. Carta, G.; Murru, E.; Banni, S.; Manca, C. Palmitic Acid: Physiological Role, Metabolism and Nutritional Implications. *Front. Physiol.* **2017**, *8*, 902. [CrossRef]
60. Bojesen, I.N.; Hansen, H.S. Membrane Transport of Anandamide through Resealed Human Red Blood Cell Membranes. *J. Lipid Res.* **2005**, *46*, 1652–1659. [CrossRef]

61. Kaczocha, M.; Glaser, S.T.; Maher, T.; Clavin, B.; Hamilton, J.; O'Rourke, J.; Rebecchi, M.; Puopolo, M.; Owada, Y.; Thanos, P.K. Fatty Acid Binding Protein Deletion Suppresses Inflammatory Pain through Endocannabinoid/N-Acylethanolamine-Dependent Mechanisms. *Mol. Pain* **2015**, *11*, 52. [CrossRef]
62. Kaczocha, M.; Vivieca, S.; Sun, J.; Glaser, S.T.; Deutsch, D.G. Fatty Acid-Binding Proteins Transport N-Acylethanolamines to Nuclear Receptors and Are Targets of Endocannabinoid Transport Inhibitors. *J. Biol. Chem.* **2012**, *287*, 3415–3424. [CrossRef]
63. Oddi, S.; Fezza, F.; Pasquariello, N.; D'Agostino, A.; Catanzaro, G.; De Simone, C.; Rapino, C.; Finazzi-Agrò, A.; Maccarrone, M. Molecular Identification of Albumin and Hsp70 as Cytosolic Anandamide-Binding Proteins. *Chem. Biol.* **2009**, *16*, 624–632. [CrossRef]
64. Korbecki, J.; Bobiński, R.; Dutka, M. Self-Regulation of the Inflammatory Response by Peroxisome Proliferator-Activated Receptors. *Inflamm. Res.* **2019**, *68*, 443–458. [CrossRef] [PubMed]
65. Lo Verme, J.; Fu, J.; Astarita, G.; La Rana, G.; Russo, R.; Calignano, A.; Piomelli, D. The Nuclear Receptor Peroxisome Proliferator-Activated Receptor-Alpha Mediates the Anti-Inflammatory Actions of Palmitoylethanolamide. *Mol. Pharmacol.* **2005**, *67*, 15–19. [CrossRef] [PubMed]
66. Lo Verme, J.; La Rana, G.; Russo, R.; Calignano, A.; Piomelli, D. The Search for the Palmitoylethanolamide Receptor. *Life Sci.* **2005**, *77*, 1685–1698. [CrossRef] [PubMed]
67. LoVerme, J.; Russo, R.; La Rana, G.; Fu, J.; Farthing, J.; Mattace-Raso, G.; Meli, R.; Hohmann, A.; Calignano, A.; Piomelli, D. Rapid Broad-Spectrum Analgesia through Activation of Peroxisome Proliferator-Activated Receptor-Alpha. *J. Pharmacol. Exp. Ther.* **2006**, *319*, 1051–1061. [CrossRef]
68. Zhou, G.; Fu, X.; Wang, L.; Cao, Y.; Zhuang, J.; Hu, J.; Li, Y.; Xu, C.; Gao, S.; Shao, A.; et al. Palmitoylethanolamide Ameliorates Neuroinflammation via Modulating PPAR-α to Promote the Functional Outcome after Intracerebral Hemorrhage. *Neurosci. Lett.* **2022**, *781*, 136648. [CrossRef]
69. Annunziata, C.; Pirozzi, C.; Lama, A.; Senzacqua, M.; Comella, F.; Bordin, A.; Monnolo, A.; Pelagalli, A.; Ferrante, M.C.; Mollica, M.P.; et al. Palmitoylethanolamide Promotes White-to-Beige Conversion and Metabolic Reprogramming of Adipocytes: Contribution of PPAR-α. *Pharmaceutics* **2022**, *14*, 338. [CrossRef]
70. Im, D.-S. GPR119 and GPR55 as Receptors for Fatty Acid Ethanolamides, Oleoylethanolamide and Palmitoylethanolamide. *Int. J. Mol. Sci.* **2021**, *22*, 1034. [CrossRef]
71. Ryberg, E.; Larsson, N.; Sjögren, S.; Hjorth, S.; Hermansson, N.-O.; Leonova, J.; Elebring, T.; Nilsson, K.; Drmota, T.; Greasley, P.J. The Orphan Receptor GPR55 Is a Novel Cannabinoid Receptor. *Br. J. Pharmacol.* **2007**, *152*, 1092–1101. [CrossRef]
72. Petrosino, S.; Schiano Moriello, A.; Cerrato, S.; Fusco, M.; Puigdemont, A.; De Petrocellis, L.; Di Marzo, V. The Anti-Inflammatory Mediator Palmitoylethanolamide Enhances the Levels of 2-Arachidonoyl-Glycerol and Potentiates Its Actions at TRPV1 Cation Channels. *Br. J. Pharmacol.* **2016**, *173*, 1154–1162. [CrossRef]
73. Di Marzo, V.; Melck, D.; Orlando, P.; Bisogno, T.; Zagoory, O.; Bifulco, M.; Vogel, Z.; De Petrocellis, L. Palmitoylethanolamide Inhibits the Expression of Fatty Acid Amide Hydrolase and Enhances the Anti-Proliferative Effect of Anandamide in Human Breast Cancer Cells. *Biochem. J.* **2001**, *358*, 249–255. [CrossRef]
74. Julius, D. TRP Channels and Pain. *Annu. Rev. Cell Dev. Biol.* **2013**, *29*, 355–384. [CrossRef] [PubMed]
75. Costa, B.; Comelli, F.; Bettoni, I.; Colleoni, M.; Giagnoni, G. The Endogenous Fatty Acid Amide, Palmitoylethanolamide, Has Anti-Allodynic and Anti-Hyperalgesic Effects in a Murine Model of Neuropathic Pain: Involvement of CB(1), TRPV1 and PPARgamma Receptors and Neurotrophic Factors. *Pain* **2008**, *139*, 541–550. [CrossRef] [PubMed]
76. De Petrocellis, L.; Davis, J.B.; Di Marzo, V. Palmitoylethanolamide Enhances Anandamide Stimulation of Human Vanilloid VR1 Receptors. *FEBS Lett.* **2001**, *506*, 253–256. [CrossRef]
77. Ambrosino, P.; Soldovieri, M.V.; Russo, C.; Taglialatela, M. Activation and Desensitization of TRPV1 Channels in Sensory Neurons by the PPARα Agonist Palmitoylethanolamide. *Br. J. Pharmacol.* **2013**, *168*, 1430–1444. [CrossRef] [PubMed]
78. Ho, W.-S.V.; Barrett, D.A.; Randall, M.D. "Entourage" Effects of N-Palmitoylethanolamide and N-Oleoylethanolamide on Vasorelaxation to Anandamide Occur through TRPV1 Receptors. *Br. J. Pharmacol.* **2008**, *155*, 837–846. [CrossRef]
79. Campora, L.; Miragliotta, V.; Ricci, E.; Cristino, L.; Di Marzo, V.; Albanese, F.; Federica Della Valle, M.; Abramo, F. Cannabinoid Receptor Type 1 and 2 Expression in the Skin of Healthy Dogs and Dogs with Atopic Dermatitis. *Am. J. Vet. Res.* **2012**, *73*, 988–995. [CrossRef]
80. Barbero, R.; Vercelli, C.; Cuniberti, B.; Della Valle, M.F.; Martano, M.; Re, G. Expression of Functional TRPV1 Receptor in Primary Culture of Canine Keratinocytes. *J. Vet. Pharmacol. Ther.* **2018**, *41*, 795–804. [CrossRef]
81. Dall'Aglio, C.; Mercati, F.; Pascucci, L.; Boiti, C.; Pedini, V.; Ceccarelli, P. Immunohistochemical Localization of CB1 Receptor in Canine Salivary Glands. *Vet. Res. Commun* **2010**, *34* (Suppl. S1), S9–S12. [CrossRef]
82. Mercati, F.; Dall'Aglio, C.; Pascucci, L.; Boiti, C.; Ceccarelli, P. Identification of Cannabinoid Type 1 Receptor in Dog Hair Follicles. *Acta Histochem.* **2012**, *114*, 68–71. [CrossRef]
83. Fernández-Trapero, M.; Espejo-Porras, F.; Rodríguez-Cueto, C.; Coates, J.R.; Pérez-Díaz, C.; de Lago, E.; Fernández-Ruiz, J. Upregulation of CB2 Receptors in Reactive Astrocytes in Canine Degenerative Myelopathy, a Disease Model of Amyotrophic Lateral Sclerosis. *Dis. Model. Mech.* **2017**, *10*, 551–558. [CrossRef]
84. Freundt-Revilla, J.; Kegler, K.; Baumgärtner, W.; Tipold, A. Spatial Distribution of Cannabinoid Receptor Type 1 (CB1) in Normal Canine Central and Peripheral Nervous System. *PLoS ONE* **2017**, *12*, e0181064. [CrossRef]

85. Freundt-Revilla, J.; Heinrich, F.; Zoerner, A.; Gesell, F.; Beyerbach, M.; Shamir, M.; Oevermann, A.; Baumgärtner, W.; Tipold, A. The Endocannabinoid System in Canine Steroid-Responsive Meningitis-Arteritis and Intraspinal Spirocercosis. *PLoS ONE* **2018**, *13*, e0187197. [CrossRef]
86. Galiazzo, G.; Giancola, F.; Stanzani, A.; Fracassi, F.; Bernardini, C.; Forni, M.; Pietra, M.; Chiocchetti, R. Localization of Cannabinoid Receptors CB1, CB2, GPR55, and PPARα in the Canine Gastrointestinal Tract. *Histochem. Cell Biol.* **2018**, *150*, 187–205. [CrossRef] [PubMed]
87. Stanzani, A.; Galiazzo, G.; Giancola, F.; Tagliavia, C.; De Silva, M.; Pietra, M.; Fracassi, F.; Chiocchetti, R. Localization of Cannabinoid and Cannabinoid Related Receptors in the Cat Gastrointestinal Tract. *Histochem. Cell Biol.* **2020**, *153*, 339–356. [CrossRef]
88. Gebremedhin, D.; Lange, A.R.; Campbell, W.B.; Hillard, C.J.; Harder, D.R. Cannabinoid CB1 Receptor of Cat Cerebral Arterial Muscle Functions to Inhibit L-Type Ca2+ Channel Current. *Am. J. Physiol.* **1999**, *276*, H2085–H2093. [CrossRef]
89. Miragliotta, V.; Ricci, P.L.; Albanese, F.; Pirone, A.; Tognotti, D.; Abramo, F. Cannabinoid Receptor Types 1 and 2 and Peroxisome Proliferator-Activated Receptor-α: Distribution in the Skin of Clinically Healthy Cats and Cats with Hypersensitivity Dermatitis. *Vet. Dermatol.* Online ahead of print. **2018**. [CrossRef]
90. Pirone, A.; Cantile, C.; Miragliotta, V.; Lenzi, C.; Giannessi, E.; Cozzi, B. Immunohistochemical Distribution of the Cannabinoid Receptor 1 and Fatty Acid Amide Hydrolase in the Dog Claustrum. *J. Chem. Neuroanat.* **2016**, *74*, 21–27. [CrossRef]
91. Pirone, A.; Lenzi, C.; Briganti, A.; Abbate, F.; Levanti, M.; Abramo, F.; Miragliotta, V. Spatial Distribution of Cannabinoid Receptor 1 and Fatty Acid Amide Hydrolase in the Cat Ovary and Oviduct. *Acta Histochem.* **2017**, *119*, 417–422. [CrossRef]
92. Ndong, C.; O'Donnell, D.; Ahmad, S.; Groblewski, T. Cloning and Pharmacological Characterization of the Dog Cannabinoid CB_2 receptor. *Eur. J. Pharmacol.* **2011**, *669*, 24–31. [CrossRef]
93. Silver, R.J. The Endocannabinoid System of Animals. *Animals* **2019**, *9*, 686. [CrossRef]
94. Iannotta, M.; Belardo, C.; Trotta, M.C.; Iannotti, F.A.; Vitale, R.M.; Maisto, R.; Boccella, S.; Infantino, R.; Ricciardi, F.; Mirto, B.F.; et al. N-Palmitoyl-D-Glucosamine, a Natural Monosaccharide-Based Glycolipid, Inhibits TLR4 and Prevents LPS-Induced Inflammation and Neuropathic Pain in Mice. *Int. J. Mol. Sci.* **2021**, *22*, 1491. [CrossRef] [PubMed]
95. Impellizzeri, D.; Di Paola, R.; Cordaro, M.; Gugliandolo, E.; Casili, G.; Morittu, V.M.; Britti, D.; Esposito, E.; Cuzzocrea, S. Adelmidrol, a Palmitoylethanolamide Analogue, as a New Pharmacological Treatment for the Management of Acute and Chronic Inflammation. *Biochem. Pharmacol.* **2016**, *119*, 27–41. [CrossRef] [PubMed]
96. Cerrato, S.; Brazis, P.; Della Valle, M.F.; Miolo, A.; Puigdemont, A. Inhibitory Effect of Topical Adelmidrol on Antigen-Induced Skin Wheal and Mast Cell Behavior in a Canine Model of Allergic Dermatitis. *BMC Vet. Res.* **2012**, *8*, 230. [CrossRef] [PubMed]
97. Cordaro, M.; Siracusa, R.; Impellizzeri, D.; D' Amico, R.; Peritore, A.F.; Crupi, R.; Gugliandolo, E.; Fusco, R.; Di Paola, R.; Schievano, C.; et al. Safety and Efficacy of a New Micronized Formulation of the ALIAmide Palmitoylglucosamine in Preclinical Models of Inflammation and Osteoarthritis Pain. *Arthritis Res. Ther.* **2019**, *21*, 254. [CrossRef]
98. Petrosino, S.; Cordaro, M.; Verde, R.; Schiano Moriello, A.; Marcolongo, G.; Schievano, C.; Siracusa, R.; Piscitelli, F.; Peritore, A.F.; Crupi, R.; et al. Oral Ultramicronized Palmitoylethanolamide: Plasma and Tissue Levels and Spinal Anti-Hyperalgesic Effect. *Front. Pharmacol.* **2018**, *9*, 249. [CrossRef] [PubMed]
99. Bilia, A.R.; Piazzini, V.; Guccione, C.; Risaliti, L.; Asprea, M.; Capecchi, G.; Bergonzi, M.C. Improving on Nature: The Role of Nanomedicine in the Development of Clinical Natural Drugs. *Planta Med.* **2017**, *83*, 366–381. [CrossRef]
100. Takano, R.; Furumoto, K.; Shiraki, K.; Takata, N.; Hayashi, Y.; Aso, Y.; Yamashita, S. Rate-Limiting Steps of Oral Absorption for Poorly Water-Soluble Drugs in Dogs; Prediction from a Miniscale Dissolution Test and a Physiologically-Based Computer Simulation. *Pharm. Res.* **2008**, *25*, 2334–2344. [CrossRef]
101. Dhiman, A.; Prabhakar, P.K. Micronization in Food Processing: A Comprehensive Review of Mechanistic Approach, Physicochemical, Functional Properties and Self-Stability of Micronized Food Materials. *J. Food Eng.* **2021**, *292*, 110248. [CrossRef]
102. Impellizzeri, D.; Bruschetta, G.; Cordaro, M.; Crupi, R.; Siracusa, R.; Esposito, E.; Cuzzocrea, S. Micronized/Ultramicronized Palmitoylethanolamide Displays Superior Oral Efficacy Compared to Nonmicronized Palmitoylethanolamide in a Rat Model of Inflammatory Pain. *J. Neuroinflammat.* **2014**, *11*, 136. [CrossRef]
103. Impellizzeri, D.; Campolo, M.; Di Paola, R.; Bruschetta, G.; de Stefano, D.; Esposito, E.; Cuzzocrea, S. Ultramicronized Palmitoylethanolamide Reduces Inflammation an a Th1-Mediated Model of Colitis. *Eur. J. Inflamm.* **2015**, *13*, 14–31. [CrossRef]
104. Cerrato, S.; Brazis, P.; Della Valle, M.F.; Miolo, A.; Petrosino, S.; Di Marzo, V.; Puigdemont, A. Effects of Palmitoylethanolamide on the Cutaneous Allergic Inflammatory Response in Ascaris Hypersensitive Beagle Dogs. *Vet. J.* **2012**, *191*, 377–382. [CrossRef] [PubMed]
105. Miragliotta, V.; Noli, C. Dermatology: Endocannabinoids and Related N-Acylethanolamines in the Skin. In *Cannabis Therapy in Veterinary Medicine: A Complete Guide*; Cital, S., Kramer, K., Hughston, L., Gaynor, J.S., Eds.; Springer International Publishing: Cham, Switzerland, 2021; pp. 207–230. ISBN 978-3-030-68317-7.
106. Abramo, F.; Salluzzi, D.; Leotta, R.; Auxilia, S.; Noli, C.; Miolo, A.; Mantis, P.; Lloyd, D.H. Mast Cell Morphometry and Densitometry in Experimental Skin Wounds Treated with a Gel Containing Adelmidrol: A Placebo Controlled Study. *Wounds* **2008**, *20*, 149–157. [PubMed]
107. Mantis, P.; Lloyd, D.H.; Pfeiffer, D.; Stevens, K.; Auxilia, S.; Noli, C.; Auxilia, S.; Noli, C.; Abramo, F.; Miolo, A. Assessment of the Effect of an Aliamide-Containing Topical Gel by Evaluation of the Reduction of Wound Volume Measured by High R. *Wounds* **2007**, *19*, 113–119. [PubMed]

108. Marsella, R.; Joyce, J.; Nicklin, C.; Lopez, J. Evaluation of the Effects of Palmitoylethanolamide on Clinical Signs in House Dust Mite Allergic High IgE Beagle Dogs Using a Randomized, Double Blinded, Placebo Controlled Design. *Vet. Dermatol.* **2005**, *16*, 202.
109. Noli, C.; Della Valle, M.F.; Miolo, A.; Medori, C.; Schievano, C. Skinalia Clinical Research Group Effect of Dietary Supplementation with Ultramicronized Palmitoylethanolamide in Maintaining Remission in Cats with Nonflea Hypersensitivity Dermatitis: A Double-Blind, Multicentre, Randomized, Placebo-Controlled Study. *Vet. Dermatol.* **2019**, *30*, 387-e117. [CrossRef]
110. Waisglass, S.; Araujio, J.; Della Valle, M.F.; Milgram, N.W. Palmitoylethanolamide in the Management of Canine Atopic Dermatitis. Randomized, Double-Blind, Placebo Controlled Study. In Proceedings of the International Congress of Scivac, Rimini, Italy, 29–31 May 2009; pp. 56–61.
111. Noli, C.; Della Valle, M.F.; Miolo, A.; Medori, C.; Schievano, C. Skinalia Clinical Research Group Efficacy of Ultra-Micronized Palmitoylethanolamide in Canine Atopic Dermatitis: An Open-Label Multi-Centre Study. *Vet. Dermatol.* **2015**, *26*, 432–440, e101. [CrossRef]
112. Fabbrini, F.; Leone, F. Applicazione Topica Di Adelmidrol (2%) per La Gestione Del Prurito Associato a Dermatite Atopica Del Cane: Studio Osservazionale—Veterinaria. *Veterinaria* **2013**, *27*.
113. Schiano Moriello, A.; Di Marzo, V.; Petrosino, S. Mutual Links between the Endocannabinoidome and the Gut Microbiome, with Special Reference to Companion Animals: A Nutritional Viewpoint. *Animals* **2022**, *12*, 348. [CrossRef]
114. Russo, R.; Cristiano, C.; Avagliano, C.; De Caro, C.; La Rana, G.; Raso, G.M.; Canani, R.B.; Meli, R.; Calignano, A. Gut-Brain Axis. Role of Lipids in the Regulation of Inflammation, Pain and CNS Diseases. *Curr. Med. Chem.* **2018**, *25*, 3930–3952. [CrossRef]
115. Hedlund, P.; Gratzke, C. The Endocannabinoid System - a Target for the Treatment of LUTS? *Nat. Rev. Urol.* **2016**, *13*, 463–470. [CrossRef]
116. Zhou, S.; Ling, X.; Meng, P.; Liang, Y.; Shen, K.; Wu, Q.; Zhang, Y.; Chen, Q.; Chen, S.; Liu, Y.; et al. Cannabinoid Receptor 2 Plays a Central Role in Renal Tubular Mitochondrial Dysfunction and Kidney Ageing. *J. Cell Mol. Med.* **2021**, *25*, 8957–8972. [CrossRef] [PubMed]
117. Chua, J.T.; Argueta, D.A.; DiPatrizio, N.V.; Kovesdy, C.P.; Vaziri, N.D.; Kalantar-Zadeh, K.; Moradi, H. Endocannabinoid System and the Kidneys: From Renal Physiology to Injury and Disease. *Cannabis Cannabinoid Res.* **2019**, *4*, 10–20. [CrossRef] [PubMed]
118. Klawitter, J.; Sempio, C.; Jackson, M.J.; Smith, P.H.; Hopp, K.; Chonchol, M.; Gitomer, B.Y.; Christians, U.; Klawitter, J. Endocannabinoid System in Polycystic Kidney Disease. *Am. J. Nephrol.* **2022**, *53*, 264–272. [CrossRef] [PubMed]
119. Britti, D.; Crupi, R.; Impellizzeri, D.; Gugliandolo, E.; Fusco, R.; Schievano, C.; Morittu, V.M.; Evangelista, M.; Di Paola, R.; Cuzzocrea, S. A Novel Composite Formulation of Palmitoylethanolamide and Quercetin Decreases Inflammation and Relieves Pain in Inflammatory and Osteoarthritic Pain Models. *BMC Vet. Res.* **2017**, *13*, 229. [CrossRef] [PubMed]
120. Gugliandolo, E.; Peritore, A.F.; Impellizzeri, D.; Cordaro, M.; Siracusa, R.; Fusco, R.; D'Amico, R.; Paola, R.D.; Schievano, C.; Cuzzocrea, S.; et al. Dietary Supplementation with Palmitoyl-Glucosamine Co-Micronized with Curcumin Relieves Osteoarthritis Pain and Benefits Joint Mobility. *Animals* **2020**, *10*, E1827. [CrossRef]
121. Annunziata, C.; Lama, A.; Pirozzi, C.; Cavaliere, G.; Trinchese, G.; Di Guida, F.; Nitrato Izzo, A.; Cimmino, F.; Paciello, O.; De Biase, D.; et al. Palmitoylethanolamide Counteracts Hepatic Metabolic Inflexibility Modulating Mitochondrial Function and Efficiency in Diet-Induced Obese Mice. *FASEB J.* **2020**, *34*, 350–364. [CrossRef] [PubMed]
122. Rahman, S.M.K.; Uyama, T.; Hussain, Z.; Ueda, N. Roles of Endocannabinoids and Endocannabinoid-Like Molecules in Energy Homeostasis and Metabolic Regulation: A Nutritional Perspective. *Annu Rev. Nutr.* **2021**, *41*, 177–202. [CrossRef]
123. Sihag, J.; Di Marzo, V. (Wh)Olistic (E)Ndocannabinoidome-Microbiome-Axis Modulation through (N)Utrition (WHEN) to Curb Obesity and Related Disorders. *Lipids Health Dis.* **2022**, *21*, 9. [CrossRef]
124. Esposito, G.; Capoccia, E.; Turco, F.; Palumbo, I.; Lu, J.; Steardo, A.; Cuomo, R.; Sarnelli, G.; Steardo, L. Palmitoylethanolamide Improves Colon Inflammation through an Enteric Glia/Toll like Receptor 4-Dependent PPAR-α Activation. *Gut* **2014**, *63*, 1300–1312. [CrossRef]
125. Karwad, M.A.; Macpherson, T.; Wang, B.; Theophilidou, E.; Sarmad, S.; Barrett, D.A.; Larvin, M.; Wright, K.L.; Lund, J.N.; O'Sullivan, S.E. Oleoylethanolamine and Palmitoylethanolamine Modulate Intestinal Permeability in Vitro via TRPV1 and PPARα. *FASEB J.* **2017**, *31*, 469–481. [CrossRef]
126. Sarnelli, G.; Seguella, L.; Pesce, M.; Lu, J.; Gigli, S.; Bruzzese, E.; Lattanzi, R.; D'Alessandro, A.; Cuomo, R.; Steardo, L.; et al. HIV-1 Tat-Induced Diarrhea Is Improved by the PPARalpha Agonist, Palmitoylethanolamide, by Suppressing the Activation of Enteric Glia. *J. Neuroinflammat.* **2018**, *15*, 94. [CrossRef] [PubMed]
127. Borrelli, F.; Romano, B.; Petrosino, S.; Pagano, E.; Capasso, R.; Coppola, D.; Battista, G.; Orlando, P.; Di Marzo, V.; Izzo, A.A. Palmitoylethanolamide, a Naturally Occurring Lipid, Is an Orally Effective Intestinal Anti-Inflammatory Agent. *Br. J. Pharmacol.* **2015**, *172*, 142–158. [CrossRef] [PubMed]
128. Capasso, R.; Orlando, P.; Pagano, E.; Aveta, T.; Buono, L.; Borrelli, F.; Di Marzo, V.; Izzo, A.A. Palmitoylethanolamide Normalizes Intestinal Motility in a Model of Post-Inflammatory Accelerated Transit: Involvement of CB_1 Receptors and TRPV1 Channels. *Br. J. Pharmacol.* **2014**, *171*, 4026–4037. [CrossRef] [PubMed]
129. Couch, D.G.; Cook, H.; Ortori, C.; Barrett, D.; Lund, J.N.; O'Sullivan, S.E. Palmitoylethanolamide and Cannabidiol Prevent Inflammation-Induced Hyperpermeability of the Human Gut In Vitro and In Vivo-A Randomized, Placebo-Controlled, Double-Blind Controlled Trial. *Inflammat. Bowel Dis.* **2019**, *25*, 1006–1018. [CrossRef] [PubMed]

130. Di Paola, R.; Impellizzeri, D.; Torre, A.; Mazzon, E.; Cappellani, A.; Faggio, C.; Esposito, E.; Trischitta, F.; Cuzzocrea, S. Effects of Palmitoylethanolamide on Intestinal Injury and Inflammation Caused by Ischemia-Reperfusion in Mice. *J. Leukoc. Biol.* **2012**, *91*, 911–920. [CrossRef]
131. Petrini, D.; Di Giuseppe, M.; Deli, G.; De Caro Carella, C. Cystolithiasis in a Syrian Hamster: A Different Outcome. *Open Vet. J.* **2016**, *6*, 135–138. [CrossRef] [PubMed]
132. Gugliandolo, E.; Crupi, R.; Peritore, A.F.; Licata, P.; Piras, C.; Cuzzocrea, S.; Britti, D. Interstitial Cystitis/Bladder Pain Syndrome (IC/BPS): Role of N-Palmitoyl-D-Glucosamine-Hesperidin. In Proceedings of the 74th SISVET, Virtual Congress, 23–26 June 2021; p. 243.
133. Vezzoni, A.; Crupi, F.; Boiocchi, S.; Boano, S. Effect of Palmitoylethanolamide Co-Ultra Micronized with Quercetin in Dogs with Osteoarthritis by Means of Dynamic Gate Analysis and Canine Brief Pain Inventory Questionnaire. In Proceedings of the 5th World Veterinary Orthopaedic Congress ESVOT-VOS, Barcelona, Spain, 12–15 October 2018; 2018; pp. 771–772.
134. Asperio, R.M. Integrazione dietetica con PGA-Cur: Indagine osservazionale su 181 cani con osteoartrite [Dietary supplementation with PGA-cur: A survey on 181 osteoarthritis dogs]. *Summa* **2020**, *8*, 39–48.
135. Bonello, D.; Squarzoni, P. Effect of a Mucoadhesive Gel and Dental Scaling on Gingivitis in Dogs. *J. Vet. Dent.* **2008**, *25*, 28–32. [CrossRef]
136. De Luca, L.; Ferracane, R.; Vitaglione, P. Food Database of N-Acyl-Phosphatidylethanolamines, N-Acylethanolamines and Endocannabinoids and Daily Intake from a Western, a Mediterranean and a Vegetarian Diet. *Food Chem.* **2019**, *300*, 125218. [CrossRef]
137. Nestmann, E.R. Safety of Micronized Palmitoylethanolamide (MicroPEA): Lack of Toxicity and Genotoxic Potential. *Food Sci. Nutr.* **2017**, *5*, 292–309. [CrossRef]

biomolecules

Review

Synaptic Effects of Palmitoylethanolamide in Neurodegenerative Disorders

Martina Assogna [1,2], Francesco Di Lorenzo [1], Alessandro Martorana [1,2] and Giacomo Koch [1,3,*]

1 Department of Clinical and Behavioural Neurology, Santa Lucia Foundation IRCCS, 00179 Rome, Italy
2 Memory Clinic, Department of Systems Medicine, University of Tor Vergata, 00133 Rome, Italy
3 Department of Neuroscience and Rehabilitation, University of Ferrara, 44121 Ferrara, Italy
* Correspondence: g.koch@hsantalucia.it

Abstract: Increasing evidence strongly supports the key role of neuroinflammation in the pathophysiology of neurodegenerative diseases, such as Alzheimer's disease, frontotemporal dementia, and amyotrophic lateral sclerosis. Neuroinflammation may alter synaptic transmission contributing to the progression of neurodegeneration, as largely documented in animal models and in patients' studies. In the last few years, palmitoylethanolamide (PEA), an endogenous lipid mediator, and its new composite, which is a formulation constituted of PEA and the well-recognized antioxidant flavonoid luteolin (Lut) subjected to an ultra-micronization process (co-ultraPEALut), has been identified as a potential therapeutic agent in different disorders by exerting potential beneficial effects on neurodegeneration and neuroinflammation by modulating synaptic transmission. In this review, we will show the potential therapeutic effects of PEA in animal models and in patients affected by neurodegenerative disorders.

Keywords: PEA; neuroinflammation; neurodegeneration; synaptic plasticity; frontotemporal dementia; Alzheimer's disease; amyotrophic lateral sclerosis; transcranial magnetic stimulation; endocannabinoids

Citation: Assogna, M.; Di Lorenzo, F.; Martorana, A.; Koch, G. Synaptic Effects of Palmitoylethanolamide in Neurodegenerative Disorders. *Biomolecules* **2022**, *12*, 1161. https://doi.org/10.3390/biom12081161

Academic Editors: Salvatore Cuzzocrea and Rosalia Crupi

Received: 17 June 2022
Accepted: 6 August 2022
Published: 22 August 2022

1. Introduction

As life expectancy is continuously rising, the global economic effect of neurodegenerative disorders, such as Alzheimer's disease (AD), amyotrophic lateral sclerosis (ALS) and frontotemporal dementia (FTD) is increasing significantly [1]. The pathogenic mechanisms driving neurodegenerative illnesses, however, are still unknown. Several factors are involved, including genetic, environmental, and endogenous influences. Pathophysiological causes include abnormal protein dynamics, oxidative stress with reactive oxygen species, mitochondrial dysfunction, DNA damage, synaptic deficits, and neuroinflammatory processes [2].

Neuroinflammation is a complex process mediated by cytokines, which are primarily generated by microglia and astrocytes and whose activation can be harmful or protective to neurons. When implicated in the induction and control of neuronal development, cell survival, and synaptic plasticity pathways, beneficial pro-inflammatory cytokines are protective. Prolonged and abnormal pro-inflammatory signaling, however, is responsible for tissue neurodegeneration [3].

From a neuropathological point of view, neurodegenerative diseases are characterized by the deposition of misfolded proteins, such as amyloid beta (Aβ) and tau aggregates for AD and TAR DNA-binding protein 43 (TDP-43) in ALS and FTD. The progressive accumulation of these proteins triggers various pathological phenomena that contribute to the pathophysiological cascade of events that lead to the onset of clinical symptoms. Especially the impairment of the synaptic efficacy and the trigger and sustenance of neuroinflammation processes are increasingly being studied in neurodegenerative disorders as it has been shown their pivotal role in the progression of neurodegeneration and their potential modulation as therapeutic target in neurodegenerative diseases.

The aim of this review is to give insights into the interplay between synaptic machinery and neuroinflammation processes in neurodegenerative disorders and to clarify how palmitoylethanolamide (PEA)—an endogenous lipid mediator with high affinity for endocannabinoid receptor—and its new composite—which is a formulation constituted of PEA and the well-recognized antioxidant flavonoid luteolin (Lut) subjected to an ultramicronization process (co-ultraPEAlut)—might be able to modulate this relationship in animal models and in patients.

1.1. *Synaptic Impairment and Neuroinflammation in Neurodegenerative Disorders*

1.1.1. Alzheimer's Disease

Synaptic Impairment

AD is macroscopically characterized by brain atrophy while microscopic hallmarks are the deposition of amyloid plaques and neurofibrillary tangles. Recently, with the introduction of biomarkers able to reflect in vivo the neuropathological alterations occurring in the disease, substantial modifications have been posed to AD definition; however, the clinical course of the disease remains unpredictable due to the scarce comprehension of pathophysiological mechanisms.

At this regard, there is strong evidence that synaptic density loss occurs before neuronal death, implying that impaired synaptic plasticity processes play a major role in AD etiology [4,5]. The loss of synaptic density has been reported to have the strongest statistical link with the degree of cognitive impairment in AD, rather than Aβ plaques, tangle formation, or neuronal death [6].

As a result, synaptic transmission impairment caused by toxic oligomeric species [7] can predict disease severity more accurately than gross neuronal death—a later occurrence—establishing synaptic dysfunction as a fundamental driver of AD-related cognitive decline rather than a byproduct [8]. Indeed, experimental studies in AD animal models have shown that Aβ peptides and tau proteins interact with physiological mechanisms of neuronal synaptic plasticity [9,10].

Moreover, N-methyl-d-aspartate receptor (NMDAr) mediated glutamatergic neurotransmission is crucial for synaptic plasticity and survival of neurons. Nevertheless, excessive NMDAr activity, mediated by excessive Ca^{2+} influx, may result in excitotoxicity and promotes cell death underlying a potential mechanism of neurodegeneration [11]. In humans, neurophysiological techniques, such as transcranial magnetic stimulation (TMS), can help in differentiating different neurodegenerative diseases [12] and forecast AD disease progression by estimating cortical functioning at a specific time [13].

TMS can be used to examine cortical plasticity mechanisms, such as long-term potentiation (LTP), one of the most important neurophysiological correlates for learning and memory [14]. We previously demonstrated that AD patients had a consistent deficit of LTP-like cortical plasticity in motor function [15,16] and the cerebellar cortex [17], with a sparing of mechanisms of long-term depression (LTD), evident also in early mild cognitive impairment (MCI) patients [18].

Moreover, in AD animal models, the synaptic dysfunction has been linked to a disorder of high-frequency neuronal oscillatory activity, in particular in the gamma range (40 Hz) [19,20]. Accordingly, in a recent work, TMS combined with EEG (TMS-EEG) recordings have shown that AD patients had more prominent decrease in gamma activity in the prefrontal cortex with a stronger impairment of LTP-like plasticity mechanisms and more prominent cognitive decline [21]. Interestingly, the optogenetic entrainment of fast-spiking parvalbumin-positive interneurons of AD animal model at gamma frequencies was able to reduce the total amyloid levels, probably acting on both neurons and microglia [22].

Similarly, intranasal administration of pro-resolving lipid mediator in a mouse model of AD was able to improve memory dysfunction and restore gamma oscillation impairment, accompanied by a modulation of microglial activation [23].

Neuroinflammation in AD

Microglial cells are a primary target of neurodegenerative disease research because they play a vital part in the inflammatory process of the central nervous system. Depending on the specific stimulus the microglia have been exposed to, it could maintain a balance between a pro-inflammatory status (M1 phenotype), characterized by the synthesis of inflammatory cytokines, such as interleukin 1 (IL-1), interleukin 6 (IL-6), and tumor necrosis factor (TNF), and the synthesis and release of anti-inflammatory cytokines (IL-4, IL-8, and IL-10) and neurotrophic factors (M2 phenotype) [24]. Thus, the complex involvement of inflammatory cytokines in both neurodegeneration and neuroprotection is far from complete in such a complex environment. Amyloid peptides, comprising both oligomeric and senile plaque forms, are thought to be the key inflammatory trigger in AD. A prolonged pro-inflammatory signaling caused by amyloid mis-metabolism, in particular, can result in an overproduction of pro-inflammatory cytokines involved in neurodegenerative pathways signaling [3]. While there is evidence that persistent neuroinflammation causes an increase in amyloid synthesis [25], a clear relationship between tau pathology and neuroinflammation is still unclear. We recently showed that human astrocytes cultures incubated with cerebrospinal fluid (CSF) samples from AD patients were vulnerable in terms of increased apoptosis only in the presence of high levels of tau protein and APOE4 genotype [26].

As a result of these observations, we hypothesized that tau proteins play a substantial role in astrocyte degradation and a proinflammatory role in APOE4 patients [27]. Surprisingly, APOE4 carriers have been found to have an imbalanced flipping of the microglial phenotype M1–M2 [28]. Furthermore, microglial apolipoprotein E (ApoE) regulates microglial homeostatic gene expression downstream, resulting in a neurodegenerative phenotypic switch that could exacerbate AD pathogenesis [29]. Consistent with this framework, we showed that during early phases of AD, in APOE4 carriers, amyloid pathology likely induces a specific cytokines pattern synthesis associated to cognitive preservation [30] (Figure 1).

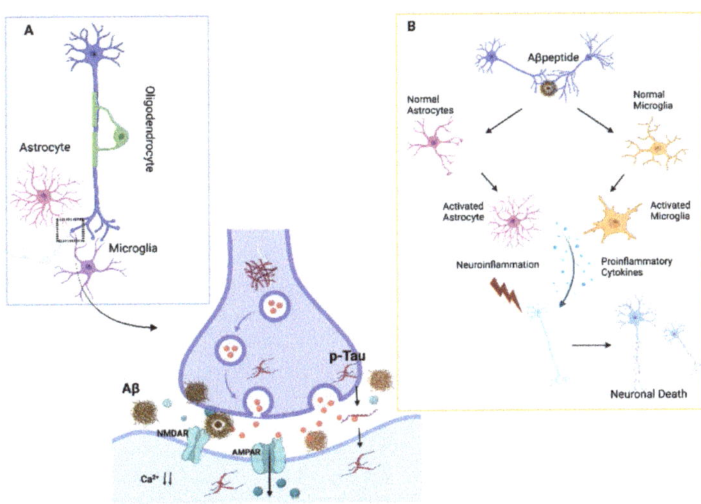

Figure 1. Neuroinflammation and synaptic impairment in AD: Panel (**A**) displays the interplay among the neuron and the glial cells involved in AD pathology, such as astrocyte, microglia, and oligodendrocyte. In the dotted square, the particular of synaptic transmission impairment induced by Aβ and p-Tau activation; panel (**B**) shows the probable mechanisms of action of the activated microglia and astrocyte induced by amyloid beta deposition and the pro-inflammatory cytokines cascade in the pathophysiology of AD. AD: Alzheimer's disease; Aβ: amyloid beta; p-Tau: phosphorylated tau.

1.1.2. Frontotemporal Lobar Degeneration

Synaptic Impairment in FTLD

Similar to AD, synaptic disruption appears to precede neuronal death also in frontotemporal lobar degeneration (FTLD). Indeed, FTLD abnormal brain connectivity (connectopathy) or synaptopathy has been described [31,32]. In the affected cortex, extensive synaptic loss and a reduction in the number of spines have been shown post-mortem [33,34]. In some [35], but not all, studies [36], a significant decrease in synaptic density evaluated with synaptophysin in the superficial layers of the prefrontal cortex of FTLD individuals compared to normal controls was described. In Pick's illness, synaptophysin immunoreactivity was likewise diminished in the hippocampus dentate gyrus' outer molecular layer [37].

Recently, in a sample of behavioral variant of FTD (bvFTD), synaptic loss was measured in vivo with synaptic vesicle glycoprotein 2A (SV2A)-PET, a metabolic marker of synaptopathy, in the anterior parahippocampal gyrus of a sample of bvFTD patients [38].

Interestingly, in a TMS study, altered mechanisms of plasticity were observed also in pre-symptomatic FTD carriers of progranulin (GRN) and C9orf72 genes mutation in comparison to age-matched healthy controls, reinforcing the notion that alteration of synaptic machinery begins years before the onset of clinical symptoms [39].

Neuroinflammation in FTLD

Neuroinflammation and immune-mediated processes have been identified as key contributors to the degenerative process of FTD [40–42]. Neuroinflammation is a hotly contested topic [43], whether it is a main or secondary event in the neurodegeneration associated with FTD or has an overall helpful or negative effect. Alternatively, the initial pathological insult (that is, aggregation and/or accumulation of amyloid-, tau-, or TDP43) induces an ongoing cytotoxic response that results in secondary chronic neuroinflammation and altered neuronal function in brain regions specific to the disease phenotype [44,45]. The buildup of aberrant conformations of tau or TDP43 signals generated by injured neurons [46] or deregulation of the systems for clearing misfolded or damaged neuronal proteins are likely to stimulate immune activation in FTD. These mechanisms eventually result in neurodegeneration [47,48].

Depending on the stage and severity of the disease, many immunological mechanisms, both innate and adaptive, are likely to be engaged. As a result, looking into neuroinflammatory and immune-mediated pathways for diagnostic biomarkers, innovative therapy targets, and disease-modifying medicines for FTD is a potential line of research.

1.1.3. Amyotrophic Lateral Sclerosis

Synaptic Impairment in ALS

Amyotrophic lateral sclerosis (ALS) is a neurodegenerative illness characterized by motor neurons loss (MNs). Misfolded proteins, glutamate excitotoxicity, mitochondrial dysfunction at distal axon terminals, and alterations in the neuronal cytoskeleton are all pathogenic characteristics of ALS. C9orf72 gene-mediated pathogenesis is aided by synergies between the loss in C9orf72 functions and the gain in function caused by toxic consequences of repeat expansions [49]. Neuropathological hallmarks of ALS are dendritic and synaptic degeneration in the cortex and corticospinal (CS) motor neurons [50].

Braak and colleagues classified ALS as a disease of big axon neurons with different stages of trans-synaptic dissemination based on the cortical synaptic degeneration hallmark [51]. According to this classification, the disease begins in corticospinal motor neurons (CSMNs), develops to MNs, and then to extra-motor regions [51,52]. Cortical hyperexcitability is an early clinical characteristic of ALS, and this pathological insight is consistent with it [53]. Furthermore, the trans-synaptic spread theory implies a prion-like mechanism for spreading misfolded protein aggregates to distant populations of neurons [51]. This supports the theory that ALS is caused by a synaptopathy [54].

Synaptopathy is a general term for disorders characterized by synaptic dysfunction, independent of the underlying causes [55]. Synaptopathy, in its broadest sense, refers to a collection of symptoms that, over time, contribute to synaptic failure. Changes in Ca^{2+} levels at synapses, glutamate excitotoxicity, structural changes in pre- and postsynaptic anchoring proteins, altered synaptic structure and function, which is frequently associated with dendritic spine loss, dysfunctional neurotransmitter release, impaired maintenance and regeneration of axons by Schwann cells, and cognitive deficits are among these characteristics [54,55]. Neuronal loss, mitochondrial dysfunction, accumulation of misfolded proteins associated with faulty proteostasis, and dysfunctional neuromuscular junctions are all symptoms of synaptopathies [55,56].

Several studies have shown aberrant synapses' shape and function in ALS, with the nature of the abnormalities varying depending on the disease's progression [54,57]. In addition, neuronal loss, alterations in dendritic spine density, and morphology in excitatory neurotransmission locations, particularly in pyramidal cells, such as CSMNs, have been documented in ALS post-mortem samples [54,57]. This trait has also been seen in many ALS animal models [58]. We will examine how alterations in multiple pathways can contribute to synaptic dysfunction in the C9orf72-ALS pathogenesis in this review.

Neuroinflammation ALS

Neuroinflammation is a pathogenic process defined by the invasion of activated microglia and astrocytes in ALS patients with and without genetic abnormalities. In post-mortem tissues of ALS patients, activated microglia and astrocytes that produce pro-inflammatory cytokines are increased [59–61]. In living ALS patients, a PET investigation revealed increases in activated microglia [11C-(R)PK11195 PET] and astrocytes (11C-DED PET) [62,63]. Furthermore, sporadic ALS patients with varied degrees of disease severity have considerably greater CSF soluble triggering receptor expressed on myeloid cells 2 (sTREM2) levels than controls [64]. CSF sTREM2 levels are highest in early-stage ALS, and higher levels of CSF sTREM2 are associated with slower disease development in late-stage ALS. High levels of sTREM2 in the CSF for a long time could indicate a neuroprotective phenotype [64].

2. PEA and PEA Combined with Luteolin Mechanisms of Action

2.1. PEA Synaptic Mechanisms of Action

PEA is an endogenous lipid mediator, which belongs to the class of acylethanolamides (AEs), produced "on demand" from phospholipids membranes with a well-recognized anti-inflammatory, analgesic, and neuroprotective function in various conditions in both central and peripheral nervous system [65–68]. The biological effect of PEA has been extensively investigated by preclinical studies, and it seems to be mediated by the direct activation of the orphan GPCR 55 receptor (GPR55), while the affinity for type-1 and type-2 cannabinoid receptors (CB1 and CB2) and for the transient receptor potential vanilloid 1 (TRPV1) channels is lower. Moreover, PEA plays a central role in the modulation of pain and inflammation pathways by the activation of peroxisome proliferator-activated receptor-alpha receptor (PPAR-α) [69,70]. In particular, the activation of PPAR-α induced by PEA, through the interaction with transcriptions factors is involved in the reduction in NF-κB activation and pro-inflammatory enzyme synthesis, thus promoting anti-inflammatory and analgesic effects [71,72]. Moreover, the protective effects of PEA in neurodegeneration and neuroinflammation preclinical models of different pathologies are reversed by the pharmacological modulation of PPAR-α with antagonists or its genetic silencing [73–76]. GPR55 receptors are broadly expressed in several brain areas, and they were identified as cannabinoid receptors with a different signaling pathway from CB1 and CB2 receptors [77]. The PEA-induced anti-inflammatory effects seems in part to be mediated by the activation of GPR55 in an experimental murine model of colitis [78], in a murine model of Parkinson's disease [79], and in chronic arterial inflammation [80]. While the immunomodulatory function of GPR55 is widely recognized, their effect on neuronal cells and their localization is still elusive [81]. A recent work from Musella et al., [82] investigated the involvement of PEA in both excitatory and inhibitory transmission in the striatum of a rodent model. The authors found for the first time that PEA can enhance GABA transmission and modulate the synthesis of 2-Arachidonoylglycerol (2-AG), which acts as inhibiting in a retrograde manner at a presynaptic site the CB1R and GABA release. Furthermore, PEA can modulate and enhance indirectly the levels of other endocannabinoids, through the so-called entourage effect [65,67]. Despite the low affinity for CB1 and CB2 receptors exerted by PEA, they can be activated indirectly in different ways. Indeed, PEA can reduce the degradation of anandamide (AEA), acting on the fatty acid amide hydrolase (FAAH), the enzyme responsible for endocannabinoids degradation, leading to an increase in the cannabinoid receptor mediated transmission. More recently, PEA exposure was found to induce changes in microglia morphology and activation through the PPAR-α activation, including increased migration and phagocytosis due to a reactive microglial phenotypes mediated by its indirect regulation of CBR2 receptors [83]. In addition, in a recent study from D'Aloia et al., the treatment with PEA inhibited the M1 microglial polarization induced by lipopolysaccharide (LPS), while the anti-inflammatory markers in microglial cells were upregulated, highlighting a possible role for PEA in inducing the anti-inflammatory M2a phenotype and as a potential therapeutic tool in neurodegenerative disorders with chronic microglial hyperactivity [84]. Finally, PEA can also indirectly modulate other endocannabinoids targets, such us the TRPV1 channel [85,86]. Interestingly, it has been recently reported that TRPV1 activation in animal models of Parkinson's and AD can exert neuroprotective effects [87–90]. Accordingly, the role of the TRVP1 activation by the endocannabinoid anandamide was found to be effective in reversing memory deficits and hippocampal function from Aβ-induced cytotoxicity, in rodents, in particular through the effect of rescues of gamma oscillations [90], which an alteration is emerging as a key mechanism for the pathophysiology of AD [91]. The administration of the TRPV1-receptor agonist capsaicin was able to restore hippocampal damage and gamma oscillations by reversing both the desynchronization of action potential firing in CA3 pyramidal cells and the shift in excitatory/inhibitory current balance, thus suggesting the pathway as a possible therapeutic target for AD [90].

Noteworthy, PEA seems to exert neuroprotective effects also through the modulation of the glutamatergic transmission and synaptic plasticity. Indeed, Lin and colleagues showed that in rat cerebrocortical nerve terminals, PEA could act on glutamate pathways exerting a presynaptic inhibition of glutamate, likely through a reduction in the Ca^{2+} influx, which might be linked to the activation of presynaptic cannabinoid CB1 receptors [92]; the inhibition of aberrant glutamatergic activity might thus result in an anti-excitotoxic effect, which could counteract the neurodegenerative process.

All these findings strongly support that PEA could be considered as a promising molecule to counteract neuroinflammation and neurodegeneration through its multiple synaptic targets.

2.2. *PEA Combined with Luteolin Effects*

Several studies have been shown a role of flavonoids, in particular luteolin, in displaying many neuroprotective and anti-inflammatory properties in chronic conditions and neurodegenerative disorders [93]. Moreover, luteolin has been shown to exert antioxidant activity and to improve glucose metabolism by potentiating insulin sensitivity and modulating Aβ deposition by the activation of the gut-microbiota-liver-brain axis in AD models [94]. The combination of the pharmacodynamic properties of PEA and those of luteolin was found to be more effective in counteracting both inflammation and oxidative stress. Accordingly, many studies have shown that using PEALut can provide better effects by stimulating both hippocampal neurogenesis and dendritic spine maturation [95,96] and that the two molecules potentiate their synergic effect when simultaneously submitted to the micronization process [67]. Additionally, the association with flavones seems to stabilize the two molecules and enhance their pharmacological activities [97], even in experimental models of AD and other neurodegenerative disorders [76,98], mainly through the modulation of the neuroinflammatory and apoptotic pathways, the cytokines release, the activation of astrocytes and microglia, and the ability to modulate the autophagic process [95,99].

3. Synaptic Effects of PEA and PEALut in Neurodegenerative Disorders

In the last few years, PEA has been identified as a potential therapeutic agent in different neurodegenerative disorders [100,101]. Based on several preclinical and clinical studies, PEA can exert potential beneficial effects on neurodegeneration and neuroinflammation by modulating synaptic transmission.

3.1. Preclinical Models

The first in vitro study of the mechanism of PEA in AD model from Scuderi et al. in 2011 demonstrated the ability of PEA to reduce Aβ-induced astrocyte activation and proinflammatory molecules and cytokine release in primary rat astrocytes through a PPAR-α-dependent mechanism [72]. In line with the same observations, a later study of the same authors showed that PEA reduced reactive gliosis and attenuated neuronal damage in rat models of Aβ-induced neurotoxicity, with a mechanism strictly depending on PPAR-α activation [102,103].

Accordingly, the neuroprotective effect of PEA was demonstrated in mice injected with amyloid-β 25–35 (Aβ 25–35) peptide intracerebroventricularly, evaluated for learning and memory deficits [73]. Surprisingly, PEA was able to reduce or prevent, in a dose-dependent manner, the Aβ induced behavioral deficits, while it failed to rescue memory impairment in peroxisome proliferator-activated receptor-α (PPAR-α) null mice, thus further highlighting the importance of PPAR-α modulation for neuroprotection and PEA efficacy against amyloid neuronal damage [73].

A following study from Scuderi et al. in 2014 investigated the systemic administration of PEA in murine models given an injection of beta-amyloid 1–42 (Aβ 1–42) in the hippocampal cortex, to further elucidate its therapeutic potential and the mechanisms underlying the behavioral effects. The authors reported significant modifications in biomarkers related to astrogliosis and amyloidogenesis, finding new evidence that PEA can restore behavioral deficits and impaired molecular pathways similar to early traits of AD by activating PPAR-α [104].

The anti-inflammatory properties of the co-ultraPEALut, a composite of PEA and the antioxidant flavonoid luteolin, has been investigated in organotypic model of AD by incubating with Aβ1-42 peptide differentiated human neuroblastoma cells and hippocampal slice cultures [75]. The authors showed that the compound exerted a protective effect on glial cells by reducing significantly apoptosis and glial fibrillary acidic protein expression and restoring neuronal nitric oxide synthase and brain-derived neurotrophic factor [75].

Accordingly, the incubation with co-ultraPEALut significantly reduced the TNF-α-induced serum amyloid A (SAA) mRNA expression in oligodendrocyte precursor cells, which is relevant since SAA has been demonstrated to localize immunohistochemically with aβ deposits in AD brain [105].

Moreover, co-ultra PEALut was able to counteract the Aβ1–42-mediated inflammation and astrocyte reactivity in an in vitro model of AD and to restore oligodendrocytes homeostasis through a mechanism that could involve PPAR-α activation [106]. Interestingly, PEA effects in neuroinflammation and AD angiogenesis were demonstrated in Aβ-treated C6 rat astroglioma cells and human umbilical vein endothelial cells (HUVEC) [107]. As expected, PEA was able to induce a dose-dependent reduction in pro-inflammatory and pro-angiogenic biomarker in the cells stimulated with Aβ, and the effect was blocked in the model by the treatment with the PPAR-α antagonist GW6471, further suggesting the mechanism on astroglial cells is proliferator-activated receptor alpha-dependent [107].

Similarly, PEA was tested for its neuroprotective effects against Aβ-induced toxicity on cell vitality and glutamatergic transmission in AD mice, particularly in primary cultures of cortical neurons and astrocytes from mice with the triple-transgenic AD model (3×Tg-AD) and the wild type mice [108]. As expected, PEA reversed the effects of the Aβ1-42 fragment on glutamatergic transmission and cell viability in cultured neurons and astrocytes isolated from wild type mice. On the contrary, PEA did not prevent the formation of Aβ-plaques and neurofibrillary tangles or early synaptic dysfunction or cognitive decline in 3×Tg-AD mice, thus suggesting a possible efficacy only in early AD [108].

Another study investigated the reactive astrogliosis process in 3×Tg-AD mice treated with PEA; the authors found that the astrocytes' reactive state and neurons' viability were improved by PEA revealing its beneficial neuro-supportive function [109].

Interestingly, PEA also has been shown to exert immunomodulatory, analgesic, and neuroprotective effects and restore cognitive dysfunction in different chronic pain conditions by restoring glutamatergic synapses' functioning deficits [110–112].

In particular, the involvement of metabotropic glutamate receptor (mGluR) 5 and 8 in ultra-micronized (um-PEA) effects on cognition and long-term potentiation (LTP) mechanisms was investigated in the entorhinal cortex (LEC)-dentate gyrus (DG) pathway in mouse models of spare nerve injury (SNI) [113].

The chronic treatment with this compound rescued discriminative memory and LTP deficits at the LEC-DG pathway in SNI mice. The authors, based on the physiological role exerted from glutamate in memory formation processes in the hippocampus hypothesized that the modulation of glutamatergic activity might be at the base of the PEA-induced restoration of LTP and cognitive behavior in SNI models of chronic pain, thus reducing glutamate excitotoxicity [111]. In particular, the administration of mGluR5 antagonist facilitated memory and plasticity mechanisms while the mGluR8 blockage prevented the protective action of the PEA on LTP, thus displaying different roles but both necessary to mediate the efficacy of PEA in neuropathic pain [111].

In a recent work from Beggiato et al. from 2020, the authors investigated the effects of ultra-micronized PEA (um-PEA) treatment in 3×Tg-AD mice, and they found that it was able to reduce the typical increase in hippocampal glutamate levels observed in the AD mouse model [114].

Accordingly, chronic treatment with PEA reversed memory deficit and LTP impairment in SNI wild type mouse models, but not in PPARα null mice, and restored glutamatergic transmission deficits, the loss in synaptic density, and the expression of phosphorylated GluR1 subunits, as well as an increase in neuroblasts [111]. Moreover, the increase in synaptogenesis induced by PEA in SNI mice was correlated to the improvement in episodic memory and LTP. Taken together, these results open new perspectives for the use of PEA for the translation of the same result in AD pathology for which an impaired or a same synaptic transmission deficit and impairment of LTP mechanisms have been demonstrated [12].

The therapeutic potential of three-months subcutaneously administration of ultra-micronized PEA (um-PEA) was investigated in 3×Tg-AD mice, and mitochondrial bioenergetics alterations, which can lead to glutamatergic neurotransmission alterations and excitotoxicity, were evaluated in the frontal cortex (FC) and hippocampus (HIPP), with the results showing that um-PEA was able to counteract mitochondrial dysfunctions and rescue brain energy metabolism in the FC but not in the HIPP [115].

Local synaptic effects of PEA on GPR 55 transmission in ventral-hippocampus (vHipp), a key region to memory function, and the consequent modulation of mesolimbic activity were investigated in in vivo rat brains from Kramar et al., 2017 [116].

As hypothesized, vHipp GPR55 activation was found to increase glutamatergic levels in the hippocampus, potentiating excitatory transmission from the vHipp to the mesolimbic cortex, including the ventral tegmental area (VTA).

According to the result of this study, PEA administration was found to increase VTA dopaminergic frequency and bursting rates through a local NMDA-receptor dependent mechanism [116].

Finally, the pharmacological properties of co-ultra PEALut were investigated in an in vivo mouse model of prodromal AD. In Aβ infused rats, the early administration of PEA reduced the astrogliosis and microgliosis and prevent the over-expression of pro-inflammatory cytokine genes and the reduction in mRNA levels BDNF and GDNF, which are fundamental neurotrophins regulating synaptic plasticity mechanisms and neuronal growth and branching [117].

3.2. Clinical Studies

Thus far, several studies reported the protective effects of PEA in neuropathic pain and peripheral conditions sustained by neuroinflammation, while there are few works aimed at investigating the efficacy of PEA administration in neurodegenerative disorders in humans.

The first case report on um-PEA oral administration in a patient with ALS showed an improvement of the clinical picture, as measured by electromyographic analysis and respiratory capacity, due probably to the ability to modulate neuroinflammation [118]. According to this previous finding, a larger study on ALS tested the clinical and molecular effects of 600 mg um-PEA administration (in 28 patients) twice daily as an addition to standard therapy alone (50 mg riluzole, 36 patients) for six months [119].

Surprisingly, the trial showed that PEA-treated ALS patients had a slower decline in respiratory function as measured by a lower decrease in their forced vital capacity (FVC) and later need for a tracheotomy, compared to the untreated patients.

Moreover, the authors, by micro-transplanting human muscle membranes from muscle biopsies of ALS patients into Xenopus oocytes, showed that PEA can reduce the desensitization of AChRs-evoked currents in both ALS and non-ALS human samples after repetitive neurotransmitter application, which is selectively effective on human e-AChRs subtype, thus providing molecular basis for PEA efficacy on muscle excitability and evidences on acetylcholine modulation exerted by the compound [119].

Until now, there are no clinical data published on the possible beneficial effects of PEA in AD patients. Only one study investigated the efficacy of nine months PEALut high-dose administration in amnestic MCI in a patient [120].

The patient at baseline underwent a neuropsychological examination, which included attentive matrices, Babcock Story recall test, Mini-Mental State examination (MMSE), Montreal Cognitive Assessment, Rey Auditory-Verbal Learning Test, Trail Making Test, and verbal fluency tests as well as a perfusion single-photon emission computed tomography, which documented a significant hypoperfusion in the parietal, inferior-temporal, and temporo-occipital areas. At the nine-months follow-up, the neuropsychological evaluation was almost normal, and the SPECT hypometabolism was normalized [120], thus opening to the design of larger clinical trials in this population.

A recent study published by our group [121] investigated, for the first time, the cognitive and neurophysiological effects of four weeks of PEALut administration in seventeen patients with FTD. For the purpose of this study, patients underwent an extensive cognitive and behavioral assessment, which included neuropsychiatric inventory (NPI), MMSE, frontal assessment battery (FAB), screening for aphasia in neurodegeneration (SAND), FTLD-modified clinical dementia rating scale sum of boxes (FTLD-SOB.) To further investigate in vivo neurophysiological synaptic effects of the compound administration, we used paired-pulse and repetitive TMS protocols assessing LTP mechanisms and long-interval intracortical inhibition. Moreover, TMS-EEG recordings were collected to evaluate changes in frontal oscillatory activity.

Surprisingly, the results showed that PEALut can improve frontal lobe function and behavioral disturbances, mainly through the modulation of GABAergic activity and high-frequency cortical oscillatory activity, which is impaired in FTD patients [121] (Figures 2 and 3).

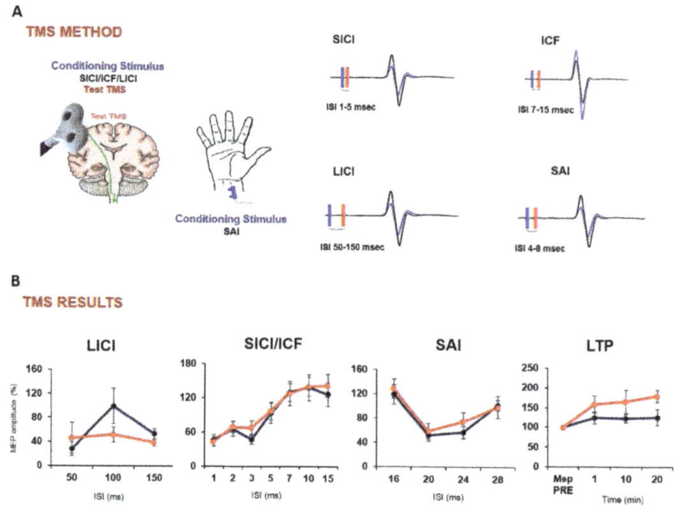

Figure 2. Neurophysiological effects of PEALut in FTD patients: paired-pulse TMS results and iTBS after effects (adapted from [121]): Panel (**A**) provides a schematic illustration of TMS protocols investigating synaptic transmission and cortical plasticity; panel (**B**) shows corticospinal measures. After one month of treatment with PEALut, we observed a significant restoration of LICI and LTP, suggesting a restoration of GABA(B) activity and cortical plasticity. No effects were found in protocols measuring cholinergic neurotransmission (SAI) and GABA(A) activity (SICI). (blue line represents the pre-treatment results, red line post-treatment). FTD: frontotemporal dementia; TMS: transcranial magnetic stimulation; PEALut: palmithoylethanolamide combined with luteolin; LICI: long-interval intracortical inhibition; LTP: long-term potentiation; SICI: short-interval intracortical inhibition; SAI: short-latency afferent inhibition; iTBS: intermittent theta burst stimulation.

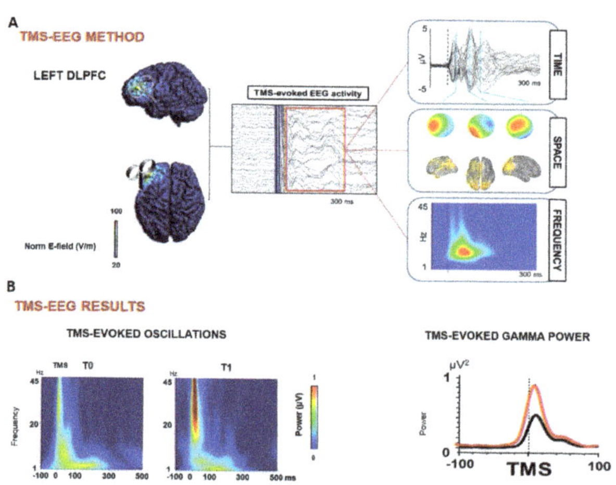

Figure 3. Effects of PEALut on cortical oscillations in FTD patients (adapted from [121]): Panel (**A**) provides a schematic illustration of TMS-EEG protocols investigating cortical reactivity, oscillatory activity, and connectivity on left DLPFC; panel (**B**) displays cortical measures results. After one month of treatment with PEALut, we observed a significant increase in high-frequency oscillations (black line represents the gamma power pre-treatment, red line post-treatment). FTD: frontotemporal dementia; TMS: transcranial magnetic stimulation; PEALut: palmithoylethanolamide combined with luteolin; DLPFC: dorsolateral prefrontal cortex.

4. Conclusions

So far, the pharmacological treatment for AD and other neurodegenerative dementias has been mostly based on symptomatic drugs enhancing cognition and reducing behavioral alterations (antipsychotic drugs, acetylcholinesterase inhibitors, and NMDA receptor antagonist) with contrasting results and no efficacy in modifying disease progression [122].

Passive immunotherapies based on the administration of exogenous antibodies targeting the two hallmarks in AD pathology, which are Aβ and tau protein, have given ambiguous results and minimal therapeutic benefit [123].

Based on this premise, a conceptual shift in the approach to neurodegenerative dementias treatment is needed urgently. In this context, new therapies focusing on synaptic dysfunction [124,125] and neurotransmitters deficits, as well as new recognized pathogenetic mechanisms, such as neuroinflammation, might be considered as a promising therapeutic agent to counteract neurodegeneration.

In this scenario and based on the results presented in this review, new compounds such as PEA, and new formulations, such as PEAlut, with its multiple pharmacological targets and mechanisms of action, fulfill the criteria for a potential promising key role in modulating neuroinflammation and synaptic neurotransmission, especially at early stages, thus modifying disease progression.

While preclinical studies data are promising, the lack of larger clinical trials is necessary to further elucidate the role of PEA and PEAlut in neurodegenerative diseases treatment.

Author Contributions: All authors designed the review, critically reviewed the contents, and contributed to manuscript preparation. All authors have read and agreed to the published version of the manuscript.

Funding: This work was partially supported by a grant received from the Italian Ministry of Health (grant GR-2019-12370435) to FDL.

Institutional Review Board Statement: Not applicable.

Informed Consent Statement: Not applicable.

Data Availability Statement: Not applicable.

Conflicts of Interest: The authors declare no conflict of interest.

References

1. Cova, I.; Markova, A.; Campini, I.; Grande, G.; Mariani, C.; Pomati, S. Worldwide Trends in the Prevalence of Dementia. *J. Neurol. Sci.* **2017**, *379*, 259–260. [CrossRef] [PubMed]
2. Jellinger, K.A. Basic Mechanisms of Neurodegeneration: A Critical Update. *J. Cell Mol. Med.* **2010**, *14*, 457–487. [CrossRef] [PubMed]
3. Zheng, C.; Zhou, X.-W.; Wang, J.-Z. The Dual Roles of Cytokines in Alzheimer's Disease: Update on Interleukins, TNF-α, TGF-β and IFN-γ. *Transl. Neurodegener.* **2016**, *5*, 7. [CrossRef] [PubMed]
4. Scheff, S.W.; Price, D.A. Synaptic Pathology in Alzheimer's Disease: A Review of Ultrastructural Studies. *Neurobiol. Aging* **2003**, *24*, 1029–1046. [CrossRef]
5. Ingelsson, M.; Fukumoto, H.; Newell, K.L.; Growdon, J.H.; Hedley-Whyte, E.T.; Frosch, M.P.; Albert, M.S.; Hyman, B.T.; Irizarry, M.C. Early Abeta Accumulation and Progressive Synaptic Loss, Gliosis, and Tangle Formation in AD Brain. *Neurology* **2004**, *62*, 925–931. [CrossRef]
6. Terry, R.D.; Masliah, E.; Salmon, D.P.; Butters, N.; DeTeresa, R.; Hill, R.; Hansen, L.A.; Katzman, R. Physical Basis of Cognitive Alterations in Alzheimer's Disease: Synapse Loss Is the Major Correlate of Cognitive Impairment. *Ann. Neurol.* **1991**, *30*, 572–580. [CrossRef]
7. Selkoe, D.J. Alzheimer's Disease Is a Synaptic Failure. *Science* **2002**, *298*, 789–791. [CrossRef]
8. Martorana, A.; Di Lorenzo, F.; Belli, L.; Sancesario, G.; Toniolo, S.; Sallustio, F.; Sancesario, G.M.; Koch, G. Cerebrospinal Fluid Aβ 42 Levels: When Physiological Become Pathological State. *CNS Neurosci. Ther.* **2015**, *21*, 921–925. [CrossRef]
9. Lasagna-Reeves, C.A.; Castillo-Carranza, D.L.; Sengupta, U.; Clos, A.L.; Jackson, G.R.; Kayed, R. Tau Oligomers Impair Memory and Induce Synaptic and Mitochondrial Dysfunction in Wild-Type Mice. *Mol. Neurodegener.* **2011**, *6*, 39. [CrossRef]
10. Llorens, F.; Schmitz, M.; Karch, A.; Cramm, M.; Lange, P.; Gherib, K.; Varges, D.; Schmidt, C.; Zerr, I.; Stoeck, K. Comparative Analysis of Cerebrospinal Fluid Biomarkers in the Differential Diagnosis of Neurodegenerative Dementia. *Alzheimers Dement.* **2016**, *12*, 577–589. [CrossRef]

11. Wang, R.; Reddy, P.H. Role of Glutamate and NMDA Receptors in Alzheimer's Disease. *J. Alzheimers Dis.* **2017**, *57*, 1041–1048. [CrossRef] [PubMed]
12. Benussi, A.; Grassi, M.; Palluzzi, F.; Koch, G.; Di Lazzaro, V.; Nardone, R.; Cantoni, V.; Dell'Era, V.; Premi, E.; Martorana, A.; et al. Classification Accuracy of Transcranial Magnetic Stimulation for the Diagnosis of Neurodegenerative Dementias. *Ann. Neurol.* **2020**, *87*, 394–404. [CrossRef] [PubMed]
13. Di Lorenzo, F.; Motta, C.; Casula, E.P.; Bonnì, S.; Assogna, M.; Caltagirone, C.; Martorana, A.; Koch, G. LTP-like Cortical Plasticity Predicts Conversion to Dementia in Patients with Memory Impairment. *Brain Stimul.* **2020**, *13*, 1175–1182. [CrossRef] [PubMed]
14. Bliss, T.V.; Collingridge, G.L. A Synaptic Model of Memory: Long-Term Potentiation in the Hippocampus. *Nature* **1993**, *361*, 31–39. [CrossRef]
15. Koch, G.; Esposito, Z.; Kusayanagi, H.; Monteleone, F.; Codecá, C.; Di Lorenzo, F.; Caltagirone, C.; Bernardi, G.; Martorana, A. CSF Tau Levels Influence Cortical Plasticity in Alzheimer's Disease Patients. *JAD* **2011**, *26*, 181–186. [CrossRef]
16. Di Lorenzo, F.; Ponzo, V.; Motta, C.; Bonnì, S.; Picazio, S.; Caltagirone, C.; Bozzali, M.; Martorana, A.; Koch, G. Impaired Spike Timing Dependent Cortico-Cortical Plasticity in Alzheimer's Disease Patients. *J. Alzheimers Dis.* **2018**, *66*, 983–991. [CrossRef]
17. Di Lorenzo, F.; Martorana, A.; Ponzo, V.; Bonnì, S.; D'Angelo, E.; Caltagirone, C.; Koch, G. Cerebellar Theta Burst Stimulation Modulates Short Latency Afferent Inhibition in Alzheimer's Disease Patients. *Front. Aging Neurosci.* **2013**, *5*, 2. [CrossRef]
18. Martorana, A.; Di Lorenzo, F.; Manenti, G.; Semprini, R.; Koch, G. Homotaurine Induces Measurable Changes of Short Latency Afferent Inhibition in a Group of Mild Cognitive Impairment Individuals. *Front. Aging Neurosci.* **2014**, *6*, 254. [CrossRef]
19. Arroyo-García, L.E.; Isla, A.G.; Andrade-Talavera, Y.; Balleza-Tapia, H.; Loera-Valencia, R.; Alvarez-Jimenez, L.; Pizzirusso, G.; Tambaro, S.; Nilsson, P.; Fisahn, A. Impaired Spike-Gamma Coupling of Area CA3 Fast-Spiking Interneurons as the Earliest Functional Impairment in the AppNL-G-F Mouse Model of Alzheimer's Disease. *Mol. Psychiatry* **2021**, *26*, 5557–5567. [CrossRef]
20. Nakazono, T.; Lam, T.N.; Patel, A.Y.; Kitazawa, M.; Saito, T.; Saido, T.C.; Igarashi, K.M. Impaired In Vivo Gamma Oscillations in the Medial Entorhinal Cortex of Knock-in Alzheimer Model. *Front. Syst. Neurosci.* **2017**, *11*, 48. [CrossRef]
21. Casula, E.P.; Pellicciari, M.C.; Bonnì, S.; Borghi, I.; Maiella, M.; Assogna, M.; Minei, M.; Motta, C.; D'Acunto, A.; Porrazzini, F.; et al. Decreased Frontal Gamma Activity in Alzheimer Disease Patients. *Ann. Neurol.* **2022**. *online ahead of print*. [CrossRef]
22. Iaccarino, H.F.; Singer, A.C.; Martorell, A.J.; Rudenko, A.; Gao, F.; Gillingham, T.Z.; Mathys, H.; Seo, J.; Kritskiy, O.; Abdurrob, F.; et al. Gamma Frequency Entrainment Attenuates Amyloid Load and Modifies Microglia. *Nature* **2016**, *540*, 230–235. [CrossRef]
23. Emre, C.; Arroyo-García, L.E.; Do, K.V.; Jun, B.; Ohshima, M.; Alcalde, S.G.; Cothern, M.L.; Maioli, S.; Nilsson, P.; Hjorth, E.; et al. Intranasal Delivery of Pro-Resolving Lipid Mediators Rescues Memory and Gamma Oscillation Impairment in AppNL-G-F/NL-G-F Mice. *Commun. Biol.* **2022**, *5*, 245. [CrossRef] [PubMed]
24. Udeochu, J.C.; Shea, J.M.; Villeda, S.A. Microglia Communication: Parallels between Aging and Alzheimer's Disease. *Clin. Exp. Neuroimmunol.* **2016**, *7*, 114–125. [CrossRef] [PubMed]
25. Alasmari, F.; Alshammari, M.A.; Alasmari, A.F.; Alanazi, W.A.; Alhazzani, K. Neuroinflammatory Cytokines Induce Amyloid Beta Neurotoxicity through Modulating Amyloid Precursor Protein Levels/Metabolism. *Biomed. Res. Int.* **2018**, *2018*, 3087475. [CrossRef] [PubMed]
26. Koch, G.; Di Lorenzo, F.; Loizzo, S.; Motta, C.; Travaglione, S.; Baiula, M.; Rimondini, R.; Ponzo, V.; Bonnì, S.; Toniolo, S.; et al. CSF Tau Is Associated with Impaired Cortical Plasticity, Cognitive Decline and Astrocyte Survival Only in APOE4-Positive Alzheimer's Disease. *Sci. Rep.* **2017**, *7*, 13728. [CrossRef]
27. Shi, Y.; Yamada, K.; Liddelow, S.A.; Smith, S.T.; Zhao, L.; Luo, W.; Tsai, R.M.; Spina, S.; Grinberg, L.T.; Rojas, J.C.; et al. ApoE4 Markedly Exacerbates Tau-Mediated Neurodegeneration in a Mouse Model of Tauopathy. *Nature* **2017**, *549*, 523–527. [CrossRef]
28. Palop, J.J.; Mucke, L. Amyloid-Beta-Induced Neuronal Dysfunction in Alzheimer's Disease: From Synapses toward Neural Networks. *Nat. Neurosci.* **2010**, *13*, 812–818. [CrossRef]
29. Klyubin, I.; Cullen, W.K.; Hu, N.-W.; Rowan, M.J. Alzheimer's Disease Aβ Assemblies Mediating Rapid Disruption of Synaptic Plasticity and Memory. *Mol. Brain* **2012**, *5*, 25. [CrossRef]
30. Motta, C.; Finardi, A.; Toniolo, S.; Di Lorenzo, F.; Scaricamazza, E.; Loizzo, S.; Mercuri, N.B.; Furlan, R.; Koch, G.; Martorana, A. Protective Role of Cerebrospinal Fluid Inflammatory Cytokines in Patients with Amnestic Mild Cognitive Impairment and Early Alzheimer's Disease Carrying Apolipoprotein E4 Genotype. *J. Alzheimers Dis.* **2020**, *76*, 681–689. [CrossRef]
31. Clare, R.; King, V.G.; Wirenfeldt, M.; Vinters, H.V. Synapse Loss in Dementias. *J. Neurosci. Res.* **2010**, *88*, 2083–2090. [CrossRef] [PubMed]
32. van der Ende, E.L.; Xiao, M.; Xu, D.; Poos, J.M.; Panman, J.L.; Jiskoot, L.C.; Meeter, L.H.; Dopper, E.G.; Papma, J.M.; Heller, C.; et al. Neuronal Pentraxin 2: A Synapse-Derived CSF Biomarker in Genetic Frontotemporal Dementia. *J. Neurol. Neurosurg. Psychiatry* **2020**, *91*, 612–621. [CrossRef] [PubMed]
33. Lipton, A.M.; Cullum, C.M.; Satumtira, S.; Sontag, E.; Hynan, L.S.; White, C.L.; Bigio, E.H. Contribution of Asymmetric Synapse Loss to Lateralizing Clinical Deficits in Frontotemporal Dementias. *Arch. Neurol.* **2001**, *58*, 1233–1239. [CrossRef] [PubMed]
34. Ferrer, I. Neurons and Their Dendrites in Frontotemporal Dementia. *Dement. Geriatr. Cogn. Disord.* **1999**, *10*, 55–60. [CrossRef]
35. Liu, X.; Erikson, C.; Brun, A. Cortical Synaptic Changes and Gliosis in Normal Aging, Alzheimer's Disease and Frontal Lobe Degeneration. *Dementia* **1996**, *7*, 128–134. [CrossRef]
36. Connelly, S.J.; Mukaetova-Ladinska, E.B.; Abdul-All, Z.; Alves da Silva, J.; Brayne, C.; Honer, W.G.; Mann, D.M.A. Synaptic Changes in Frontotemporal Lobar Degeneration: Correlation with MAPT Haplotype and APOE Genotype. *Neuropathol. Appl. Neurobiol.* **2011**, *37*, 366–380. [CrossRef]

37. Lippa, C.F. Synaptophysin Immunoreactivity in Pick's Disease: Comparison with Alzheimer's Disease and Dementia with Lewy Bodies. *Am. J. Alzheimers Dis. Other Dement.* **2004**, *19*, 341–344. [CrossRef]
38. Salmon, E.; Bahri, M.A.; Plenevaux, A.; Becker, G.; Seret, A.; Delhaye, E.; Degueldre, C.; Balteau, E.; Lemaire, C.; Luxen, A.; et al. In Vivo Exploration of Synaptic Projecti.ions in Frontotemporal Dementia. *Sci. Rep.* **2021**, *11*, 16092. [CrossRef]
39. Benussi, A.; Cosseddu, M.; Filareto, I.; Dell'Era, V.; Archetti, S.; Sofia Cotelli, M.; Micheli, A.; Padovani, A.; Borroni, B. Impaired Long-Term Potentiation-like Cortical Plasticity in Presymptomatic Genetic Frontotemporal Dementia. *Ann. Neurol.* **2016**, *80*, 472–476. [CrossRef]
40. Broce, I.; Karch, C.M.; Wen, N.; Fan, C.C.; Wang, Y.; Tan, C.H.; Kouri, N.; Ross, O.A.; Höglinger, G.U.; Muller, U.; et al. Immune-Related Genetic Enrichment in Frontotemporal Dementia: An Analysis of Genome-Wide Association Studies. *PLoS Med.* **2018**, *15*, e1002487. [CrossRef]
41. Miller, Z.A.; Rankin, K.P.; Graff-Radford, N.R.; Takada, L.T.; Sturm, V.E.; Cleveland, C.M.; Criswell, L.A.; Jaeger, P.A.; Stan, T.; Heggeli, K.A.; et al. TDP-43 Frontotemporal Lobar Degeneration and Autoimmune Disease. *J. Neurol. Neurosurg. Psychiatry* **2013**, *84*, 956–962. [CrossRef] [PubMed]
42. Miller, Z.A.; Sturm, V.E.; Camsari, G.B.; Karydas, A.; Yokoyama, J.S.; Grinberg, L.T.; Boxer, A.L.; Rosen, H.J.; Rankin, K.P.; Gorno-Tempini, M.L.; et al. Increased Prevalence of Autoimmune Disease within C9 and FTD/MND Cohorts. *Neurol. Neuroimmunol. Neuroinflamm.* **2016**, *3*, e301. [CrossRef] [PubMed]
43. Pasqualetti, G.; Brooks, D.J.; Edison, P. The Role of Neuroinflammation in Dementias. *Curr. Neurol. Neurosci. Rep.* **2015**, *15*, 17. [CrossRef] [PubMed]
44. Mrak, R.E.; Griffin, W.S.T. Common Inflammatory Me.echanisms in Lewy Body Disease and Alzheimer Disease. *J. Neuropathol. Exp. Neurol.* **2007**, *66*, 683–686. [CrossRef]
45. Imamura, K.; Hishikawa, N.; Ono, K.; Suzuki, H.; Sawada, M.; Nagatsu, T.; Yoshida, M.; Hashizume, Y. Cytokine Production of Activated Microglia and Decrease in Neurotrophic Factors of Neurons in the Hippocampus of Lewy Body Disease Brains. *Acta Neuropathol.* **2005**, *109*, 141–150. [CrossRef]
46. Wyss-Coray, T.; Mucke, L. Inflammation in Neurodegenerative Disease—A Double-Edged Sword. *Neuron* **2002**, *35*, 419–432. [CrossRef]
47. Heneka, M.T.; Kummer, M.P.; Latz, E. Innate Immune Activation in Neurodegenerative Disease. *Nat. Rev. Immunol.* **2014**, *14*, 463–477. [CrossRef]
48. Sochocka, M.; Diniz, B.S.; Leszek, J. Inflammatory Response in the CNS: Friend or Foe? *Mol. Neurobiol.* **2017**, *54*, 8071–8089. [CrossRef]
49. Nishimura, A.L.; Arias, N. Synaptopathy Mechanisms in ALS Caused by C9orf72 Repeat Expansion. *Front. Cell. Neurosci.* **2021**, *15*, 660693. [CrossRef]
50. Fogarty, M.J. Amyotrophic Lateral Sclerosis as a Synaptopathy. *Neural Regen Res.* **2019**, *14*, 189–192. [CrossRef]
51. Braak, H.; Brettschneider, J.; Ludolph, A.C.; Lee, V.M.; Trojanowski, J.Q.; Del Tredici, K. Amyotrophic Lateral Sclerosis–a Model of Corticofugal Axonal Spread. *Nat. Rev. Neurol.* **2013**, *9*, 708–714. [CrossRef] [PubMed]
52. Brettschneider, J.; Del Tredici, K.; Toledo, J.B.; Robinson, J.L.; Irwin, D.J.; Grossman, M.; Suh, E.; Van Deerlin, V.M.; Wood, E.M.; Baek, Y.; et al. Stages of PTDP-43 Pathology in Amyotrophic Lateral Sclerosis. *Ann. Neurol.* **2013**, *74*, 20–38. [CrossRef] [PubMed]
53. Vucic, S.; Rothstein, J.D.; Kiernan, M.C. Advances in Treating Amyotrophic Lateral Sclerosis: Insights from Pathophysiological Studies. *Trends Neurosci.* **2014**, *37*, 433–442. [CrossRef] [PubMed]
54. Fogarty, M.J.; Mu, E.W.H.; Lavidis, N.A.; Noakes, P.G.; Bellingham, M.C. Motor Areas Show Altered Dendritic Structure in an Amyotrophic Lateral Sclerosis Mouse Model. *Front. Neurosci.* **2017**, *11*, 609. [CrossRef]
55. Lepeta, K.; Lourenco, M.V.; Schweitzer, B.C.; Martino Adami, P.V.; Banerjee, P.; Catuara-Solarz, S.; de La Fuente Revenga, M.; Guillem, A.M.; Haidar, M.; Ijomone, O.M.; et al. Synaptopathies: Synaptic Dysfunction in Neurological Disorders—A Review from Students to Students. *J. Neurochem.* **2016**, *138*, 785–805. [CrossRef]
56. Wishart, T.M.; Parson, S.H.; Gillingwater, T.H. Synaptic Vulnerability in Neurodegenerative Disease. *J. Neuropathol. Exp. Neurol.* **2006**, *65*, 733–739. [CrossRef] [PubMed]
57. Genç, B.; Jara, J.H.; Lagrimas, A.K.B.; Pytel, P.; Roos, R.P.; Mesulam, M.M.; Geula, C.; Bigio, E.H.; Özdinler, P.H. Apical Dendrite Degeneration, a Novel Cellular Pathology for Betz Cells in ALS. *Sci. Rep.* **2017**, *7*, 41765. [CrossRef]
58. Jara, J.H.; Villa, S.R.; Khan, N.A.; Bohn, M.C.; Özdinler, P.H. AAV2 Mediated Retrograde Transduction of Corticospinal Motor Neurons Reveals Initial and Selective Apical Dendrite Degeneration in ALS. *Neurobiol. Dis.* **2012**, *47*, 174–183. [CrossRef]
59. Oksanen, M.; Lehtonen, S.; Jaronen, M.; Goldsteins, G.; Hämäläinen, R.H.; Koistinaho, J. Astrocyte Alterations in Neurodegenerative Pathologies and Their Modeling in Human Induced Pluripotent Stem Cell Platforms. *Cell. Mol. Life Sci.* **2019**, *76*, 2739–2760. [CrossRef]
60. Henkel, J.S.; Engelhardt, J.I.; Siklós, L.; Simpson, E.P.; Kim, S.H.; Pan, T.; Goodman, J.C.; Siddique, T.; Beers, D.R.; Appel, S.H. Presence of Dendritic Cells, MCP-1, and Activated Microglia/Macrophages in Amyotrophic Lateral Sclerosis Spinal Cord Tissue. *Ann. Neurol.* **2004**, *55*, 221–235. [CrossRef]
61. Kawamata, T.; Akiyama, H.; Yamada, T.; McGeer, P.L. Immunologic Reactions in Amyotrophic Lateral Sclerosis Brain and Spinal Cord Tissue. *Am. J. Pathol.* **1992**, *140*, 691–707. [PubMed]
62. Turner, M.R.; Cagnin, A.; Turkheimer, F.E.; Miller, C.C.J.; Shaw, C.E.; Brooks, D.J.; Leigh, P.N.; Banati, R.B. Evidence of Widespread Cerebral Microglial Activation in Amyotrophic Lateral Sclerosis: An [11C](R)-PK11195 Positron Emission Tomography Study. *Neurobiol. Dis.* **2004**, *15*, 601–609. [CrossRef] [PubMed]

63. Johansson, A.; Engler, H.; Blomquist, G.; Scott, B.; Wall, A.; Aquilonius, S.-M.; Långström, B.; Askmark, H. Evidence for Astrocytosis in ALS Demonstrated by [11C](L)-Deprenyl-D2 PET. *J. Neurol. Sci.* **2007**, *255*, 17–22. [CrossRef] [PubMed]
64. Cooper-Knock, J.; Green, C.; Altschuler, G.; Wei, W.; Bury, J.J.; Heath, P.R.; Wyles, M.; Gelsthorpe, C.; Highley, J.R.; Lorente-Pons, A.; et al. A Data-Driven Approach Links Microglia to Pathology and Prognosis in Amyotrophic Lateral Sclerosis. *Acta Neuropathol. Commun.* **2017**, *5*, 23. [CrossRef]
65. Mattace Raso, G.; Russo, R.; Calignano, A.; Meli, R. Palmitoylethanolamide in CNS Health and Disease. *Pharmacol. Res.* **2014**, *86*, 32–41. [CrossRef]
66. Tsuboi, K.; Uyama, T.; Okamoto, Y.; Ueda, N. Endocannabinoids and Related N-Acylethanolamines: Biological Activities and Metabolism. *Inflamm. Regener.* **2018**, *38*, 28. [CrossRef]
67. Petrosino, S.; Di Marzo, V. The Pharmacology of Palmitoylethanolamide and First Data on the Therapeutic Efficacy of Some of Its New Formulations: Palmitoylethanolamide and Its New Formulations. *Br. J. Pharmacol.* **2017**, *174*, 1349–1365. [CrossRef]
68. Iannotti, F.A.; Di Marzo, V.; Petrosino, S. Endocannabinoids and Endocannabinoid-Related Mediators: Targets, Metabolism and Role in Neurological Disorders. *Prog. Lipid Res.* **2016**, *62*, 107–128. [CrossRef]
69. Lo Verme, J.; Fu, J.; Astarita, G.; La Rana, G.; Russo, R.; Calignano, A.; Piomelli, D. The Nuclear Receptor Peroxisome Proliferator-Activated Receptor-α Mediates the Anti-Inflammatory Actions of Palmitoylethanolamide. *Mol. Pharmacol.* **2005**, *67*, 15–19. [CrossRef]
70. Pertwee, R.G. GPR55: A New Member of the Cannabinoid Receptor Clan? Commentary. *Br. J. Pharmacol.* **2009**, *152*, 984–986. [CrossRef]
71. D'Agostino, G.; La Rana, G.; Russo, R.; Sasso, O.; Iacono, A.; Esposito, E.; Mattace Raso, G.; Cuzzocrea, S.; Loverme, J.; Piomelli, D.; et al. Central Administration of Palmitoylethanolamide Reduces Hyperalgesia in Mice via Inhibition of NF-KappaB Nuclear Signalling in Dorsal Root Ganglia. *Eur. J. Pharmacol.* **2009**, *613*, 54–59. [CrossRef] [PubMed]
72. Daynes, R.A.; Jones, D.C. Emerging Roles of PPARs in Inflammation and Immunity. *Nat. Rev. Immunol.* **2002**, *2*, 748–759. [CrossRef]
73. Scuderi, C.; Esposito, G.; Blasio, A.; Valenza, M.; Arietti, P.; Steardo Jr, L.; Carnuccio, R.; De Filippis, D.; Petrosino, S.; Iuvone, T.; et al. Palmitoylethanolamide Counteracts Reactive Astrogliosis Induced by β-Amyloid Peptide. *J. Cell Mol. Med.* **2011**, *15*, 2664–2674. [CrossRef] [PubMed]
74. D'Agostino, G.; Russo, R.; Avagliano, C.; Cristiano, C.; Meli, R.; Calignano, A. Palmitoylethanolamide Protects Against the Amyloid-B25-35-Induced Learning and Memory Impairment in Mice, an Experimental Model of Alzheimer Disease. *Neuropsychopharmacology* **2012**, *37*, 1784–1792. [CrossRef] [PubMed]
75. Esposito, E.; Impellizzeri, D.; Mazzon, E.; Paterniti, I.; Cuzzocrea, S. Neuroprotective Activities of Palmitoylethanolamide in an Animal Model of Parkinson's Disease. *PLoS ONE* **2012**, *7*, e41880. [CrossRef]
76. Paterniti, I.; Cordaro, M.; Campolo, M.; Siracusa, R.; Cornelius, C.; Navarra, M.; Cuzzocrea, S.; Esposito, E. Neuroprotection by Association of Palmitoylethanolamide with Luteolin in Experimental Alzheimer's Disease Models: The Control of Neuroinflammation. *CNS Neurol. Disord. Drug Targets* **2014**, *13*, 1530–1541. [CrossRef]
77. Lauckner, J.E.; Jensen, J.B.; Chen, H.-Y.; Lu, H.-C.; Hille, B.; Mackie, K. GPR55 Is a Cannabinoid Receptor That Increases Intracellular Calcium and Inhibits M Current. *Proc. Natl. Acad. Sci. USA* **2008**, *105*, 2699–2704. [CrossRef]
78. Borrelli, F.; Romano, B.; Petrosino, S.; Pagano, E.; Capasso, R.; Coppola, D.; Battista, G.; Orlando, P.; Di Marzo, V.; Izzo, A.A. Palmitoylethanolamide, a Naturally Occurring Lipid, Is an Orally Effective Intestinal Anti-Inflammatory Agent. *Br. J. Pharmacol.* **2015**, *172*, 142–158. [CrossRef]
79. Celorrio, M.; Rojo-Bustamante, E.; Fernández-Suárez, D.; Sáez, E.; Estella-Hermoso de Mendoza, A.; Müller, C.E.; Ramírez, M.J.; Oyarzábal, J.; Franco, R.; Aymerich, M.S. GPR55: A Therapeutic Target for Parkinson's Disease? *Neuropharmacology* **2017**, *125*, 319–332. [CrossRef]
80. Rinne, P.; Guillamat-Prats, R.; Rami, M.; Bindila, L.; Ring, L.; Lyytikäinen, L.-P.; Raitoharju, E.; Oksala, N.; Lehtimäki, T.; Weber, C.; et al. Palmitoylethanolamide Promotes a Proresolving Macrophage Phenotype and Attenuates Atherosclerotic Plaque Formation. *Arterioscler. Thromb. Vasc. Biol.* **2018**, *38*, 2562–2575. [CrossRef]
81. Marichal-Cancino, B.A.; Fajardo-Valdez, A.; Ruiz-Contreras, A.E.; Mendez-Díaz, M.; Prospero-García, O. Advances in the Physiology of GPR55 in the Central Nervous System. *Curr. Neuropharmacol.* **2017**, *15*, 771–778. [CrossRef] [PubMed]
82. Musella, A.; Fresegna, D.; Rizzo, F.R.; Gentile, A.; Bullitta, S.; De Vito, F.; Guadalupi, L.; Centonze, D.; Mandolesi, G. A Novel Crosstalk within the Endocannabinoid System Controls GABA Transmission in the Striatum. *Sci. Rep.* **2017**, *7*, 7363. [CrossRef] [PubMed]
83. Guida, F.; Luongo, L.; Boccella, S.; Giordano, M.E.; Romano, R.; Bellini, G.; Manzo, I.; Furiano, A.; Rizzo, A.; Imperatore, R.; et al. Palmitoylethanolamide Induces Microglia Changes Associated with Increased Migration and Phagocytic Activity: Involvement of the CB2 Receptor. *Sci. Rep.* **2017**, *7*, 375. [CrossRef] [PubMed]
84. D'Aloia, A.; Molteni, L.; Gullo, F.; Bresciani, E.; Artusa, V.; Rizzi, L.; Ceriani, M.; Meanti, R.; Lecchi, M.; Coco, S.; et al. Palmitoylethanolamide Modulation of Microglia Activation: Characterization of Mechanisms of Action and Implication for Its Neuroprotective Effects. *Int. J. Mol. Sci.* **2021**, *22*, 3054. [CrossRef] [PubMed]
85. Zygmunt, P.M.; Ermund, A.; Movahed, P.; Andersson, D.A.; Simonsen, C.; Jönsson, B.A.G.; Blomgren, A.; Birnir, B.; Bevan, S.; Eschalier, A.; et al. Monoacylglycerols Activate TRPV1—A Link between Phospholipase C and TRPV1. *PLoS ONE* **2013**, *8*, e81618. [CrossRef]
86. Ross, R.A. Anandamide and Vanilloid TRPV1 Receptors. *Br. J. Pharmacol.* **2003**, *140*, 790–801. [CrossRef]
87. Nam, J.H.; Park, E.S.; Won, S.-Y.; Lee, Y.A.; Kim, K.I.; Jeong, J.Y.; Baek, J.Y.; Cho, E.J.; Jin, M.; Chung, Y.C.; et al. TRPV1 on Astrocytes Rescues Nigral Dopamine Neurons in Parkinson's Disease via CNTF. *Brain* **2015**, *138*, 3610–3622. [CrossRef]

88. Jayant, S.; Sharma, B.M.; Sharma, B. Protective Effect of Transient Receptor Potential Vanilloid Subtype 1 (TRPV1) Modulator, against Behavioral, Biochemical and Structural Damage in Experimental Models of Alzheimer's Disease. *Brain Res.* **2016**, *1642*, 397–408. [CrossRef]
89. Zhao, N.; Liu, C.-C.; Van Ingelgom, A.J.; Martens, Y.A.; Linares, C.; Knight, J.A.; Painter, M.M.; Sullivan, P.M.; Bu, G. Apolipoprotein E4 Impairs Neuronal Insulin Signaling by Trapping Insulin Receptor in the Endosomes. *Neuron* **2017**, *96*, 115–129.e5. [CrossRef]
90. Balleza-Tapia, H.; Crux, S.; Andrade-Talavera, Y.; Dolz-Gaiton, P.; Papadia, D.; Chen, G.; Johansson, J.; Fisahn, A. TrpV1 Receptor Activation Rescues Neuronal Function and Network Gamma Oscillations from Aβ-Induced Impairment in Mouse Hippocampus In Vitro. *eLife* **2018**, *7*, e37703. [CrossRef]
91. Palop, J.J.; Mucke, L. Network Abnormalities and Interneuron Dysfunction in Alzheimer Disease. *Nat. Rev. Neurosci.* **2016**, *17*, 777–792. [CrossRef] [PubMed]
92. Lin, T.-Y.; Lu, C.-W.; Wu, C.-C.; Huang, S.-K.; Wang, S.-J. Palmitoylethanolamide Inhibits Glutamate Release in Rat Cerebrocortical Nerve Terminals. *Int. J. Mol. Sci.* **2015**, *16*, 5555–5571. [CrossRef] [PubMed]
93. Grosso, C.; Valentão, P.; Ferreres, F.; Andrade, P.B. The Use of Flavonoids in Central Nervous System Disorders. *Curr. Med. Chem.* **2013**, *20*, 4694–4719. [CrossRef]
94. Daily, J.W.; Kang, S.; Park, S. Protection against Alzheimer's Disease by Luteolin: Role of Brain Glucose Regulation, Anti-Inflammatory Activity, and the Gut Microbiota-Liver-Brain Axis. *Biofactors* **2021**, *47*, 218–231. [CrossRef]
95. Crupi, R.; Paterniti, I.; Ahmad, A.; Campolo, M.; Esposito, E.; Cuzzocrea, S. Effects of Palmitoylethanolamide and Luteolin in an Animal Model of Anxiety/Depression. *CNS Neurol. Disord. Drug Targets* **2013**, *12*, 989–1001. [CrossRef]
96. Impellizzeri, D.; Esposito, E.; Di Paola, R.; Ahmad, A.; Campolo, M.; Peli, A.; Morittu, V.M.; Britti, D.; Cuzzocrea, S. Palmitoylethanolamide and Luteolin Ameliorate Development of Arthritis Caused by Injection of Collagen Type II in Mice. *Arthritis Res. Ther.* **2013**, *15*, R192. [CrossRef]
97. Bertolino, B.; Crupi, R.; Impellizzeri, D.; Bruschetta, G.; Cordaro, M.; Siracusa, R.; Esposito, E.; Cuzzocrea, S. Beneficial Effects of Co-Ultramicronized Palmitoylethanolamide/Luteolin in a Mouse Model of Autism and in a Case Report of Autism. *CNS Neurosci. Ther.* **2017**, *23*, 87–98. [CrossRef] [PubMed]
98. Siracusa, R.; Paterniti, I.; Impellizzeri, D.; Cordaro, M.; Crupi, R.; Navarra, M.; Cuzzocrea, S.; Esposito, E. The Association of Palmitoylethanolamide with Luteolin Decreases Neuroinflammation and Stimulates Autophagy in Parkinson's Disease Model. *CNS Neurol. Disord. Drug Targets* **2015**, *14*, 1350–1365. [CrossRef]
99. Paterniti, I.; Impellizzeri, D.; Di Paola, R.; Navarra, M.; Cuzzocrea, S.; Esposito, E. A New Co-Ultramicronized Composite Including Palmitoylethanolamide and Luteolin to Prevent Neuroinflammation in Spinal Cord Injury. *J. Neuroinflamm.* **2013**, *10*, 91. [CrossRef]
100. Beggiato, S.; Tomasini, M.C.; Ferraro, L. Palmitoylethanolamide (PEA) as a Potential Therapeutic Agent in Alzheimer's Disease. *Front. Pharmacol.* **2019**, *10*, 821. [CrossRef]
101. Landolfo, E.; Cutuli, D.; Petrosini, L.; Caltagirone, C. Effects of Palmitoylethanolamide on Neurodegenerative Diseases: A Review from Rodents to Humans. *Biomolecules* **2022**, *12*, 667. [CrossRef] [PubMed]
102. Scuderi, C.; Valenza, M.; Stecca, C.; Esposito, G.; Carratù, M.R.; Steardo, L. Palmitoylethanolamide Exerts Neuroprotective Effects in Mixed Neuroglial Cultures and Organotypic Hippocampal Slices via Peroxisome Proliferator-Activated Receptor-α. *J. Neuroinflamm.* **2012**, *9*, 49. [CrossRef] [PubMed]
103. Scuderi, C.; Steardo, L. Neuroglial Roots of Neurodegenerative Diseases: Therapeutic Potential of Palmitoylethanolamide in Models of Alzheimer's Disease. *CNS Neurol. Disord. Drug Targets* **2013**, *12*, 62–69. [CrossRef]
104. Scuderi, C.; Stecca, C.; Valenza, M.; Ratano, P.; Bronzuoli, M.R.; Bartoli, S.; Steardo, L.; Pompili, E.; Fumagalli, L.; Campolongo, P.; et al. Palmitoylethanolamide Controls Reactive Gliosis and Exerts Neuroprotective Functions in a Rat Model of Alzheimer's Disease. *Cell Death Dis.* **2014**, *5*, e1419. [CrossRef] [PubMed]
105. Barbierato, M.; Borri, M.; Facci, L.; Zusso, M.; Skaper, S.D.; Giusti, P. Expression and Differential Responsiveness of Central Nervous System Glial Cell Populations to the Acute Phase Protein Serum Amyloid A. *Sci. Rep.* **2017**, *7*, 12158. [CrossRef] [PubMed]
106. Facchinetti, R.; Valenza, M.; Gomiero, C.; Mancini, G.F.; Steardo, L.; Campolongo, P.; Scuderi, C. Co-Ultramicronized Palmitoylethanolamide/Luteolin Restores Oligodendrocyte Homeostasis via Peroxisome Proliferator-Activated Receptor-α in an In Vitro Model of Alzheimer's Disease. *Biomedicines* **2022**, *10*, 1236. [CrossRef]
107. Cipriano, M.; Esposito, G.; Negro, L.; Capoccia, E.; Sarnelli, G.; Scuderi, C.; De Filippis, D.; Steardo, L.; Iuvone, T. Palmitoylethanolamide Regulates Production of Pro-Angiogenic Mediators in a Model of β Amyloid-Induced Astrogliosis In Vitro. *CNS Neurol. Disord. Drug Targets* **2015**, *14*, 828–837. [CrossRef]
108. Tomasini, M.C.; Borelli, A.C.; Beggiato, S.; Ferraro, L.; Cassano, T.; Tanganelli, S.; Antonelli, T. Differential Effects of Palmitoylethanolamide against Amyloid-β Induced Toxicity in Cortical Neuronal and Astrocytic Primary Cultures from Wild-Type and 3×Tg-AD Mice. *J. Alzheimers Dis.* **2015**, *46*, 407–421. [CrossRef]
109. Bronzuoli, M.R.; Facchinetti, R.; Steardo, L.; Romano, A.; Stecca, C.; Passarella, S.; Steardo, L.; Cassano, T.; Scuderi, C. Palmitoylethanolamide Dampens Reactive Astrogliosis and Improves Neuronal Trophic Support in a Triple Transgenic Model of Alzheimer's Disease: In Vitro and In Vivo Evidence. *Oxid. Med. Cell Longev.* **2018**, *2018*, 4720532. [CrossRef]
110. Guida, F.; Luongo, L.; Marmo, F.; Romano, R.; Iannotta, M.; Napolitano, F.; Belardo, C.; Marabese, I.; D'Aniello, A.; De Gregorio, D.; et al. Palmitoylethanolamide Reduces Pain-Related Behaviors and Restores Glutamatergic Synapses Homeostasis in the Medial Prefrontal Cortex of Neuropathic Mice. *Mol. Brain* **2015**, *8*, 47. [CrossRef]

111. Boccella, S.; Marabese, I.; Iannotta, M.; Belardo, C.; Neugebauer, V.; Mazzitelli, M.; Pieretti, G.; Maione, S.; Palazzo, E. Metabotropic Glutamate Receptor 5 and 8 Modulate the Ameliorative Effect of Ultramicronized Palmitoylethanolamide on Cognitive Decline Associated with Neuropathic Pain. *Int. J. Mol. Sci.* **2019**, *20*, 1757. [CrossRef] [PubMed]
112. al-Ghoul, W.M.; Li Volsi, G.; Weinberg, R.J.; Rustioni, A. Glutamate Immunocytochemistry in the Dorsal Horn after Injury or Stimulation of the Sciatic Nerve of Rats. *Brain Res. Bull.* **1993**, *30*, 453–459. [CrossRef]
113. Boccella, S.; Cristiano, C.; Romano, R.; Iannotta, M.; Belardo, C.; Farina, A.; Guida, F.; Piscitelli, F.; Palazzo, E.; Mazzitelli, M.; et al. Ultra-Micronized Palmitoylethanolamide Rescues the Cognitive Decline-Associated Loss of Neural Plasticity in the Neuropathic Mouse Entorhinal Cortex-Dentate Gyrus Pathway. *Neurobiol. Dis.* **2019**, *121*, 106–119. [CrossRef]
114. Beggiato, S.; Tomasini, M.C.; Cassano, T.; Ferraro, L. Chronic Oral Palmitoylethanolamide Administration Rescues Cognitive Deficit and Reduces Neuroinflammation, Oxidative Stress, and Glutamate Levels in A Transgenic Murine Model of Alzheimer's Disease. *JCM* **2020**, *9*, 428. [CrossRef] [PubMed]
115. Bellanti, F.; Bukke, V.N.; Moola, A.; Villani, R.; Scuderi, C.; Steardo, L.; Palombelli, G.; Canese, R.; Beggiato, S.; Altamura, M.; et al. Effects of Ultramicronized Palmitoylethanolamide on Mitochondrial Bioenergetics, Cerebral Metabolism, and Glutamatergic Transmission: An Integrated Approach in a Triple Transgenic Mouse Model of Alzheimer's Disease. *Front. Aging Neurosci.* **2022**, *14*, 890855. [CrossRef] [PubMed]
116. Kramar, C.; Loureiro, M.; Renard, J.; Laviolette, S.R. Palmitoylethanolamide Modulates GPR55 Receptor Signaling in the Ventral Hippocampus to Regulate Mesolimbic Dopamine Activity, Social Interaction, and Memory Processing. *Cannabis Cannabinoid Res.* **2017**, *2*, 8–20. [CrossRef]
117. Facchinetti, R.; Valenza, M.; Bronzuoli, M.R.; Menegoni, G.; Ratano, P.; Steardo, L.; Campolongo, P.; Scuderi, C. Looking for a Treatment for the Early Stage of Alzheimer's Disease: Preclinical Evidence with Co-Ultramicronized Palmitoylethanolamide and Luteolin. *Int. J. Mol. Sci.* **2020**, *21*, 3802. [CrossRef]
118. Clemente, S. Amyotrophic Lateral Sclerosis Treatment with Ultramicronized Palmitoylethanolamide: A Case Report. *CNS Neurol. Disord. Drug Targets* **2012**, *11*, 933–936. [CrossRef]
119. Palma, E.; Reyes-Ruiz, J.M.; Lopergolo, D.; Roseti, C.; Bertollini, C.; Ruffolo, G.; Cifelli, P.; Onesti, E.; Limatola, C.; Miledi, R.; et al. Acetylcholine Receptors from Human Muscle as Pharmacological Targets for ALS Therapy. *Proc. Natl. Acad. Sci. USA* **2016**, *113*, 3060–3065. [CrossRef]
120. Calabrò, R.S.; Naro, A.; De Luca, R.; Leonardi, S.; Russo, M.; Marra, A.; Bramanti, P. PEALut Efficacy in Mild Cognitive Impairment: Evidence from a SPECT Case Study! *Aging Clin. Exp. Res.* **2016**, *28*, 1279–1282. [CrossRef]
121. Assogna, M.; Casula, E.P.; Borghi, I.; Bonnì, S.; Samà, D.; Motta, C.; Di Lorenzo, F.; D'Acunto, A.; Porrazzini, F.; Minei, M.; et al. Effects of Palmitoylethanolamide Combined with Luteoline on Frontal Lobe Functions, High Frequency Oscillations, and GABAergic Transmission in Patients with Frontotemporal Dementia. *J. Alzheimers Dis.* **2020**, *76*, 1297–1308. The publication is available at IOS Press through http://doi.org/10.3233/JAD-200426. [PubMed]
122. Cummings, J. New Approaches to Symptomatic Treatments for Alzheimer's Disease. *Mol. Neurodegener.* **2021**, *16*, 2. [CrossRef] [PubMed]
123. Yiannopoulou, K.G.; Anastasiou, A.I.; Zachariou, V.; Pelidou, S.-H. Reasons for Failed Trials of Disease-Modifying Treatments for Alzheimer Disease and Their Contribution in Recent Research. *Biomedicines* **2019**, *7*, 97. [CrossRef] [PubMed]
124. Toniolo, S.; Sen, A.; Husain, M. Modulation of Brain Hyperexcitability: Potential New Therapeutic Approaches in Alzheimer's Disease. *Int. J. Mol. Sci.* **2020**, *21*, 9318. [CrossRef]
125. Francesco, D.L.; Koch, G. Synaptic Impairment: The New Battlefield of Alzheimer's Disease. *Alzheimers Dement.* **2021**, *17*, 314–315. [CrossRef]

biomolecules

MDPI

Review

Classical and Unexpected Effects of Ultra-Micronized PEA in Neuromuscular Function

Pierangelo Cifelli [1,*], Gabriele Ruffolo [2,3], Marco Ceccanti [4], Chiara Cambieri [4], Laura Libonati [4], Eleonora Palma [2] and Maurizio Inghilleri [4,*]

1 Department of Applied Clinical and Biotechnological Sciences, University of L'Aquila, 67100 L'Aquila, Italy
2 Department of Physiology and Pharmacology, Istituto Pasteur-Fondazione Cenci Bolognetti, University of Rome Sapienza, 00185 Rome, Italy; gabriele.ruffolo@uniroma1.it (G.R.); eleonora.palma@uniroma1.it (E.P.)
3 IRCCS San Raffaele Roma, 00163 Rome, Italy
4 Department of Human Neuroscience, University of Rome Sapienza, 00185 Rome, Italy; marco.ceccanti@uniroma1.it (M.C.); chiara.cambieri@uniroma1.it (C.C.); laura.libonati@uniroma1.it (L.L.)
* Correspondence: pierangelo.cifelli@univaq.it (P.C.); maurizio.inghilleri@uniroma1.it (M.I.)

Abstract: Recently, the endocannabinoid system has attracted growing attention from the scientific community for its involvement in homeostatic and pathological processes as they pertains to human physiology. Among the constituents of the endocannabinoid system, the molecule palmitoyl ethanolamide has particularly been studied for its ability to reduce several inflammatory processes involving the central nervous system. Here, we reviewed published literature and summarized the main targets of the palmitoyl ethanolamide, along with its unique possible mechanisms for restoring correct functioning of the central nervous system. Moreover, we have highlighted a less-known characteristic of palmitoyl ethanolamide, namely its ability to modulate the function of the neuromuscular junction by binding to acetylcholine receptors in different experimental conditions. Indeed, there are several studies that have highlighted how ultra-micronized palmitoyl ethanolamide is an interesting nutraceutical support for the treatment of pathological neuromuscular conditions, specifically when the normal activity of the acetylcholine receptor is altered. Although further multicentric clinical trials are needed to confirm the efficacy of ultra-micronized palmitoyl ethanolamide in improving symptoms of neuromuscular diseases, all the literature reviewed here strongly supports the ability of this endocannabinoid-like molecule to modulate the acetylcholine receptors thus resulting as a valid support for the treatment of human neuromuscular diseases.

Keywords: palmitoyl ethanolamide; neuromuscular junction; neurophysiology

Citation: Cifelli, P.; Ruffolo, G.; Ceccanti, M.; Cambieri, C.; Libonati, L.; Palma, E.; Inghilleri, M. Classical and Unexpected Effects of Ultra-Micronized PEA in Neuromuscular Function. *Biomolecules* **2022**, *12*, 758. https://doi.org/10.3390/biom12060758

Academic Editors: Salvatore Cuzzocrea and Rosalia Crupi

Received: 9 May 2022
Accepted: 27 May 2022
Published: 29 May 2022

1. Introduction

Palmitoyl ethanolamide [PEA] represents the best-known endocannabinoid-like molecule of the "autacoid local injury antagonist amide" [ALIAmide] family. ALIAmides are endogenous N-acyl ethanol amines [NAEs] with homeostatic functions able to modulate different physiological pathways involved in inflammation, pain, and metabolism [1]. Among the different ALIAmides, discovered up to now, PEA, oleoyl ethanol amide (OEA), stearoyl ethanol amide (SEA), and anandamide (ANA) represent the most-studied compounds, while others such as eicosatrienoyl ethanol amide (ETEA) still need to be investigated further. In this review we will highlight the different mechanisms of action of PEA, an endocannabinoid [eCB]-like bioactive lipid mediator, and its involvement in different human pathologies. Finally, we will place under the spotlight PEA's ability to modulate the synaptic function in the neuromuscular system and at neuromuscular junction [NMJ].

2. What, Where and Why?

NAEs are endogenous bioactive lipids characterized by common precursors, namely the N-acylated ethanolaminephospholipids such as N-acylphosphatidylethanolamines

(NAPEs) and N-acylplasmenylethanolamines (pNAPEs). The biological targets and the specific functions of the different NAEs depend upon the origin of the N-acyl group, which can arise from a variety of different fatty acids [2]. The endogenous levels of NAEs in different tissues are mainly regulated by enzymatic reactions responsible for both their biosynthesis and degradation. Two main enzyme reactions are involved in NAEs biosynthesis. The first is the process of N-acylation of ethanolamine phospholipids mediated by a Ca^{2+}-dependent N-acyltransferase, while the second is represented by the release of the different NAEs from NAPEs, mediated by NAPE-hydrolyzing phospholipase (NAPE-PLD) [3]. Catabolism of NAEs is rapidly put in effect by the action of the free acid amide hydroxylase [FAAH], an integral membrane enzyme able to hydrolyze them in their corresponding fatty acids and ethanolamine [4].

Among NAEs, PEA was first identified in mammalian tissues in 1965 [5]. In central nervous system [CNS] it is produced and released both from neurons and glial cells as an "on-demand molecule" to promote endogenous neuroprotection following tissue damage and the subsequent processes of neuroinflammation. PEA is part of the endocannabinoid system (ECS), a regulatory signaling network made up of lipidic molecules, the endocannabinoids (eCBs), their receptors, and different enzymes involved in the metabolism and catabolism of the lipidic signals. The ECS is strongly involved in different homeostatic processes of our body, both in physiological and pathological conditions [6]. Indeed, the ECS is implicated in the regulation of several functions, such as the cycle of sleep, the regulation of food intake, and the perception of painful stimuli and in high cognitive functions such as mood regulation, memory, and reward [7].

As mentioned, PEA can carry out a plethora of actions, and its tissue levels are often up regulated in several pathological conditions [8]. Its capacity to modulate divergent pathways is largely due to its ability to interact, directly or indirectly, with different targets [9]. The main target of PEA is represented by the nuclear peroxisome proliferator-activated receptor-α (PPAR-α) [10]. This receptor, part of the larger family of nuclear receptor proteins, has the capability, when activated, to modulate and regulate the expression of different genes involved in inflammation [11]. Indeed, the activation of PPAR-α receptors induces and promotes a cascade of events leading to inhibition and suppression of the release of pro-inflammatory cytokines, such as Interleukin-1β (IL-1β), Interleukin 6 (IL6), and the tumor-necrosis factor (TNF-α) [12]. All these actions were further confirmed in several pre-clinical studies, where the anti-inflammatory effect of PEA was totally abolished when applied on PPAR-α knockout models of disease [12]. Another target activated by PEA is represented by GPR55 and GPR119, two orphan receptors [13], considered up to now as novel cannabinoid receptors [14]. PEA is also able to activate and modulate the transient receptor potential vanilloid receptor 1 (TRPV1) channels, accounting thus, at least in part, for its anti-nociceptive effect [15]. Furthermore, PEA can indirectly activate the canonical cannabinoid receptors, namely CB1 and CB2, increasing the level of anandamide (AEA) by inhibiting its enzymatic-mediated hydrolysis carried out by fatty acid amide hydrolase (FAAH) [16].

One of the main targets of PEA is represented by the resident immune cells of CNS, namely microglial cells. These latter are responsible for CNS homeostasis, being able to modify their function following CNS injuries [17]. Indeed, after multiple insults, [e.g., inflammation, tumors, and traumas], they can switch their phenotype between the M1, considered unanimously the pro-inflammatory, and the M2, considered instead as anti-inflammatory [18]. The imbalance between the two states is considered as one of the main events underlying many neurogenerative diseases, where M1 phenotypes are found in a condition of continuous activation [19]. In this state, M1-activated microglia releases many inflammatory cytokines, such as t TNF-α, IL-1β, and inducible reactive oxygen species (iROS), which lead the neuronal population to death [19]. Recent studies demonstrated that PEA can modulate microglia polarization, reducing the release of pro-inflammatory cytokines and promoting migration and phagocytic activity [20]. In detail, PEA can reduce the lipopolysaccharide (LPS)-induced microglial activation by switching their phenotypic constructs from M1 to M2, thus inhibiting the release of pro-inflammatory elements in the extracellular milieu and reducing the hyperexcitability of cultured primary cortical cells together with LPS-activated microglial cells [21]. Notably, these effects were all mediated by cannabinoid receptor type 2 (CB2), thus confirming the strong involvement of the ECS during inflammatory processes [22].

3. Pharmacology of PEA

PEA is a lipophilic molecule with low solubility rate. Hence, specific formulations, such as micronized and ultra-micronized PEA, have been used to increase its solubility and enhance its bioavailability. Indeed, with these techniques PEA particles became smaller, thus increasing surface areas to improve absorption [23]. Different studies highlight that ultra-micronized PEA has greater absorbability than micronized and non-micronized PEA, also showing better results in terms of reducing pain perception [24–26] Another issue about PEA is that its bioavailability is related to its possible pre-systemic metabolism. Since enzymes involved in its metabolism are expressed both in the small intestine and in the liver [27], it is quite difficult to calculate its range of distribution and how this parameter may change in disparate individuals.

Due to its partially unclear mechanisms of absorption and pre-systemic metabolism, the distribution mechanism of PEA is still far from being elucidated. However, due to its lipophilic nature, several studies demonstrated that, even with its low level of bioavailability, PEA its able to reach tissues to target its receptors effectively [28,29]. Currently, the only way to characterise its absorption is by measuring its blood levels after administration [28].

Once PEA reaches the tissues, there is a fast cellular uptake in order to expose PEA to its intracellular PPAR-α receptors. This crucial step needs to be further deciphered, as research pertaining to its mechanism is still lacking. It appears that the intracellular process of uptake is strictly regulated by PEA hydrolysis, to balance its extracellular/intracellular ratio [30]. In contrast with data about absorption and availability, studies about PEA metabolism are abundant in literature. Indeed, PEA is hydrolysed in two compounds, namely palmitic acid and ethanolamine by the fatty acid amide hydrolase [FAAH] enzyme, located on endoplasmic reticulum membranes [31,32]. Subsequently, another enzyme involved in PEA hydrolysis was found and described: the FAAH-2 [33]. Notably, both FAAH and FAAH-2 are able to hydrolase AEA faster than PEA, while the third enzyme involved in PEA catabolism, namely N-acylethanolamine acid amidase (NAAA), is able to hydrolase more efficiently PEA than AEA [34,35].

The reason why three distinct enzymes are involved in PEA catabolism is still obscure. Nevertheless, recent works have highlighted the possibility that it may depend on the source of PEA, with differences between endogenous and exogenous PEA. Endogenous PEA is hydrolysed by means of both FAAH and NAAA, but several studies indicate that the prevalence of one or the other is strongly tissue-specific and may change also in different pathological conditions [36].

Alterations were described when catabolism of exogenous PEA was scrutinized; indeed, the fatty acid binding proteins do not deliver exogenous PEA to lysosomal NAAA, thus suggesting a prevalent role of FAAH for exogenous PEA metabolism [37].

The catabolism of the main metabolite of PEA, palmitic acid, was fully described by Carta et al. in 2017 [38]. However, data describing the excretion of unmetabolized PEA is still lacking, even though the kidney route seems to be prominent [5].

4. PEA, a Multi Target Drug for Different Clinical Applications

Since, as mentioned above, PEA can bind different receptor targets and is thus able to modulate different signalling pathways, its rationale for clinical uses embraces a wide variety of pathological conditions, from chronic inflammations to perseverant pain conditions. Here we will address the main applications in routinely clinical practice with a particular focus on neurological conditions.

5. PEA and Neuroinflammation

Nowadays, neuroinflammation is considered a key element in the pathogenesis of a broad spectrum of neurological diseases, from neurodegenerative conditions, such as Alzheimer's disease [39–43] and Parkinson's disease [44], to traumatic diseases, such as traumatic brain injuries [TBI] [45–49], strokes [50], and other conditions where neuronal excitability is increased, such as epilepsy [51]. Moreover, recently, attention has been called to the significant effect of PEA in clinical conditions involving the higher functions of CNS, such as cognitive impairment [52] and mood disorders [53]. In all these conditions, microglial cells, together with mast cells and astrocytes, play a crucial role in inducing and maintaining the inflammatory processes, leading to neuronal cell death, and thus impairing the CNS functions. PEA capacity to dampen neuroinflammation could represent a viable tool to preserve neuronal populations and to retain physiological brain functions [54,55]. Most of PEA's effects are due, as already mentioned, to its ability to modulate and modify microglial, astrocyte and mast cells activation [56,57]. Indeed, PEA can enhance microglial migration without switching their phenotype towards the pro-inflammatory M1 state, thus increasing the resistance to infections without activation of inflammatory cascade [57]. PEA is able to carry out its neuroprotective effects on CNS also with other mechanisms, like for example, inhibiting the apoptosis processes and modulating different pathways, such as the bax/bcl-2 and Akt/mTOR/p70S6K pathways [58]. PEA is also able to target and modulate NMDA receptors, thus protecting cells for glutamate toxicity [59]. PEA neuroprotective actions are also carried out by promoting and modulating synaptic homeostasis and favouring neurogenesis [60–62]. Notably, all these PEA-mediated actions were confirmed using different animal models of disease characterized by neuroinflammation. Chronic administration of PEA at different concentrations (ranging from 10 to 100 mg/Kg/day) was able to significantly reduce neuroinflammation [63–65], to protect neurons from death [21,66], to reduce iROS productions [66], and to improve behavioural, motor and cognitive deficits [67–69].

6. PEA in Central Nervous System Diseases

It is well known that both ECS and NAEs are strongly involved in regulation and modulation of behaviour, cognition and in mood regulation [70]. Plasmatic PEA levels are significantly altered in conditions as post-traumatic stress disease (PTSD) [71], depression [72,73], and in autism spectrum disorders (ASD) [74]. The mechanisms by which PEA can improve these conditions are different and include increased hippocampal neurogenesis [59–61] and increased maturation of oligodendrocyte precursor cells [59]. PEA was able to prevent the plasmatic reduction of the brain-derived neurotrophic factor (BDNF) and glial cell line-derived neurotrophic factor (GDNF) in a murine model of stroke [75], thus explaining, at least in part, its neuroprotective effects. Interestingly, PEA is also strongly involved in the neuromodulation of GABAergic transmission. As a matter of fact, PEA by binding GPR55 receptors is able to enhance GABAergic transmission in the *corpus striatum* and to increase the synthesis of the endocannabinoid 2-AG at postsynaptic level thus modulating GABA release through the stimulation of presynaptic CB1Rs [76]. These biological effects may partially explain the neuroprotective action of PEA and its potential use to treat these patients in clinical activities. To confirm the pre-clinical results obtained by using PEA, human studies were performed on different pathological diseases. In a yearlong study, PD patients used ultra-micronized PEA as add-on therapy along with levodopa, showing a significant improvement on both motor- and non-motor-symptoms, ameliorating the typical mood deficits of Parkinsonian patients as well as reducing fatigue and improving sleep-cycle coupled with improved responses to different mental tasks [77]. Notably, in a double-blind, randomized, placebo-controlled study performed on patients with major depressive disorder [MDD], 600 mg of ultra-micronized PEA used as add-on therapy with typical selective serotonin reuptake inhibitors [SSRI] for 6 weeks was able to significantly improve the depressive scores, ameliorating symptoms when compared to SSRI plus placebo group [78]. It is important to note that, also if further clinical studies are required in all the human studies performed to date, PEA was well tolerated and displayed only few and minor adverse effects [79].

7. Analgesic Properties of PEA

The direct link between inflammation and pain kicked off a series of studies with the aim to emphasize PEA analgesic properties. Taking advantage of pre-clinical models of inflammation and neuropathic pain, it was possible to demonstrate the strong analgesic properties of PEA [80–82]. When inflammation processes persist, PEA levels increase as a protective mechanism to inhibit inflammatory pathways leading to tissue damage and loss of functions. If the stimulus is protracted, PEA levels start to decrease, thus indicating that PEA's increased concentration is not enough to restore the physiological conditions when inflammation becomes chronic [83]. The PEA's analgesic properties are guaranteed by several mechanisms. For instance, its ability to directly bind to PPAR-α and GPR55 receptors or to indirectly bind to CB1, CB2, and TRPV1 receptors may partially account for its properties [15]. The analgesic properties of PEA also take into account its capacity to suppress inflammation by reducing mast cells activation and to reduce the production and the release of different inflammatory mediators such as nerve growth factor (NGF), cyclo-oxygenase-2 (COX-2), TNF-α, and iROS [84]. All these pathways are responsible for the preservation of peripheral nerve morphology and for reducing inflammation-coupled oedema and microglial activation [85,86]. In addition, several preclinical and clinical studies confirm the anti-nociceptive properties of PEA in neuropathic pain [87,88]. PEA treatment was also able to preserve Langerhans islets' morphology in the pancreatic gland of diabetic mice [89]. Orefice et al., in a randomized, double-blind, placebo-controlled study, demonstrated that ultra-micronized PEA as add-on therapy in patients afflicted by multiple sclerosis was able to significantly reduce the pain sensation and to improve their quality of life [90].

Notably, currently available drugs used to treat inflammation and pain-related conditions often display several adverse effects, especially when used chronically, such gastrointestinal, hepatic, renal, and cardiovascular disorders. PEA thus represents a promising alternative treatment, since it shows a high profile of safety and tolerability [78].

8. PEA in Neuromuscular Alterations and Diseases

The first study that hypothesized a potential use of ultra-micronized PEA in pathological conditions involving neuromuscular transmission was published in 2012 by Clemente who described the improvement of different vital parameters in a single case of a patient affected by amyotrophic lateral sclerosis (ALS), a neurodegenerative disorder able to disrupt the physiological function of the neuromuscular junction leading, in a few years, to muscle paralysis and death of the patients [85]. Up to now, there are several studies that highlight some of the possible mechanisms involved in this pathology, but unfortunately, ALS is still a disease with a poor prognosis lacking effective therapies and treatments in spite of recent efforts to develop new therapeutic strategies [91,92]. The degeneration of the second motor neuron is considered a hallmark of this disease, but different studies performed on animal models of ALS highlighted that the first alterations are localized at the neuromuscular junction (NMJ) [93,94]. Indeed, it was described that, even during the pre-symptomatic period, changes in NMJ function and muscular functions are present, thus leading to the "dying-back" theory [95]. According to this hypothesis, the first pathogenic step in ALS is represented by strong functional alterations at the NMJ level leading subsequently to the progression of the disease involving motor neuron degeneration [96], muscle atrophy, mitochondrial dysfunction, microgliosis, and neuronal degeneration [97]. To strengthen this hypothesis, it was demonstrated that muscular acetylcholine receptors (AChRs) obtained from muscle biopsy of ALS patients are functionally impaired when compared with AChRs obtained from non-ALS denervated muscle, showing also a reduction of sensitivity to ACh 98]. In detail, it is well known that after denervation, muscular AChRs subunits' composition undergoes modification, with a wide up-regulation of the $\alpha1\beta1\gamma\delta$ (γ-AChR) on the muscular sarcolemma. Indeed, in ALS muscles where both denervation and aberrant reinnervation are present, both the γ-AChR and the ε-AChRs are expressed. It was demonstrated that PEA can significantly reduce the rate of AChR desensitization of receptors obtained from ALS patients, during repetitive application of ACh [98]. In detail PEA was able to reduce this functional property predominantly on the ε-AChRs form, while this effect was to a lesser extent on γ-AChR [98]. To further strengthen this observation in the same study, an observational double-blind clinical test was performed on a cohort of ALS patients recruited by the neurological unit at the Policlinico Umberto I in Rome headed by Inghilleri. In detail, 28 patients were co-treated with riluzole (at the time of the study, the only officially approved treatment) [99] plus ultra-micronized PEA twice a day (50 mg and 600 mg, respectively) and compared with 36 patients receiving only riluzole. The two groups of patients were monitored, and clinical and electrophysiological test were assessed. Interestingly, ultra-micronized PEA-treated patients showed a slow respiratory impairment and a delayed need of tracheotomy. This clinical improvement was associated with a reduced decay of the forced vital capacity (FVC) in comparison with patients treated only with riluzole [100]. Therefore, the authors suggested that the PEA's effect on FVC is due to its ability to reduce the muscle AChRs desensitization induced by repetitive stimulation, as in respiratory muscles. Taken together, data obtained from this study showed that PEA add-on therapy slows the disease progression, suggesting that PEA could represent a valid aid to slow respiratory impairment in these patients, thus increasing their life expectancy. In another open-label pilot study, also carried out by Inghilleri's group at the Policlinico Umberto I in Rome, oral ultra-micronized PEA was administered to a cohort of patients afflicted by myasthenia gravis (MG) [101]. This autoimmune disease is characterized by the presence of auto-antibodies able to disrupt the physiological function of NMJ. Symptoms are often variable, from mild deficits limited to single muscle groups to generalized weakness involving vital functions. However, the main symptom is muscular fatigue, which causes characteristic fluctuations in symptoms that worsen in the evening and improve after periods of rest. The progression of the disease can be sneaky, with long periods of spontaneous remission or rapid progression [102]. In the aforementioned study [101], after only one week of treatment, patients displayed a significant improvement of all the neurophysiological parameters taken in account. In detail, ultra-micronized PEA induced a statistically significant improvement of the quantitative myasthenia gravis score (QMG), thus reducing the level of disability of the treated patients, and also a significant effect in improving the pathological muscle responses (measured as "decremental muscle responses", RNS). Since these positive effects appeared after only one week of treatment, it is not surprising that they disappeared one week after the withdrawal of the ultra-micronized PEA. Notably, the antibody titre did not significantly change following PEA treatment [101], suggesting a possible direct effect of ultra-micronized PEA on nAChRs as already shown in ALS patients. The capacity of PEA to reduce the release of pro-inflammatory cytokines could also be exploited to treat diseases such as sarcopenia, a condition characterized by progressive and generalized loss of skeletal muscle mass and strength combined with low physical performance [103]. Recently, new efforts have been made to improve PEA bioavailability, in order to better reach peripheral cells such as muscle cells [104], opening thus new scenarios in the treatments of diseases characterized by strong inflammation. On the other hand, new insights on the use of ultra-micronized PEA on inflammatory pathways modulation came from studies using PEA in association with antioxidant agents, such as phycocyanin extract (PC) from spirulina algae [66], that further enhanced PEA's beneficial properties. Indeed, a multi-center, double-blinded, randomized placebo-controlled clinical trial showed that co-administration of PEA plus Luteolin was able to significantly improve the recovery of olfactory function in patients after SARS-CoV-2 infection [105].

9. Conclusions

PEA signifies a valid support in different pathological conditions, since it displays a wide range of positive effects without noteworthy side effects. Until now, PEA has been extensively used to treat and ameliorate conditions characterized by neuroinflammation and to treat and reduce the perception of neuropathic pain. It is possible to find separate commercially available PEA formulations, (micronized, ultra-micronized), which increase its solubility and bioavailability, and in doses ranging from 200 mg to up to 1000 mg. While its neuroprotective effect is well known, its capacity to interact with the NMJs deserves further consideration, as it could pave the way for novel therapeutic implications. In detail, the ability to modulate the AChR function could open up new perspectives for its use in different pathological conditions characterized by alterations of NMJ, such as ALS, but also in other diseases such as MG where AChRs function is strongly altered. Further studies including larger double-blind multicentre clinical studies are required in order to find the best medical approach for any condition taken into consideration.

Author Contributions: Conceptualization, E.P. and M.I.; writing—original draft preparation, P.C., G.R., M.C., E.P. and M.I.; writing—review and editing, P.C., G.R., M.C., C.C., L.L., E.P. and M.I.; funding acquisition, P.C., E.P. and M.I. All authors have read and agreed to the published version of the manuscript.

Funding: This work was supported by the Fondazione Italiana di Ricerca per la Sclerosi Laterale Amiotrofica (AriSLA Grant "MUSALS-AChR") to E.P. and M.I. This research was supported by intramural "DISCAB" GRANT 2022 code 07_DG_2022_05 to P.C. G.R. was supported by the Italian Ministry of Health "Ricerca Corrente". The work was supported by grants from the Ateneo Project (Sapienza University)", grant n° RM11916B84D24429 to E.P.

Acknowledgments: The authors thank Epitech group SpA for the ongoing support to this research. Authors also thank Anwesha Ghosh for English proofreading.

Conflicts of Interest: The authors declare no conflict of interest.

References

1. Aloe, L.; Leon, A.; Levi-Montalcini, R. A proposed autacoid mechanism controlling mastocyte behaviour. *Agents Actions* **1993**, *39*, C145–C147. [CrossRef] [PubMed]
2. Wellner, N.; Diep, T.A.; Janfelt, C.; Hansen, H.S. N-acylation of phosphatidylethanolamine and its biological functions in mammals. *Biochim. Biophys Acta.* **2013**, *1831*, 652–662. [CrossRef] [PubMed]
3. Rahman, I.A.S.; Tsuboi, K.; Uyama, T.; Ueda, N. New players in the fatty acyl ethanolamide metabolism. *Pharmacol. Res.* **2014**, *86*, 1–10. [CrossRef] [PubMed]
4. Deutsch, D.G.; Ueda, N.; Yamamoto, S. The fatty acid amide hydrolase (FAAH). *Prostaglandins Leukot Essent Fat. Acids* **2002**, *66*, 201–210. [CrossRef] [PubMed]
5. Rankin, L.; Fowler, C.J. The Basal Pharmacology of Palmitoylethanolamide. *Int. J. Mol. Sci.* **2020**, *21*, 7942. [CrossRef]
6. Di Marzo, V.; Piscitelli, F. The endocannabinoid system and its modulation by phytocannabinoids. *Neurotherapeutics* **2015**, *12*, 692–698. [CrossRef]
7. PPacher, P.; Bátkai, S.; Kunos, G. The endocannabinoid system as an emerging target of pharmacotherapy. *Pharmacol. Rev.* **2006**, *58*, 389–462. [CrossRef]
8. Raso, G.M.; Russo, R.; Calignano, A.; Meli, R. Palmitoylethanolamide in CNS health and disease. *Pharmacol. Res.* **2014**, *86*, 32–41. [CrossRef]
9. Di Marzo, V.; Melck, D.; Bisogno, T.; de Petrocellis, L. Endocannabinoids: Endogenous cannabinoid receptor ligands with neuromodulatory action. *Trends Neurosci.* **1998**, *21*, 521–528. [CrossRef]
10. Verme, J.L.; Fu, J.; Astarita, G.; la Rana, G.; Russo, R.; Calignano, A.; Piomelli, D. The Nuclear Receptor Peroxisome Proliferator-Activated Receptor-α Mediates the Anti-Inflammatory Actions of Palmitoylethanolamide. *Mol. Pharmacol.* **2004**, *67*, 15–19. [CrossRef]
11. Bougarne, N.; Weyers, B.; Desmet, S.J.; Deckers, J.; Ray, D.W.; Staels, B.; de Bosscher, K. Molecular actions of PPARα in lipid metabolism and inflammation. *Endocr. Rev.* **2018**, *39*, 760–802. [CrossRef] [PubMed]
12. Verme, J.L.; Russo, R.; la Rana, G.; Fu, J.; Farthing, J.; Mattace-Raso, G.; Meli, R.; Hohmann, A.; Calignano, A.; Piomelli, D. Rapid broad-spectrum analgesia through activation of peroxisome proliferator-activated receptor-alpha. *J. Pharmacol. Exp. Ther.* **2006**, *319*, 1051–1061.
13. Ryberg, E.; Larsson, N.; Sjögren, S.; Hjorth, S.; Hermansson, N.-O.; Leonova, J.; Elebring, T.; Nilsson, K.; Drmota, T.; Greasley, P.J. The orphan receptor GPR55 is a novel cannabinoid receptor. *Br. J. Pharmacol.* **2009**, *152*, 1092–1101. [CrossRef] [PubMed]

14. Brown, A.J. Novel cannabinoid receptors. *Br. J. Pharmacol.* **2007**, *152*, 567–575. [CrossRef]
15. Costa, B.; Comelli, F.; Bettoni, I.; Colleoni, M.; Giagnoni, G. The endogenous fatty acid amide, palmitoylethanolamide, has antiallodynic and anti-hyperalgesic effects in a murine model of neuropathic pain: Involvement of CB1, TRPV1 and PPARγ receptors and neurotrophic factors. *Pain* **2008**, *139*, 541–550. [CrossRef]
16. Deutsch, D.; Glaser, S.; Howell, J.; Kunz, J.; Puffenbarger, R.; Hillard, C.; Abumrad, N. The cellular uptake of anandamide is coupled to its breakdown by fatty-acid amide hydrolase. *J. Biol. Chem.* **2001**, *276*, 6967–6973. [CrossRef]
17. Kettenmann, H.; Hanisch, U.-K.; Noda, M.; Verkhratsky, A. Physiology of microglia. *Physiol. Rev.* **2011**, *91*, 461–553. [CrossRef]
18. Kigerl, K.A.; Gensel, J.C.; Ankeny, D.P.; Alexander, J.K.; Donnelly, D.J.; Popovich, P.G. Identification of two distinct macrophage subsets with divergent effects causing either neurotoxicity or regeneration in the injured mouse spinal cord. *J. Neurosci. Off. J. Soc. Neurosci.* **2009**, *29*, 13435–13444. [CrossRef]
19. Ransohoff, R.M.; Perry, V.H. Microglial physiology: Unique stimuli, specialized responses. *Annu Rev. Immunol.* **2009**, *27*, 119–145. [CrossRef]
20. Cordaro, M.; Cuzzocrea, S.; Crupi, R. An Update of Palmitoylethanolamide and Luteolin Effects in Preclinical and Clinical Studies of Neuroinflammatory Events. *Antioxidants* **2020**, *9*, 216. [CrossRef]
21. D'Aloia, A.; Molteni, L.; Gullo, F.; Bresciani, E.; Artusa, V.; Rizzi, L.; Ceriani, M.; Meanti, R.; Lecchi, M.; Coco, S.; et al. Palmitoylethanolamide Modulation of Microglia Activation: Characterization of Mechanisms of Action and Implication for Its Neuroprotective Effects. *Int. J. Mol. Sci.* **2021**, *22*, 3054. [CrossRef] [PubMed]
22. Guida, F.; Luongo, L.; Boccella, S.; Giordano, M.E.; Romano, R.; Bellini, G.; Manzo, I.; Furiano, A.; Rizzo, A.; Imperatore, R.; et al. Palmitoylethanolamide induces microglia changes associated with increased migration and phagocytic activity: Involvement of the CB2 receptor. *Sci. Rep.* **2017**, *7*, 375. [CrossRef] [PubMed]
23. Petrosino, S.; di Marzo, V. The pharmacology of palmitoylethanolamide and first data on the therapeutic efficacy of some of its new formulations. *Br. J. Pharmacol.* **2017**, *174*, 1349–1365. [CrossRef] [PubMed]
24. Petrosino, S.; Cordaro, M.; Verde, R.; Moriello, A.S.; Marcolongo, G.; Schievano, C.; Siracusa, R.; Piscitelli, F.; Peritore, A.; Crupi, R.; et al. Oral Ultramicronized Palmitoylethanolamide: Plasma and Tissue Levels and Spinal Anti-hyperalgesic Effect. *Front Pharmacol.* **2018**, *9*, 249. [CrossRef]
25. Impellizzeri, D.; Bruschetta, G.; Cordaro, M.; Crupi, R.; Siracusa, R.; Esposito, E.; Cuzzocrea, S. Micronized/ultramicronized palmitoylethanolamide displays superior oral efficacy compared to nonmicronized palmitoylethanolamide in a rat model of inflammatory pain. *J. Neuroinflamm.* **2014**, *11*, 136. [CrossRef]
26. Fusco, R.; Gugliandolo, E.; Campolo, M.; Evangelista, M.; di Paola, R.; Cuzzocrea, S. Effect of a new formulation of micronized and ultramicronized N-palmitoylethanolamine in a tibia fracture mouse model of complex regional pain syndrome. *PLoS ONE* **2017**, *12*, e0178553. [CrossRef]
27. Schmid, P.; Zuzarte-Augustin, M.; Schmid, H. Properties of rat liver N-acylethanolamine amidohydrolase. *J. Biol. Chem.* **1985**, *260*, 14145–14149. [CrossRef]
28. Vacondio, F.; Bassi, M.; Silva, C.; Castelli, R.; Carmi, C.; Scalvini, L.; Lodola, A.; Vivo, V.; Flammini, L.; Barocelli, E.; et al. Amino acid derivatives as palmitoylethanolamide prodrugs: Synthesis, in vitro metabolism and in vivo plasma profile in rats. *PLoS ONE* **2015**, *10*, e0128699. [CrossRef]
29. Bracey, M.; Hanson, M.; Masuda, K.; Stevens, R.; Cravatt, B. Structural adaptions in a membrane enzyme that terminates endocannabinoid signaling. *Science* **2002**, *298*, 1793–1796. [CrossRef]
30. Boger, D.; Fecik, R.; Patterson, J.; Miyauchi, H.; Patricelli, M.; Cravatt, B. Fatty acid amide hydrolase substrate specificity. *Bioorg. Med. Chem. Lett.* **2000**, *10*, 2613–2616. [CrossRef]
31. Tsuboi, K.; Sun, Y.-X.; Okamoto, Y.; Araki, N.; Tonai, T.; Ueda, N. Molecular characterization of N-acylethanolamine-hydro lyzing acid amidase, a novel member of the choloylglycine hydrolase family with structural and functional similarity to acid ceramidase. *J. Biol. Chem.* **2005**, *280*, 11082–11092. [CrossRef] [PubMed]
32. Ueda, N.; Yamanaka, K.; Yamamoto, S. Purification and characterization of an acid amidase selective for N-palmitoylethan olamine, a putative endogenous anti-inflammatory substance. *J. Biol. Chem.* **2001**, *276*, 35552–35557. [CrossRef] [PubMed]
33. Wei, B.; Mikkelsen, T.; McKinney, M.; Lander, E.; Cravatt, B. A second fatty acid amide hydrolase with variable distribution among placental mammals. *J. Biol. Chem.* **2006**, *281*, 36569–36578. [CrossRef] [PubMed]
34. Bonezzi, F.T.; Sasso, O.; Pontis, S.; Realini, N.; Romeo, E.; Ponzano, S.; Nuzzi, A.; Fiasella, A.; Bertozzi, F.; Piomelli, D. An important role for N-acylethanolamine acid amidase in the complete Freund's adjuvant rat model of arthritis. *J. Pharmacol. Exp. Ther.* **2016**, *356*, 656–663. [CrossRef]
35. Artamonov, M.; Zhukov, O.; Shuba, I.; Storozhuk, L.; Khmel, T.; Klimashevsky, V.; Mikosha, A.; Gula, N. Incorporation of labelled N-acylethanolamine (NAE) into rat brain regions in vivo and adaptive properties of saturated NAE under X-ray irradiation (1999). *Ukr. Biokhimicheskii Zhurnal* **2005**, *77*, 51–62.
36. Jhaveri, M.; Richardson, D.; Robinson, I.; Garle, M.; Patel, A.; Sun, Y.; Sagar, D.; Bennett, A.; Alexander, S.; Kendall, D.; et al. Inhibition of fatty acid amide hydrolase and cyclooxygenase-2 increases levels of endocannabinoid related molecules and pro duces analgesia via peroxisome proliferator-activated receptor-alpha in a model of inflammatory pain. *Neuropharmacology* **2008**, *55*, 85–93. [CrossRef]
37. Kaczocha, M.; Glaser, S.T.; Chae, J.; Brown, D.A.; Deutsch, D.G. Lipid droplets are novel sites of N-acylethanolamine inacti vation by fatty acid amide hydrolase-2. *J. Biol. Chem.* **2010**, *285*, 2796–2806. [CrossRef]

38. Carta, G.; Murru, E.; Bani, S.; Manca, C. Palmitic acid: Physiological role, metabolism and nutritional implications. *Front. Physiol.* **2017**, *8*, 902. [CrossRef]
39. Scuderi, C.; Stecca, C.; Valenza, M.; Ratano, P.; Bronzuoli, M.R.; Bartoli, S.; Steardo, L.; Pompili, E.; Fumagalli, L.; Campolongo, P.; et al. Palmitoylethanolamide controls reactive gliosis and exerts neuroprotective functions in a rat model of Alzheimer's dis ease. *Cell Death Dis.* **2014**, *11*, e1419. [CrossRef]
40. Bronzuoli, M.R.; Facchinetti, R.; Steardo, L.; Romano, A.; Stecca, C.; Passarella, S.; Cassano, T.; Scuderi, C. Palmitoylethanola mide Dampens Reactive Astrogliosis and Improves Neuronal Trophic Support in a Triple Transgenic Model of Alzheimer's Disease: In Vitro and In Vivo Evidence. *Oxid Med. Cell Longev.* **2018**, *2018*, 4720532. [CrossRef]
41. Beggiato, S.; Tomasini, M.C.; Ferraro, L. Palmitoylethanolamide (PEA) as a Potential Therapeutic Agent in Alzheimer's Disease. *Front. Pharmacol.* **2019**, *24*, 821. [CrossRef] [PubMed]
42. D'Agostino, G.; Russo, R.; Avagliano, C.; Cristiano, C.; Meli, R.; Calignano, A. Palmitoylethanolamide protects against the amyloidβ25-35-induced learning and memory impairment in mice, an experimental model of Alzheimer disease. *Neuropsy Chopharmacol.* **2012**, *37*, 1784–1792. [CrossRef] [PubMed]
43. Scuderi, C.; Steardo, L. Neuroglial roots of neurodegenerative diseases: Therapeutic potential of palmitoylethanolamide in models of Alzheimer's disease. *CNS. Neurol. Disord. Drug. Targets* **2013**, *12*, 62–69. [CrossRef] [PubMed]
44. Esposito, E.; Impellizzeri, D.; Mazzon, E.; Paterniti, I.; Cuzzocrea, S. Neuroprotective activities of palmitoylethanolamide in an animal model of Parkinson's disease. *PLoS ONE* **2012**, *7*, e41880. [CrossRef]
45. Esposito, E.; Cuzzocrea, S. Palmitoylethanolamide in homeostatic and traumatic central nervous system injuries. *CNS Neu. rol. Disord. Drug Targets* **2013**, *12*, 55–61. [CrossRef]
46. De Palma, L.; Santamato, A.; Montillo, V.; Fiore, P. Co-ultra micronized palmitoylethanolamide/luteolin treatment as add- on to intensive neuro-rehabilitation in a young patient with traumatic brain injury. *Gazz Med. Ital. Arch. Sci. Med.* **2018**, *177*, 237–242. [CrossRef]
47. Ordaro, M.; Impellizzeri, D.; Paterniti, I.; Bruschetta, G.; Siracusa, R.; de Stefano, D.; Cuzzocrea, S.; Esposito, E. Neuroprotective Effects of Co-UltraPEALut on Secondary Inflammatory Process and Autophagy Involved in Traumatic Brain Injury. *J. Neuro Trauma* **2016**, *33*, 132–146.
48. Ahmad, A.; Crupi, R.; Impellizzeri, D.; Campolo, M.; Marino, A.; Esposito, E.; Cuzzocrea, S. Administration of palmitoyleth anolamide (PEA) protects the neurovascular unit and reduces secondary injury after traumatic brain injury in mice. *Brain. Behav. Immun.* **2012**, *26*, 1310–1321. [CrossRef]
49. Guida, F.; Boccella, S.; Iannotta, M.; de Gregorio, D.; Giordano, C.; Belardo, C.; Romano, R.; Palazzo, E.; Scafuro, M.A.; Serra, N.; et al. Palmitoylethanolamide Reduces Neuropsychiatric Behaviors by Restoring Cortical Electrophysiological Activity in a Mouse Model of Mild Traumatic Brain Injury. *Front. Pharmcol.* **2017**, *6*, 95. [CrossRef]
50. Caltagirone, C.; Cisari, C.; Schievano, C.; di Paola, R.; Cordaro, M.; Bruschetta, G.; Esposito, E.; Cuzzocrea, S. Stroke Study Group. Co-ultramicronized Palmitoylethanolamide/Luteolin in the Treatment of Cerebral Ischemia: From Rodent to Man. *Transl. Stroke. Res.* **2016**, *7*, 54–69. [CrossRef]
51. Citraro, R.; Russo, E.; Scicchitano, F.; van Rijn, C.M.; Cosco, D.; Avagliano, C.; Russo, R.; D'Agostino, G.; Petrosino, S.; Guida, F.; et al. Antiepileptic action of N-palmitoylethanolamine through CB1 and PPAR-α receptor activation in a genetic model of absence epilepsy. *Neuropharmacology* **2013**, *69*, 115–126. [CrossRef] [PubMed]
52. Kruk-Slomka, M.; Dzik, A.; Budzynska, B.; Biala, G. Endocannabinoid System: The Direct and Indirect Involvement in the Memory and Learning Processes—A Short Review. *Mol. Neurobiol.* **2017**, *54*, 8332–8347. [CrossRef] [PubMed]
53. Hou, R.; Garner, M.; Holmes, C.; Osmond, C.; Teeling, J.; Lau, L.; Baldwin, D.S. Peripheral inflammatory cytokines and im mune balance in generalised anxiety disorder: Case-controlled study. *Brain Behav. Immun.* **2017**, *62*, 212–218. [CrossRef] [PubMed]
54. Klein, T.W. Cannabinoid-based drugs as anti-inflammatory therapeutics. *Nat. Rev. Immunol.* **2005**, *5*, 400–411. [CrossRef]
55. Skaper, S.D.; Facci, L. Mast cell-glia axis in neuroinflammation and therapeutic potential of the anandamide congener pal mitoylethanolamide. *Philos. Trans. R. Soc. Lond. B Biol. Sci.* **2012**, *367*, 3312–3325. [CrossRef]
56. Skaper, S.D.; Facci, L.; Giusti, P. Mast cells, glia and neuroinflammation: Partners in crime? *Immunology* **2014**, *141*, 314–327. [CrossRef]
57. Nau, R.; Ribes, S.; Djukic, M.; Eiffert, H. Strategies to increase the activity of microglia as efficient protectors of the brain against infections. *Front. Cell Neurosci.* **2014**, *22*, 138. [CrossRef]
58. D'Agostino, G.; la Rana, G.; Russo, R.; Sasso, O.; Iacono, A.; Esposito, E.; Raso, G.M.; Cuzzocrea, S.; Verme, J.L.; Peiomelli, D. Central administration of palmitoylethanolamide reduces hyperalgesia in mice via inhibition of NF-κB nuclear signalling in dorsal root ganglia. *Eur. J. Pharmcol.* **2009**, *613*, 54–59. [CrossRef]
59. Zeise, M.; Lehmann, A. Phosphoethanolamine transiently enhances excitability of rat hippocampal neurons in vitro. *J. Comp. Physiol. A.* **1987**, *161*, 461–467. [CrossRef]
60. Barbierato, M.; Facci, L.; Marinelli, C.; Zusso, M.; Argentini, C.; Skaper, S.D.; Giusti, P. Co-ultramicronized Palmitoylethano lamide/Luteolin Promotes the Maturation of Oligodendrocyte Precursor Cells. *Sci. Rep.* **2015**, *5*, 16676. [CrossRef]
61. Fidaleo, M.; Fanelli, F.; Ceru, M.P.; Moreno, S. Neuroprotective properties of peroxisome proliferator-activated receptor alpha (PPARα) and its lipid ligands. *Curr. Med. Chem.* **2014**, *21*, 2803–2821. [CrossRef] [PubMed]

62. Boccella, S.; Cristiano, C.; Romano, R.; Iannotta, M.; Belardo, C.; Farina, A.; Guida, F.; Piscitelli, F.; Palazzo, E.; Mazzitelli, M.; et al. Ultra-micronized pal mitoylethanolamide rescues the cognitive decline-associated loss of neural plasticity in the neuropathic mouse entorhinal cor tex-dentate gyrus pathway. *Neurobiol. Dis.* **2019**, *121*, 106–119. [CrossRef] [PubMed]
63. Scuderi, C.; Esposito, G.; Blasio, A.; Valenza, M.; Arietti, P.L.; Steardo, J.; Carnuccio, R.; de Filippis, D.; Petrosino, S.; Iuvone, T.; et al. Palmitoylethanolamide counteracts reactive astrogliosis induced by β-amyloid peptide. *J. Cell Mol. Med.* **2011**, *15*, 2664–2674. [CrossRef] [PubMed]
64. Keppel Hesselink, J.M.R. Levi-Montalcini on Nerve Growth Factor, mast cells and palmitoylethanolamide, an endogenous anti-inflammatory and analgesic compound. *Pain Relief.* **2013**, *2*, 1. [CrossRef]
65. Beggiato, S.; Cassano, T.; Ferraro, L.; Tomasini, M.C. Astrocytic palmitoylethanolamide pre-exposure exerts neuroprotective effects in astrocyte-neuron co-cultures from a triple transgenic mouse model of Alzheimer's disease. *Life Sci.* **2020**, *257*, 118037. [CrossRef]
66. Bergandi, L.; Apprato, G.; Silvagno, F. Antioxidant and Anti-Inflammatory Activity of Combined Phycocyanin and Pal mitoylethanolamide in Human Lung and Prostate Epithelial Cells. *Antioxidants* **2022**, *11*, 201. [CrossRef]
67. Antonucci, N.; Cirillo, A.; Siniscalco, D. Beneficial Effects of Palmitoylethanolamide on Expressive Language, Cognition, and Behaviors in Autism: A Report of Two Cases. *Case Rep. Psychiatry* **2015**, *2015*, 325061. [CrossRef]
68. Müller-Vahl, K.R.; Bindila, L.; Lutz, B.; Musshoff, F.; Skripuletz, T.; Baumgaertel, C.; Sühs, K.W. Cerebrospinal fluid endocan nabinoid levels in Gilles de la Tourette syndrome. *Neuropsychopharmacology* **2020**, *45*, 1323–1329. [CrossRef]
69. Altamura, C.; Ventriglia, M.; Martini, M.G.; Montesano, D.; Errante, Y.; Piscitelli, F.; Scrascia, F.; Quattrocchi, C.; Palazzo, P.; Seccia, S.; et al. Elevation of Plasma 2-Arachidonoylglycerol Levels in Alzheimer's Disease Patients as a Po tential Protective Mechanism against Neurodegenerative Decline. *J. Alzheimers Dis.* **2015**, *46*, 497–506. [CrossRef]
70. Alcaraz-Silva, J.; Feingold, D.; Viana-Torre, G.; Budde, H.; Imperatori, C.; Machado, S.; Murillo-Rodríguez, E. The Endocanna binoid System as a Biomarker for Diagnostic and Therapeutic Applications in Depression and Anxiety. *CNS Neurol. Disord. Drug Targets* **2022**. [CrossRef]
71. Hauer, D.; Schelling, G.; Gola, H.; Campolongo, P.; Morath, J.; Roozendaal, B.; Hamuni, G.; Karabatsiakis, A.; Atsak, P.; Vogeser, M.; et al. Plasma concentrations of endocannabinoids and related primary fatty acid amides in patients with post-trau matic stress disorder. *PLoS ONE* **2013**, *8*, e62741. [CrossRef] [PubMed]
72. Hill, M.N.; Miller, G.E.; Carrier, E.J.; Gorzalka, B.B.; Hillard, C.J. Circulating endocannabinoids and N-acyl ethanolamines are differentially regulated in major depression and following exposure to social stress. *Psychoneuroendocrinology* **2009**, *34*, 1257–1262. [CrossRef] [PubMed]
73. Hill, M.N.; Miller, G.E.; Ho, W.S.; Gorzalka, B.B.; Hillard, C.J. Serum endocannabinoid content is altered in females with depressive disorders: A preliminary report. *Pharmacopsychiatry* **2008**, *41*, 48–53. [CrossRef] [PubMed]
74. Pietropaolo, S.; Marsicano, G. The role of the endocannabinoid system as a therapeutic target for autism spectrum disorder: Lessons from behavioral studies on mouse models. *Neurosci. Biobehav. Rev.* **2022**, *132*, 664–678. [CrossRef]
75. Impellizzeri, D.; Cordaro, M.; Bruschetta, G.; Siracusa, R.; Crupi, R.; Esposito, E.; Cuzzocrea, S. *N*-Palmitoylethanolamine-Oxazoline as a New Therapeutic Strategy to Control Neuroinflammation: Neuroprotective Effects in Experimental Models of Spinal Cord and Brain Injury. *J. Neurotrauma.* **2017**, *34*, 2609–2623. [CrossRef]
76. Musella, A.; Fresegna, D.; Rizzo, F.R.; Gentile, A.; Bullitta, S.; de Vito, F.; Guadalupi, L.; Centonze, D.; Mandolesi, G. A novel crosstalk within the endocannabinoid system controls GABA transmission in the striatum. *Sci. Rep.* **2017**, *7*, 7363. [CrossRef]
77. Ghazizadeh-Hashemi, M.; GhajarBrotini, S.; Schievano, C.; Guidi, L. Ultra-micronized Palmitoylethanolamide: An Efficacious Adjuvant Therapy for Parkinson's Disease. *CNS Neurol. Disord Drug Targets* **2017**, *16*, 705–713.
78. Ghazizadeh-Hashemi, M.; Ghajar, A.; Shalbafan, M.R.; Ghazizadeh-Hashemi, F.; Afarideh, M.; Malekpour, F.; Ghaleiha, A.; Ardebili, M.E.; Akhondzadeh, S. Palmitoylethanolamide as adjunctive therapy in major depressive disorder: A double-blind, randomized and placebo-controlled trial. *J. Affect. Disord.* **2018**, *232*, 127–133. [CrossRef]
79. Gabrielsson, L.; Mattson, S.; Fowler, C.J. Palmitoylethanolamide for the treatment of pain: Pharmacokinetics, safety and efficacy. *Br. J. Clin. Pharmacol.* **2016**, *82*, 932–942. [CrossRef]
80. Bettoni, I.; Comelli, F.; Colombo, A.; Bonfanti, P.; Costa, B. Non-neuronal cell modulation relieves neuropathic pain: Efficacy of the endogenous lipid palmitoylethanolamide. *CNS Neurol. Disord. Drug Targets* **2013**, *12*, 34–44. [CrossRef]
81. Mannelli, L.D.C.; D'Agostino, G.; Pacini, A.; Russo, R.; Zanardelli, M.; Ghelardini, C.; Calignano, A. Palmitoylethanolamide is a disease-modifying agent in peripheral neuropathy: Pain relief and neuroprotection share a PPAR-alpha-mediated mecha nism. *Mediat. Inflamm.* **2013**, *2013*, 328797.
82. Petrosino, S.; Moriello, A.S.; Cerrato, S.; Fusco, M.; Puigdemont, A.; de Petrocellis, L.; di Marzo, V. The anti-inflamma tory mediator palmitoylethanolamide enhances the levels of 2-arachidonoyl-glycerol and potentiates its actions at TRPV1 cat ion channels. *Br. J. Pharmacol.* **2016**, *173*, 1154–1162. [CrossRef] [PubMed]
83. Cerrato, S.; Brazis, P.; della Valle, M.F.; Miolo, A.; Puigdemont, A. Effects of palmitoylethanolamide on immunologically induced histamine, PGD2 and TNFα release from canine skin mast cells. *Vet. Immunol. Immunopathol.* **2010**, *133*, 9–15. [CrossRef] [PubMed]
84. Davis, A. The dangers of NAIDS: Look both ways. *Br. J. Gen. Pract.* **2016**, *66*, 172–173. [CrossRef] [PubMed]
85. Clemente, S. Amyotrophic lateral sclerosis treatment with ultramicronized palmitoylethanolamide: A case report. *CNS Neurol Disord Drug Targets* **2012**, *11*, 933–936. [CrossRef]

86. Impellizzeri, D.; Peritore, A.F.; Cordaro, M.; Gugliandolo, E.; Siracusa, R.; Crupi, R.; D'Amico, R.; Fusco, R.; Evangelista, M.; Cuzzocrea, S.; et al. The neuroprotective effects of micronized PEA (PEA-m) formulation on diabetic peripheral neuropa thy in mice. *FASEB J.* **2019**, *33*, 11364–11380. [CrossRef]
87. Hesselink, J.M.K.; Hekker, T.A. Therapeutic utility of palmitoylethanolamide in the treatment of neuropathic pain associated with various pathological conditions: A case series. *J. Pain Res.* **2012**, *5*, 437–442. [CrossRef]
88. Schifilliti, C.; Cucinotta, L.; Fedele, V.; Ingegnosi, C.; Luca, S.; Leotta, C. Micronized palmitoylethanolamide reduces the symp toms of neuropathic pain in diabetic patients. *Pain Res. Treat* **2014**, *2014*, 849623. [CrossRef]
89. Donvito, G.; Bettoni, I.; Comelli, F.; Colombo, A.; Costa, B. Palmitoylethanolamide relieves pain and preserves pancreatic islet cells in a murine model of diabetes. *CNS Neurol. Disord Drug Targets.* **2015**, *14*, 452–462. [CrossRef]
90. Orefice, N.S.; Alhouayek, M.; Carotenuto, A.; Montella, S.; Barbato, F.; Comelli, A.; Calignano, A.; Muccioli, G.G.; Orefice, G. Oral Palmitoylethanolamide Treatment Is Associated with Reduced Cutaneous Adverse Effects of Interferon-β1a and Circu lating Proinflammatory Cytokines in Relapsing-Remitting Multiple Sclerosis. *Neurotherapeutics* **2016**, *13*, 428–438. [CrossRef]
91. Taylor, J.P.; Brown, R.H.; Cleveland, D.W. Decoding ALS: From genes to mechanism. *Nature* **2016**, *539*, 197–206. [CrossRef] [PubMed]
92. Musarò, A.; Dobrowolny, G.; Cambieri, C.; Onesti, E.; Ceccanti, M.; Frasca, V.; Pisano, A.; Cerbelli, B.; Lepore, E.; Ruffolo, G.; et al. Neuromuscular magnetic stimulation counteracts muscle decline in ALS patients: Results of a randomized, double-blind, controlled study. *Sci. Rep.* **2019**, *9*, 2837. [CrossRef]
93. Dupuis, L.; Loefer, J.-P. Neuromuscular junction destruction during amyotrophic lateral sclerosis: Insights from trans genic models. *Curr. Opin. Pharmacol.* **2009**, *9*, 341–346. [CrossRef] [PubMed]
94. Rocha, M.C.; Pousinha, P.A.; Correia, A.M.; Sebastião, A.M.; Ribeiro, J.A. Early changes of neuromuscular transmission in the SOD1(G93A) mice model of ALS start long before motor symptoms onset. *PLoS ONE* **2013**, *8*, e73846. [CrossRef] [PubMed]
95. Dadon-Nachum, M.; Melamed, E.; Ofen, D. The "Dying-Back" Phenomenon of Motor Neurons in ALS. *J. Mol. Neurosci.* **2011**, *43*, 470–477. [CrossRef] [PubMed]
96. Wong, M.; Martin, L.J. Skeletal muscle-restricted expression of human SOD1 causes motor neuron degeneration in trans genic mice. *Hum. Mol. Genet.* **2010**, *19*, 2284–2302. [CrossRef] [PubMed]
97. Musarò, A. Understanding ALS: New therapeutic approaches. *FEBS J.* **2013**, *280*, 4315–4322. [CrossRef]
98. Palma, E.; Inghilleri, M.; Conti, L.; Deflorio, C.; Frasca, V.; Manteca, A.; Pichiorri, F.; Roseti, C.; Torchia, G.; Limatola, C.; et al. Physiological characterization of human muscle acetylcholine receptors from ALS patients. *Proc. Natl. Acad. Sci. USA* **2011**, *108*, 20184–20188. [CrossRef]
99. Deflorio, C.; Palma, E.; Conti, L.; Roseti, C.; Manteca, A.; Giacomelli, E.; Catalano, M.; Limatola, C.; Inghilleri, M.; Grassi, F. Riluzole blocks human muscle acetylcholine receptors. *J. Physiol.* **2012**, *590*, 2519–2528. [CrossRef]
100. Palma, E.; Reyes-Ruiz, J.M.; Lopergolo, D.; Roseti, C.; Bertollini, C.; Ruffolo, G.; Cifelli, P.; Onesti, E.; Limatola, C.; Miledi, R.; et al. Acetylcholine receptors from human muscle as pharmacological target for ALS therapy. *Proc. Natl. Acad. Sci. USA* **2016**, *113*, 3060–3065. [CrossRef]
101. Onesti, E.; Frasca, V.; Ceccanti, M.; Tartaglia, G.; Gori, M.C.; Cambieri, C.; Libonati, L.; Palma, E.; Inghilleri, M. Short-Term Ultramicronized Palmitoylethanolamide Therapy in Patients with Myasthenia Gravis: A Pilot Study to Possible Future Impli cations of Treatment. *CNS Neurol. Disord. Drug Targets.* **2019**, *18*, 232–238. [CrossRef] [PubMed]
102. Hehir, M.K.; Silvestri, N.J. Generalized Myasthenia Gravis: Classification, Clinical Presentation, Natural History, and Ep idemiology. *Neurol. Clin.* **2018**, *36*, 253–260. [CrossRef] [PubMed]
103. Cruz-Jentoft, A.J.; Bahat, G.; Bauer, J.; Boirie, Y.; Bruyère, O.; Cederholm, T.; Cooper, C.; Landi, F.; Rolland, Y.; Sayer, A.A.; et al. Sarcopenia: Revised European consensus on definition and diagnosis. *Age Ageing.* **2019**, *48*, 16–31. [CrossRef] [PubMed]
104. Maretti, E.; Molinari, S.; Battini, R.; Rustichelli, C.; Truzzi, E.; Iannuccelli, V.; Leo, E. Design, Characterization, and In Vitro Assays on Muscle Cells of Endocannabinoid-like Molecule Loaded Lipid Nanoparticles for a Therapeutic Anti-Inflammatory Approach to Sarcopenia. *Pharmaceutics* **2022**, *14*, 648. [CrossRef]
105. Di Stadio, A.; D'Ascanio, L.; Vaira, L.A.; Cantone, E.; de Luca, P.; Cingolani, C.; Motta, G.; de Riu, G.; Vitelli, F.; Spriano, G.; et al. Ultramicronized Palmitoylethanolamide and Luteolin Supplement Combined with Olfactory Training to Treat Post-COVID-19 Olfactory Impairment: A Multi-Center Double- Blinded Randomized Placebo-Controlled Clinical Trial. *Curr. Neuropharmacol.* **2022**, *20*, 1. [CrossRef]

biomolecules

Review

Effects of Palmitoylethanolamide on Neurodegenerative Diseases: A Review from Rodents to Humans

Eugenia Landolfo [1], Debora Cuturi [1,2], Laura Petrosini [1,*] and Carlo Caltagirone [1]

1 Fondazione Santa Lucia IRCCS, 00143 Rome, Italy; eugenia.landolfo@uniroma1.it (E.L.); debora.cutuli@uniroma1.it (D.C.); c.caltagirone@hsantalucia.it (C.C.)
2 Department of Psychology, University of Rome "Sapienza", 00185 Rome, Italy
* Correspondence: laura.petrosini@uniroma1.it

Abstract: Palmitoylethanolamide (PEA) stands out among endogenous lipid mediators for its neuroprotective, anti-inflammatory, and analgesic functions. PEA belonging to the N-acetylanolamine class of phospholipids was first isolated from soy lecithin, egg yolk, and peanut flour. It is currently used for the treatment of different types of neuropathic pain, such as fibromyalgia, osteoarthritis, carpal tunnel syndrome, and many other conditions. The properties of PEA, especially of its micronized or ultra-micronized forms maximizing bioavailability and efficacy, have sparked a series of innovative research to evaluate its possible application as therapeutic agent for neurodegenerative diseases. Neurodegenerative diseases are widespread throughout the world, and although they are numerous and different, they share common patterns of conditions that result from progressive damage to the brain areas involved in mobility, muscle coordination and strength, mood, and cognition. The present review is aimed at illustrating in vitro and in vivo research, as well as human studies, using PEA treatment, alone or in combination with other compounds, in the presence of neurodegeneration. Namely, attention has been paid to the effects of PEA in counteracting neuroinflammatory conditions and in slowing down the progression of diseases, such as Alzheimer's disease, Parkinson's disease, Huntington's disease, Frontotemporal dementia, Amyotrophic Lateral Sclerosis, and Multiple Sclerosis. Literature research demonstrated the efficacy of PEA in addressing the damage typical of major neurodegenerative diseases.

Keywords: PEA; ALIAmides; Alzheimer's disease; Parkinson's disease; Huntington's disease; Frontotemporal dementia; Amyotrophic Lateral Sclerosis; Multiple Sclerosis; neuroinflammation

Citation: Landolfo, E.; Cutuli, D.; Petrosini, L.; Caltagirone, C. Effects of Palmitoylethanolamide on Neurodegenerative Diseases: A Review from Rodents to Humans. *Biomolecules* **2022**, *12*, 667. https://doi.org/10.3390/biom12050667

Academic Editor: Marcus O. W. Grimm

Received: 12 April 2022
Accepted: 2 May 2022
Published: 5 May 2022

1. Introduction

1.1. PEA, an Anti-Inflammatory and Neuroprotective Substance

Lipid molecules may play a primary role essential to fight, or at least delay, chronic neuroinflammation, a phenomenon underlying many neurodegenerative diseases. A class of anti-inflammatory molecules are the Autacoid Local Injury Antagonist (ALIA) amides [1]. This acronym, coined by the research group of Rita Levi Montalcini, describes a group of endogenous bioactive acyl ethanolamides with anti-inflammatory properties [2], generally referred to as N-acylethanolamines (NAEs). NAEs include PEA, an anti-inflammatory and analgesic substance, oleoylethanolamide (OEA), an anorectic substance, and anandamide (AEA), an endocannabinoid (eCB) substance with autocrine and paracrine signaling properties [3]. PEA cannot strictly be considered a classic eCB, because it has a low affinity for the cannabinoid receptors CB1 and CB2 [4,5]. However, the presence of PEA enhances the AEA activity, likely through an "entourage effect". PEA is endowed with important anti-inflammatory, neuroprotective, and analgesic actions, and some of its effects are mediated by the peroxisome proliferator-activated receptor (PPAR)-α. PEA anti-inflammatory and neuroprotective functions have been attributed in particular to eCBs belonging to the acyl ethanolamide family, as well as to their congeners, since their production is significantly increased in the sites of neuronal damage [6]. PEA is naturally found in some foods, such as egg yolk, peanut flour, soybean oil, and corn [1,7]. In animal cells, PEA is synthesized from palmitic acid, the most common fatty acid present in many foods including palm oil, meats, cheeses, butter, and other dairy products [8]. Because of its high safety and tolerability [9–12], PEA is often used as an analgesic, anti-inflammatory, and neuroprotective mediator in the treatment of acute and chronic inflammatory diseases, alone or combined with antioxidant or analgesic molecules acting on molecular targets of central and peripheral nervous system and immune cells [11,13,14]. In the brain, PEA is produced "on demand" by neurons, microglia, and astrocytes, and thus plays a pleiotropic and pro-homeostatic role, when faced with external stressors provoking inflammation. PEA exerts a local anti-injury function by down-modulating mast cell activation and protecting neurons from excitotoxicity [15–17]. The synthesis of PEA takes place in the membranes of various cell populations and mainly involves the class of N-acylphosphatidylethanolamines (NAPEs). Similar to its eCB congeners, PEA acts as local neuroprotective mediator and its physiological tone depends on the finely regulated balance between biosynthesis (mainly catalyzed by NAPE-selective phospholipase D) and degradation (mainly catalyzed by fatty acid amide hydrolase (FAAH) and N-acylethanolamine-hydrolyzing acid amidase) [18–20].

It was proposed that PEA exerts its effects through three mechanisms, which are not mutually exclusive. The first mechanism advances that PEA acts by down-regulating mast-cell degranulation, via an ALIA effect [21–23]; the second one, the entourage effect, postulates that PEA acts by enhancing the anti-inflammatory and anti-nociceptive effects exerted by AEA [4,24,25]; and finally, the third one, the "receptor mechanism", is based on PEA's capability to directly stimulate either PPAR-α or the orphan receptor G-protein coupling, GPR55, which mediates many anti-inflammatory effects [26,27].

PEA lacks a direct antioxidant capacity to prevent the formation of free radicals and counteract the damage of DNA, lipids, and proteins [1]. With its lipid structure and the large size of heterogeneous particles in the naïve state, PEA has limitations in terms of solubility and bioavailability. To overcome these problems, PEA has been micronized (m-PEA) or ultra-micronized (um-PEA) [28]. Several in vitro and in vivo preclinical studies attest that PEA, especially in its micrometer-sized crystalline forms, may be a therapeutic agent for the effective treatment of neuroinflammatory pathologies [29]. m-PEA and um-PEA show enhanced rate of dissolution and absorption [30], better bioavailability, pharmacokinetics, and efficacy when compared to its naïve form [31,32]. Since, as already mentioned, PEA has no antioxidant effects per se, the combination of PEA's ultra-micronized forms with an antioxidant agent, such as a flavonoid, results in more efficacious forms than either molecule alone, potentiating the pharmacological effects of both compounds [33]. In fact, among the natural molecules with excellent antioxidant and antimicrobial functions there are flavonoids, as firstly luteolin, and also polydatin, quercetin, and silymarin. These compounds possess marked antioxidant and neuroprotective pharmacological actions, by modulating apoptosis and release of cytokines and free radicals (reactive oxygen and nitrogen species), suppressing the production of tumor necrosis factor alpha, inhibiting autophagy, and controlling signal transduction pathways [1,34]. In particular, luteolin is able to improve the PEA morphology: while naïve PEA has a morphology featured by large flat crystals, very small quantities of luteolin stabilize the microparticles by inhibiting the PEA crystallization process [35]. The combination of PEA and luteolin makes co-um-PEALut a product able to tackle several neuroinflammatory conditions, and to have protective effects [33].

1.2. PEA Action in the Presence of Aging and Neurodegeneration

Aging is the result of a continuous interaction between biological mechanisms and environmental factors, such as life events, health conditions, and lifestyle habits. Although aging is not necessarily synonymous with disease, the deterioration in cell function that increases with advancing age progressively increments the risk of developing disease and disability, because bodily and brain cellular responses become less and less efficient [36]. Namely, aging is characterized by gradual and permanent accumulation of cellular and molecular damage (such as abnormal protein dynamics, mitochondrial dysfunction, DNA damage, oxidative stress, neurotrophin dysfunction), progressive structural changes of neurons (deregulation of neurotransmitters and neuro-signals), loss of tissue and organ function, and neuroinflammatory processes [37–39]. Unlike the normally beneficial acute inflammatory response, chronic neuroinflammation can lead to damage and destruction of tissues, and often results from inappropriate immune responses [40]. A fundamental principle behind neuroinflammation is the existence of numerous signaling pathways between glial cells and immune system. Notably, despite different triggering events, a common feature of several central and peripheral neuropathologies is chronic immune activation, particularly of the microglia, the resident macrophages of the central nervous system [41]. Individual neurodegenerative disorders are heterogeneous in etiopathogenesis and symptomatology, but biomedical research has revealed many similarities among them at the subcellular level. These similarities suggest that therapeutic advances against one neurodegenerative disease might ameliorate other diseases as well [42].

The most common neurodegenerative diseases encompass a wide range of conditions which impair mobility, muscle coordination and strength, mood, and cognition. They are amyloidosis, tauopathies, α-synucleinopathies, proteopathies (TAR DNA-binding protein 43, TDP-43), and include Alzheimer's Disease (AD), Parkinson's Disease (PD), Huntington's Disease (HD), Frontotemporal Dementia (FTD), Amyotrophic Lateral Sclerosis (ALS), and Multiple Sclerosis (MS) (Figure 1) [43].

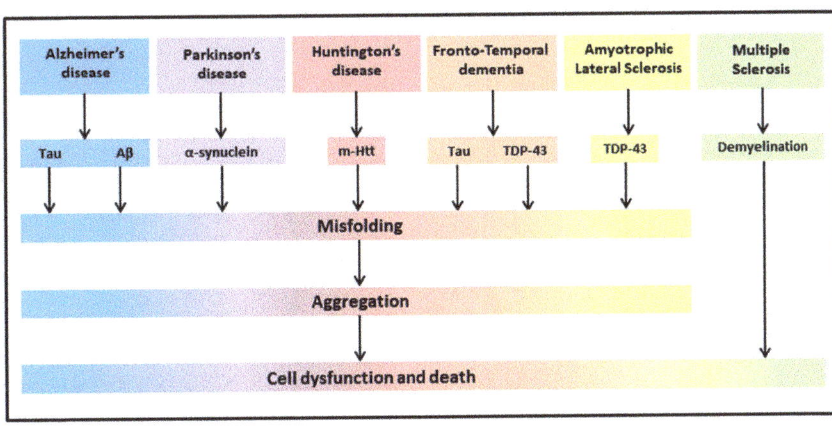

Figure 1. Neurodegenerative diseases share common pathological hallmarks leading to cell dysfunction and death.

Up to now, the treatment of most of these neurodegenerative diseases was mainly symptomatic (dopaminergic treatment for PD, inhibitors of acetylcholinesterase for cognitive disorders, antipsychotics for dementia), despite significant attempts to find drugs reducing or rescuing the debilitating symptoms [44–46]. In this context, integrative treatments of these neurodegenerative diseases have been investigated through a number of in vitro and in vivo animal models of disease, and, when combined with classical drug therapies, are in the frontline of research in an attempt to protect against neuroinflammation and oxidative stress, and thereby improve symptomatology of the neurodegenerative patients [45]. Since most clinical studies on PEA are related to neuropathic pain or inflammation-related peripheral conditions, and there are fewer studies evaluating the possible beneficial effects of PEA on neurodegenerative diseases, we were interested to offer a general overview of the effects of PEA on different symptoms of neurodegeneration, taking into account both human (Table 1) and rodent (Table 2) studies.

Table 1. Summary of human studies using PEA in the presence of neurodegeneration.

Study	Disease	Sample	um PEA (Alone or In Combination)	Dosage	Duration	Main Outcomes of PEA Treatment
[47]	MCI	1 patient	co-um-PEALut	700/70 mg daily	T3: 3 months treatment T9: 9 months follow-up	T3: mild (though not significant) cognitive improvement; T9: near-normal neuropsychological assessment; improvement in test scores; brain SPECT near-normal.
[48]	PD	30 patients	PEA added to regular levodopa	600 mg daily	12 months	Progressive reduction in the total MDS-UPDRS score; reduction in most nonmotor and motor symptoms.
[49]	PD	1 patient	co-um-PEALut added to regular car-bidopa/levodopa	700/70 mg daily	4 months	Complete resolution of leg and trunk dyskinesia and marked reduction in the onset of camptocormia during the "off" state.
[50]	FTD	17 patients	co-um-PEALut	700 mg/2 daily	4 weeks	Improvement in test scores and neurophysiological evaluation; increase in TMS-evoked frontal lobe activity and of high-frequency oscillations in the beta/gamma range.
[51]	ALS	1 patient	PEA	600 mg/2 daily	~40 days	Improvement in clinical picture.
[52]	ALS	28 treated and 36 untreated patients	PEA + 50 mg riluzole or 50 mg riluzole only	600 mg/2 daily	6 months	Lower decrease in forced vital capacity over time as compared with untreated ALS patients.
[53]	MS	24 patients 17 healthy controls	eCBs levels in blood	–	–	eCB system is altered in MS.
[54]	MS	1 patient	PEA	600 mg/2 daily	~9 months	Pain reduction; increased interval between acupuncture sessions.
[55]	MS	29 patients	PEA added to IFN-β1a or placebo	600 mg daily	12 months	Improvement in pain sensation, no reduction of erythema at the injection site, improved evaluation of quality of life, increase in PEA, AEA and OEA plasma levels, reduction of interferon-γ, tumor necrosis factor-α, and interleukin-17 serum profile.
[56]	Myasthenia gravis	22 patients	PEA	600 mg/2 daily	1 week	Reduced level of disability and decremental muscle response.

AEA-Anandamide; ALS-Amyotrophic Lateral Sclerosis; co-um-PEALut-combined ultra-micronized PEA/Lutein; eCB-endocannabinoid; FTD-Frontotemporal Dementia; IFN-β1-Interferon-beta-1; MCI-Mild Cognitive Impairment; MDS-UPDRS-Movement Disorder Society-Unified Parkinson's Disease Rating Scale; MS-Multiple Sclerosis; OEA-Oleoylethanolamide; PD-Parkinson Disease; um-ultra-micronized.

Table 2. Summary of experimental studies using PEA in the presence of neurodegeneration.

Study	Disease	Sample	um PEA (Alone or In Combination)	Dosage	Duration	Main Outcomes of PEA Treatment
[57]	AD model (Aβ 1–42 intra-hippocampal injection)	Male adult Sprague-Dawley rats (9–12/group)	i.p. PEA PEA added to GW6471	PEA:10 mg/kg; GW647: 2 mg/kg	7 days	Restoration of Aβ 1–42-induced alterations; reduced mnestic deficits.
[58]	AD model (Aβ 25–35 i.c.v. injection)	Male PPAR-α/(B6.129S4-SvJaePparatm 1Gonz) and WT mice (9–10/group)	s.c. PEA and GW7647	PEA: 3–30 mg/kg daily, GW7647: 5 mg/kg daily	1–2 weeks or a single dose	Reduction (10 mg/kg) or prevention (30 mg/kg) of behavioral impairments. No rescue of memory deficits. PEA acute treatment was ineffective.
[59]	AD model	3-month-old male 3 × Tg-AD and WT mice (9–10/group)	s.c. PEA or vehicle	10 mg/kg daily	90 days	Counteraction of disease progression, improvement of trophic support to neurons, in the absence of astrocytes and neuronal toxicity.
[60]	AD model	3-month-old or 9-month-old male 3 × Tg-AD or WT mice (7–11/group)	s.c. PEA or vehicle	10 mg/kg daily	90 days	Improvement of learning and memory, amelioration of depressive and anhedonia-like symptoms, reduced Aβ formation, tau protein phosphorylation, promotion of hippocampal neuronal survival and astrocytic function, rebalancing of glutamatergic transmission, restraint of neuroinflammation.
[61]	AD model	2-month-old male 3 × Tg-AD or WT mice (7–11/group)	oral PEA or vehicle	single dose/sub-chronic/chronic: 100 mg/kg daily	1–8–90 days	Rescue of cognitive deficit, restraint of neuroinflammation and oxidative stress, reduced increase in hippocampal glutamate levels.
[62]	PD model (MPTP)	6–7-week-old male PPAR-αKO PPAR-αWT mice (10/group)	i.p. PEA	10 mg/kg	8 days	Reduction of MPTP-induced microglial activation, glial fibrillary acidic protein positive expression astrocyte numbers, overexpression of S100b; protection against alterations in microtubule-associated protein 2a,b, dopamine transporter, nNOS-positive cells in the substantia nigra. Reversal of motor deficits.
[63]	PD model (MPTP)	3/21-month-old male CD1 mice (10/group)	oral PEA	10 mg/kg	60 days	Amelioration of behavioral deficits and of reduction of tyrosine hydroxylase and dopamine transporter in substantia nigra. Reduction of hippocampal proinflammatory cytokines and pro-neurogenic effects.

Table 2. *Cont.*

Study	Disease	Sample	um PEA (Alone or In Combination)	Dosage	Duration	Main Outcomes of PEA Treatment
[64]	PD model (6-OHDA)	Ten-week-old male Swiss CD1 mice (6 × group)	s.c. PEA or GW7647	PEA 3–30 mg/kg/day; GW7647 5 mg/kg/day	28 days	Improvement of behavioral impairment. Increased tyrosine hydroxylase expression at striatal level. Reduction in the expression of pro-inflammatory enzymes, protective scavenging effect.
[65]	PD model (MPTP)	8-week-old male C57BL/6 (10/group)	i.p. co-um-PEALut	1 mg/kg daily	8 days	Reduction of motor impairment, cataleptic response, immobility and anxiety levels. Reduction of neuronal degeneration and of specific PD markers, attenuation of inflammatory processes (activation of astrocytes, pro-inflammatory cytokines, and nitric oxide synthase), stimulation of autophagy.
[66]	PD model (MPTP)	8-week-old male C57BL/6 (10/group)	oral PEA-OXA or vehicle	10 mg/kg daily	8 days	Prevention of MPTP-induced bradykinesia and anxiety, and neuronal degeneration of the dopaminergic tract, prevention of dopamine depletion, modulation of microglia and astrocyte activation.
[67]	HD model	~32-day-old-R6/2 10-week-old R6/2 mice and WT mice (4/group)	Measurement of PEA, AEA and 2-AG endogenous levels	–	–	Alteration of the eCB system, decreased levels of PEA in the striatum
[68]	MS model (EAE)	12-week-old female C57BL/6 (8/group)	i.p. PEA or CBD or in combination	PEA 5 mg/kg CBD 5 mg/kg	3 days	Reduced severity of EAE neurobehavioral scores, diminished inflammation, demyelination, axonal damage and inflammatory cytokine expression.
[69]	MS model (chronic relapsing EAE)	Biozzi ADH mice (>6/group)	i.v. or i.p. PEA	1–10 mg/kg	Single injection	Amelioration of spasticity
[70]	MS model (EAE)	C57BL/6 mice (8/group)	i.p. co-um-PEALut or vehicle	0.1, 1, and 5 mg/kg	16 days	Dose-dependent improvement of clinical signs through anti-inflammatory signals and pro-resolving circuits.
[71]	MS model (TMEV-IDD)	Four-week female SJL/J mice	i.p. PEA or vehicle	5 mg/kg	10 days	Reduction of motor disability, anti-inflammatory effect.

Table 2. *Cont.*

Study	Disease	Sample	um PEA (Alone or In Combination)	Dosage	Duration	Main Outcomes of PEA Treatment
[72]	Vascular dementia	CD1 mice	Oral PEA-OXA or vehicle	10 mg/kg daily	15 days	Improvement of behavioral deficits, reduction of histological alterations, decrease of markers of astrocyte and microglia activation and oxidative stress, modulation of antioxidant response, inhibition of apoptotic process.

2-AG-2-Arachidonoylglycerol; 6-OHDA-6-hydroxydopamine; Aβ-amyloid beta; CBD-cannabidiol; EAE-Experimental Autoimmune Encephalomyelitis; i.c.v.-intracerebroventricular; i.p.-intraperitoneal; KO-knockout; MPTP-1-methyl-4-phenyl-1,2,3,6-tetrahydropyridine; nNOS-neuronal Nitric Oxide Synthase; PEA-OXA-2-pentadecyl-2-oxazoline; PPAR-α-peroxisome proliferator-activated receptor-α; s.c.-subcutaneous; TMEV-IDD-Theiler's Murine Encephalomyelitis Virus-Induced Demyelinating Disease; WT-Wild Type.

2. Mild Cognitive Impairment (MCI)

MCI, a neurocognitive disorder often prodromic to dementia, is a cognitive impairment beyond that expected for individual's age and education, but which does not significantly interfere with activities of daily living [73,74]. There are several subtypes of MCI, which differ based on the type and number of impaired cognitive abilities. When only memory is affected, MCI is defined "amnestic" (a-MCI). When one or more cognitive domains other than memory are affected, MCI is defined "non-amnestic" [75,76].

Only one study evaluated PEA's effects on MCI symptomatology, describing the efficacy of a 9-month supplementation of co-um-PEALut in an a-MCI patient. While, after 3-months of supplementation, a mild (although non-significant) cognitive amelioration was recorded, at the end of treatment the neuropsychological evaluation reached values almost typical for age and education level, and even the brain SPECT was almost within the normal range [47]. The findings of this case report allow hypothesizing that the use of co-um-PEALut may be a valuable option in the management of MCI-associated neuroinflammation.

3. Alzheimer's Disease (AD)

AD is characterized by progressive and irreversible brain atrophy that spreads unevenly throughout the brain, encompassing particularly vulnerable regions, such as entorhinal cortex, hippocampus, frontal cortex, temporal lobe, and forebrain basal nuclei [77–79]. The clinical manifestations are memory loss, cognitive decline, behavioral dysfunction, and inability to carry out the activities of daily living, and seem to be related to the impairment of cholinergic transmission [79,80]. Histopathologically, the main features of AD include extracellular accumulation of amyloid beta (Aβ) fibrils in senile plaques and intraneuronal neurofibrillary tangles (aggregates of the microtubule-associated hyperphosphorylated protein tau) [77].

Several in vitro studies have demonstrated that PEA is able to reduce Aβ-evoked neuroinflammation and attenuate its neurodegenerative consequences. To this purpose, rat primary astrocytes, rat primary mixed neuroglial co-cultures, and organotypic hippocampal slices were challenged with Aβ1-42, and then treated with PEA [81–83]. PEA was able to dampen Aβ-induced astrocyte activation and improve neuronal survival through selective PPAR-α activation. Data from organotypic cultures confirmed PEA's anti-inflammatory properties, implicating PPAR-α mediation, and revealing that the reduction of reactive gliosis induced a marked rebound neuroprotective effect on neurons [82]. Subsequently, the anti-inflammatory and neuroprotective activities of systemically administered PEA were analyzed in adult rats given inflammatory Aβ1-42 intrahippocampal injections mimicking some early traits of AD. By activating PPAR-α, PEA rescued altered molecular pathways and behavioral impairment [57]. Such a capacity of PEA to modulate the protective responses during inflammation suggests that endogenous PEA may be part of the complex homeostatic apparatus controlling the basal threshold of inflammatory processes. Moreover, by blunting Aβ42-induced astrocyte activation, PEA improved neuronal survival [84], cell viability, and glutamatergic transmission in mouse astrocyte-neuron co-cultures [85]. Neuroprotective activities of PEA were studied in mice intracerebroventricularly injected with Aβ25-35, and tested on spatial and non-spatial memory tasks [58]. PEA reduced or even prevented, in a dose-dependent manner, learning, and memory dysfunction. Furthermore, PEA treatment reduced molecular and biochemical inflammatory markers (lipid peroxidation, protein nitrosylation, induction of inducible nitric oxide synthase, and caspase3 activation) [58], counteracted neurodegeneration, and enhanced neuronal viability by dampening reactive astrogliosis and promoting the glial neuro-support function [59]. An additional study provided evidence that subcutaneous treatment with um-PEA is able to exert anti-inflammatory and neuroprotective effects in AD-like mice. Specifically, um-PEA improved learning and memory functions, ameliorated depression and anhedonia, reduced Aβ formation and tau protein phosphorylation, promoted neuronal survival in CA1 hippocampal subregion, normalized astrocytic functions, rebalanced glutamatergic transmission, and restrained neuroinflammation. The um-PEA efficacy was particularly evident in younger mice (6- vs. 12-month-old), suggesting its potential as an early treatment of AD [60]. In the same AD-like model, um-PEA chronic oral administration was capable of rescuing cognitive deficit and decreasing hippocampal level of extracellular glutamate. Once again, the significant neuroinflammatory effect of PEA was found, as shown by the almost complete inhibition of an increase in IL-6 in the hippocampus, by the reduced oxidative stress, and by lower signs of neuronal distress [61,86]. The use of co-ultra-PEALut in an in vitro study also showed interesting results. Pre-treatment significantly reduced inducible nitric oxide synthase and glial fibrillary acidic protein expression, restored neuronal nitric oxide synthase and brain-derived neurotrophic factor, and reduced apoptosis [87]. These results suggest that PEA alone and in combination with lutein may provide an effective strategy for AD.

4. Parkinson's Disease (PD)

PD is a common neurodegenerative disorder primarily characterized by shaking, stiffness, difficulties with balance, walking, and coordination due to the impairment of the dopaminergic nigrostriatal system [88]. Specifically, the loss of dopaminergic neurons projecting from the substantia nigra pars compacta to the caudate-putamen in the striatum results in the loss of dopaminergic neurotransmission, causing the primary motor symptoms [89,90]. Although PD was initially described as a movement disorder without dementia, it is now accepted that the progression of PD affects other dopaminergic, cholinergic, or serotonergic extra-nigral traits, leading to the appearance of cognitive and psychiatric symptoms (i.e., depression and dementia) as well as other non-motor symptoms (i.e., anosmia, sleep disturbances, and constipation) [91]. The ultimate underlying cause(s) of dopaminergic impairment remain(s) unknown [92]. While 5–10% of PD cases are of genetic origin, leading to early PD onset, most cases remain idiopathic and are associated with aging [93], and consequently with neuroinflammation [94]. However, PD must be differentiated from other parkinsonian disorders, including vascular (i.e., striatal infarction), drug-induced (i.e., neuroleptics), metabolic (i.e., Wilson's disease), infectious (i.e., HIV), toxic (i.e., carbon monoxide), normal pressure hydrocephalus, essential tremor, and other forms of neurodegenerative disease [95].

Given its anti-inflammatory effect, the efficacy of PEA in controlling neurodegeneration associated with neuroinflammation has been evaluated in the animal model of PD induced by 1-methyl-4-phenyl-1,2,3,6-tetrahydropyridine (MPTP) injections. In this model, that exhibits biochemical and cellular changes similar to those observed in idiopathic PD [96], PEA treatment protected against MPTP-induced neurotoxicity. Namely,

PEA reduced microglial and astrocyte activation as well as oxidative stress, protected against alterations in protein dynamics in the substantia nigra, and reversed motor deficits. Interestingly, in PPAR-αKO mice treated with MPTP, the lack of PPAR-α exacerbated toxicity of MPTP, indicating that PEA neuroprotection was, at least partially, dependent on PPAR-α [62]. In a more recent study, m-PEA pre-treatment in the MPTP model improved behavioral deficits (measured in open field, pole test, and elevated plus maze), improved tyrosine hydroxylase and dopamine transporter expression (index of PEA protective effect on dopaminergic neurons), reduced pro-inflammatory cytokines, and enhanced hippocampal neurogenesis [63]. These studies reveal that PEA may represent a potential therapeutic candidate to prevent neurodegeneration in the pathogenesis of PD, and propose PEA as a valuable nutraceutical approach to prevent neurodegenerative diseases associated with age. Furthermore, PEA treatment improved the motor disturbances induced by unilateral intra-striatal 6-hydroxydopamine (6-OHDA) injections, increased tyrosine hydroxylase expression at striatal level, reduced expression of pro-inflammatory enzymes, and modulated pro- and anti-apoptotic markers (a valuable index of PEA's ability to control neuroinflammation and cell death). In addition, PEA provoked a marked protective scavenging and dampened unfolding protein response. Similar data were found in the in vitro studies, where PEA inhibited the damaging stress response of the endoplasmic reticulum [64]. As previously pinpointed, the association of um-PEA and lutein has a strong neuroprotective activity. In fact, co-um-PEALut reduced the specific PD markers (immune-positive tyrosine hydroxylase) and the increased levels in activated astrocytes and pro-inflammatory cytokines, as well as inducible nitric oxide synthase, and modulated the autophagy pathway [65]. More recently, Cordaro and colleagues explored a novel neuroprotective approach based on the use of 2-pentadecyl-2-oxazoline (PEA-OXA). In the MPTP-induced PD model, daily oral PEA-OXA supplementation reduced the behavioral disturbances and neurodegeneration of the dopaminergic tract. Namely, PEA-OXA prevented dopamine depletion, increased tyrosine hydroxylase and dopamine transporter activity, decreased α-synuclein neuronal aggregation, expression of pro-inflammatory enzymes, and inducible expression of nitric oxide synthase. Furthermore, treatment with PEA-OXA significantly limited MPTP-induced microtubule-associated protein-2 alteration, and strongly reduced the increased activation of astrocytes and microglia [66].These preclinical studies on experimental models of PD indicate that m-PEA and um-PEA (alone or in association with luteolin or oxazoline) are effective in improving motor functionality through mechanisms aimed at controlling neuroinflammation and neuroprotection, based on an increase in tyrosine hydroxylase and dopamine activity, as well as on the reduction of reactive glia.

As for the human studies, two recent studies highlighted the neuroprotective PEA effect in subjects affected by PD. In particular, in an observational study, Brotini and colleagues demonstrated that oral supplementation with um-PEA slowed disease and disability score progression, significantly reducing most motor and non-motor symptoms [48]. Recently, Brotini has reported the potential efficacy of co-um-PEALut treatment as adjuvant therapy for patients with PD receiving carbidopa/levodopa in treating camptocormia, a postural deformity common in PD in which the spine bends forward while walking or standing. In particular, a 4-month treatment program resulted in complete resolution of the legs and trunk dyskinesia, and marked reduction in camptocormia onset [49]. Therefore, PEA, alone or co-administered and possibly integrating classical treatments, shows therapeutic potential in PD, by correcting dopaminergic deficits and motor dysfunctions, and contrasting pathogenetic aspects involved in the development of the disease.

5. Huntington's Disease (HD)

HD is a rare devastating autosomal dominant disorder, caused by a CAG trinucleotide repeat expansion in the *huntingtin* gene on chromosome 4, which leads to the production of the mutant huntingtin (m-Htt) protein. The degree of symptom severity, disease stage, and markers of neuronal damage correlate with levels of the m-Htt protein in the cerebrospinal fluid in patients with HD [97]. HD is characterized by loss of medium spiny neurons and astrogliosis. The first region to be substantially affected is the striatum, followed by frontal and temporal cortices. The reduced signals from the striatum to the subthalamic nucleus (which sends signals to the globus pallidus) provoke a reduced modulation of movement that results in the characteristic HD movements, called chorea. Besides motor disturbances, HD is characterized by dementia, psychiatric symptoms, and early death. The average age of HD diagnosis is about 40 years, even if the timing of onset is partially determined by the CAG repeats number [98]. Despite its well-defined genetic origin, its molecular and cellular mechanisms are still unclear [99]. There is evidence for a strong correlation between HD and the eCB system, given that the dramatic reduction in cannabinoid receptors in all regions of the basal ganglia is one of the earliest and severe alterations in HD [100].

In R6/2 mice, the transgenic model of HD exhibiting 150 CAG repeats and severe signs of HD, a progressive decline in CB1 receptor expression and abnormal sensitivity to CB1 receptor stimulation are reported. In this model, the relationship was determined between HD symptoms and changes in eCB signaling, measuring eCB levels in brain areas involved in HD (striatum, cortex, and hippocampus) at pre-symptomatic and symptomatic disease stages. Only in symptomatic R6/2 mice were the AEA, 2-Arachidonoylglycerol (2-AG) and PEA levels significantly reduced in the striatum, where cell bodies and dendrites of CB1 receptor-containing neurons are located, indicating that the eCB levels changed according to disease stages and brain regions. Thus, the impairment of the eCB system may represent a hallmark of symptomatic HD. Taking into account its anti-inflammatory and neuroprotective effects, the decreased PEA levels in the striatum of symptomatic R6/2 mice may contribute to the progression of the disease [67]. As mentioned before, with PEA being a substrate for FAAH, it was expected that the use of FAAH blockers could improve HD symptoms in R6/2 mice, suggesting that drugs that inhibit eCB degradation could be used to treat HD.

6. Frontotemporal Dementia (FTD)

FTD is a term used for a group of brain disorders characterized by atrophy of the frontal and temporal lobes, and by subsequent behavioral and language disturbances [101]. FTD patients have a characteristic histopathology with cytoplasmic inclusions containing aggregated TDP-43 or tau protein in neurons and glial cells. Signs and symptoms vary depending on which part of the brain is affected. FTD often begins between the ages of 40 and 65 [102].

The only study present in literature on PEA-treated FTD patients reported that a 4-week-treatment with co-um-PEALut reduced behavioral disturbances and improved frontal lobe function in FTD patients through modulation of cortical oscillatory activity and GABAergic transmission [50].

7. Amyotrophic Lateral Sclerosis (ALS)

ALS is a fatal neurodegenerative disease characterized by the degeneration of both upper and lower motoneurons and spinal cord which leads to muscle weakness, progressive paralysis, and death within 3–5 years of diagnosis [103,104]. ALS is accompanied by remarkable spinal inflammation mediated in particular by microglia and mast cells. ALS may share some pathological features with FTD, given that ALS patients often show widespread atrophy of the frontotemporal cortices, including premotor cortex [105,106]. ALS is associated with TDP-43 mutations that are observed in more than 90% of ALS patients.

Oral supplementation with um-PEA, in a patient affected by ALS, led to improvement in the clinical picture, muscle tone, respiratory, and motor functions, due to PEA-control of neuroinflammation [51]. More recently, a large clinical trial on ALS patients demonstrated that 6-month-administration of um-PEA, in addition to standard therapy (riluzole), significantly slowed the decline in lung functionality and the worsening of ALS symptoms. This selective PEA effect on diaphragmatic muscle is probably determined by a direct action on the neuromuscular junctions, rather than a nerve-mediated effect leading to regeneration and sprouting of peripheral nerve fibers [52].

8. Multiple Sclerosis (MS)

MS is a demyelinating disease with a prominent neurodegenerative component, caused by an autoimmune attack provoking the progressive loss of myelin sheath of neuronal axons [107]. Such a damage decreases the speed of signal transduction, resulting in cognitive and motor impairment depending on lesion's location. MS progression is caused by episodes of increasing inflammation due to the release of antigens (i.e., myelin oligodendrocyte glycoprotein) [108–110]. These antigens elicit a clear autoimmune response that, in turn, sets off a cascade of signaling molecules which result in T cells, B cells, and macrophages crossing the blood–brain barrier and attacking axonal myelin. The attack on myelin starts inflammatory processes that trigger other immune cells and the release of soluble factors, such as pro-inflammatory cytokines. Further release of antigens drives subsequent degeneration, causing further inflammation contributing to the loss of the grey matter. Apart from grey matter and white matter lesions, another distinguishing feature of MS is the presence of undifferentiated oligodendrocyte progenitor cells (OPCs), as a consequence of their inability to progress to a myelin-producing phenotype [111]. Thus, an attractive therapeutic strategy may be to replace lost oligodendrocytes and/or promote their maturation.

Studies indicate that co-um-PEALut may improve the symptoms induced by experimental autoimmune encephalomyelitis (EAE), a MS model. Treatment with co-um-PEALut facilitated the development of undifferentiated and differentiating OPCs, as indicated by the increase in myelin basic protein, proposing this co-compound as a novel approach in the treatment of inflammatory demyelinating disorders, such as MS [40,112,113]. Increasing evidence from animal studies suggests that cannabinoids could efficiently fight demyelination, inflammation, and autoimmune processes occurring in pathologies such as MS [68]. In connection with this, it is reported that in MS the human plasma levels of ECs are altered, the EC system is dynamically modulated depending on the disease subtype [53], and the exogenous administration of eCBs, PEA, and selective inhibitors of eCB re-uptake and hydrolysis improves spasticity, through the enhancement of AEA and possibly 2-AG levels [69]. These studies provide definitive evidence of the tonic control of spasticity exerted by the eCB system, and open new horizons for MS therapy. In the same EAE model of MS, the non-psychoactive effects of cannabidiol (CBD) and PEA, in combination or separately, have been evaluated. Results showed that, whereas CBD-PEA concurrent administration was not as effective as treatment, each drug per se, i.e., single PEA (or CBD) intraperitoneal administration, reduced the severity of neurobehavioral impairment, decreased inflammation, demyelination, cytokine expression, and axonal damage, thus suggesting the effectiveness of PEA as no-psychotropic CB against MS [68]. In the same animal model, intraperitoneal administration of co-um-PEALut significantly reduced the severity of clinical signs. The dose-dependent clinical improvement was associated with the reduced expression of genes coding for inflammatory proteins and the receptors involved in inflammation in the brainstem and cerebellum [70]. In another MS model, the Theiler's Murine Encephalomyelitis Virus-Induced Demyelinating Disease (TMEV-IDD), PEA administration counteracted motor deficits and exerted an anti-neuroinflammatory effect by reducing the expression of pro-inflammatory cytokines and decreasing microglial activation [71].

As for human studies, a case report described a MS patient with chronic central neuropathic pain who, after adding PEA to their acupuncture treatment, was able to increase the intervals between acupuncture sessions [54]. In a clinical trial on 29 patients with MS, um-PEA oral supplementation relieved pain at the interferon injection site and significantly reduced plasma concentrations of cytokines. Quality of life of the treated patients also improved, when compared with the placebo-treated group [55]. These findings propose that PEA and co-um-PEALut may be the starting point in a novel approach in the treatment of inflammatory demyelinating disorders, as well as in other CNS pathologies classically viewed as primarily neuronal diseases but where myelin and oligodendrocyte loss are relevant.

9. Other Diseases (Vascular Dementia, Myasthenia Gravis)

Recent studies revealed that PEA-OXA co-treatment reduced histological changes and neuronal death in vascular dementia induced by bilateral carotid artery occlusion. In particular, this treatment decreased astrocyte markers and microglia activation, increased neuronal development markers, reduced oxidative stress, modulated antioxidant response, and inhibited the apoptotic process [72]. One-week of um-PEA supplementation improved the severity of disability and muscle response to fatigue in patients with myasthenia gravis [56].

10. Conclusions

Taken together, the results reported here strongly suggest that by activating multifactorial pharmacological targets and different cellular mediators, PEA could play a promising protective role in counteracting neuroinflammation related to major neurodegenerative diseases. Despite the fact that a simplistic extrapolation of data from the animal model to human condition should be avoided, the results of both preclinical and clinical studies propose PEA as a potential therapeutic agent against neurodegeneration. Mentioned studies emphasize that PEA, especially in its ultra-micronized form, significantly impacts on the progression of some neurodegenerative diseases, acting on specific symptoms and when the pathology is at an early stage. In the final analysis, there is still much to be investigated regarding the effect of PEA, alone or in combination with other compounds, both at the preclinical level and, especially, on subjects with PD, MS, ALS, and FTD. More controversial and less clear is its role in HD. Given the numerous preclinical results regarding the efficacy of PEA in in vitro studies of AD, it would be very interesting to investigate how effective this compound really could be in AD and MCI patients.

In conclusion, through the activation of specific receptors PEA plays a protective role against neuroinflammation related to neurodegeneration, but it is necessary to continue to study the properties of this extraordinary ALIAmide to have a clearer picture on its possible use, also because, to date, no side effects have been reported.

Author Contributions: All authors designed the review, critically reviewed the contents, and contributed to manuscript preparation. E.L. performed the literature review. C.C. approved the final version of the manuscript. All authors have read and agreed to the published version of the manuscript.

Funding: This research was partially funded by Italian Ministry of Health: Ricerca Corrente (to C.C.) and Ricerca Finalizzata (Grant number RF 2018-12365391 to L.P.).

Institutional Review Board Statement: Not applicable.

Informed Consent Statement: Not applicable.

Data Availability Statement: Not applicable.

Conflicts of Interest: The authors declare no conflict of interest.

References

1. Peritore, A.F.; Siracusa, R.; Crupi, R.; Cuzzocrea, S. Therapeutic Efficacy of Palmitoylethanolamide and Its New Formulations in Synergy with Different Antioxidant Molecules Present in Diets. *Nutrients* **2019**, *11*, 2175. [CrossRef] [PubMed]
2. Chiurchiù, V.; Leuti, A.; Smoum, R.; Mechoulam, R.; Maccarrone, M. Bioactive lipids ALIAmides differentially modulate inflammatory responses of distinct subsets of primary human T lymphocytes. *FASEB J. Off. Publ. Fed. Am. Soc. Exp. Biol.* **2018**, *32*, 5716–5723. [CrossRef] [PubMed]
3. Rahman, I.A.; Tsuboi, K.; Uyama, T.; Ueda, N. New players in the fatty acyl ethanolamide metabolism. *Pharmacol. Res.* **2014**, *86*, 1–10. [CrossRef] [PubMed]

4. Ho, W.S.; Barrett, D.A.; Randall, M.D. 'Entourage' effects of N-palmitoylethanolamide and N-oleoylethanolamide on vasorelaxation to anandamide occur through TRPV1 receptors. *Br. J. Pharmacol.* **2008**, *155*, 837–846. [CrossRef]
5. Musella, A.; Fresegna, D.; Rizzo, F.R.; Gentile, A.; Bullitta, S.; De Vito, F.; Guadalupi, L.; Centonze, D.; Mandolesi, G. A novel crosstalk within the endocannabinoid system controls GABA transmission in the striatum. *Sci. Rep.* **2017**, *7*, 7363. [CrossRef]
6. Costa, B.; Conti, S.; Giagnoni, G.; Colleoni, M. Therapeutic effect of the endogenous fatty acid amide, palmitoylethanolamide, in rat acute inflammation: Inhibition of nitric oxide and cyclo–oxygenase systems. *Br. J. Pharmacol.* **2002**, *137*, 413–420. [CrossRef]
7. Carta, G.; Murru, E.; Banni, S.; Manca, C. Palmitic Acid: Physiological Role, Metabolism and Nutritional Implications. *Front. Physiol.* **2017**, *8*, 902. [CrossRef]
8. Keppel Hesselink, J.M.; de Boer, T.; Witkamp, R.F. Palmitoylethanolamide: A Natural Body–Own Anti–Inflammatory Agent, Effective and Safe against Influenza and Common Cold. *Int. J. Inflam.* **2013**, *2013*, 151028. [CrossRef]
9. Nestmann, E.R. Safety of micronized palmitoylethanolamide (microPEA): Lack of toxicity and genotoxic potential. *Food Sci. Nutr.* **2016**, *5*, 292–309. [CrossRef]
10. Gabrielsson, L.; Mattsson, S.; Fowler, C.J. Palmitoylethanolamide for the treatment of pain: Pharmacokinetics, safety and efficacy. *Br. J. Clin. Pharmacol.* **2016**, *82*, 932–942. [CrossRef]
11. Petrosino, S.; Di Marzo, V. The pharmacology of palmitoylethanolamide and first data on the therapeutic efficacy of some of its new formulations. *Br. J. Pharmacol.* **2017**, *174*, 1349–1365. [CrossRef] [PubMed]
12. Petrosino, S.; Schiano Moriello, A. Palmitoylethanolamide: A Nutritional Approach to Keep Neuroinflammation within Physiological Boundaries–A Systematic Review. *Int. J. Mol. Sci.* **2020**, *21*, 9526. [CrossRef] [PubMed]
13. Petrosino, S.; Cristino, L.; Karsak, M.; Gaffal, E.; Ueda, N.; Tüting, T.; Bisogno, T.; De Filippis, D.; D'Amico, A.; Saturnino, C.; et al. Protective role of palmitoylethanolamide in contact allergic dermatitis. *Allergy* **2010**, *65*, 698–711. [CrossRef] [PubMed]
14. Tsuboi, K.; Uyama, T.; Okamoto, Y.; Ueda, N. Endocannabinoids and related N-acylethanolamines: Biological activities and metabolism. *Inflamm. Regener.* **2018**, *38*, 28. [CrossRef]
15. Esposito, E.; Cuzzocrea, S. Palmitoylethanolamide is a new possible pharmacological treatment for the inflammation associated with trauma. *Mini Rev. Med. Chem.* **2013**, *13*, 237–255.
16. Mattace Raso, G.; Russo, R.; Calignano, A.; Meli, R. Palmitoylethanolamide in CNS health and disease. *Pharmacol. Res.* **2014**, *86*, 32–41. [CrossRef]
17. Skaper, S.D.; Facci, L.; Zusso, M.; Giusti, P. An Inflammation–Centric View of Neurological Disease: Beyond the Neuron. *Front. Cell. Neurosci.* **2018**, *12*, 72. [CrossRef]
18. Solorzano, C.; Zhu, C.; Battista, N.; Astarita, G.; Lodola, A.; Rivara, S.; Mor, M.; Russo, R.; Maccarrone, M.; Antonietti, F.; et al. Selective N-acylethanolamine-hydrolyzing acid amidase inhibition reveals a key role for endogenous palmitoylethanolamide in inflammation. *Proc. Natl. Acad. Sci. USA* **2009**, *106*, 20966–20971. [CrossRef]
19. Tsuboi, K.; Ikematsu, N.; Uyama, T.; Deutsch, D.G.; Tokumura, A.; Ueda, N. Biosynthetic pathways of bioactive N–acylethanolamines in brain. *CNS Neurol. Disord.* **2013**, *12*, 7–16. [CrossRef]
20. Rankin, L.; Fowler, C.J. The Basal Pharmacology of Palmitoylethanolamide. *Int. J. Mol. Sci.* **2020**, *21*, 7942. [CrossRef]
21. Aloe, L.; Leon, A.; Levi-Montalcini, R. A proposed autacoid mechanism controlling mastocyte behaviour. *Agents Actions* **1993**, *39*, C145–C147. [CrossRef] [PubMed]
22. Paladini, A.; Varrassi, G.; Bentivegna, G.; Carletti, S.; Piroli, A.; Coaccioli, S. Palmitoylethanolamide in the Treatment of Failed Back Surgery Syndrome. *Pain Res. Treat.* **2017**, *2017*, 1486010. [CrossRef] [PubMed]
23. Roviezzo, F.; Rossi, A.; Caiazzo, E.; Orlando, P.; Riemma, M.A.; Iacono, V.M.; Guarino, A.; Ialenti, A.; Cicala, C.; Peritore, A.; et al. Palmitoylethanolamide Supplementation during Sensitization Prevents Airway Allergic Symptoms in the Mouse. *Front. Pharmacol.* **2017**, *8*, 857. [CrossRef] [PubMed]
24. Conti, S.; Costa, B.; Colleoni, M.; Parolaro, D.; Giagnoni, G. Antiinflammatory action of endocannabinoid palmitoylethanolamide and the synthetic cannabinoid nabilone in a model of acute inflammation in the rat. *Br. J. Pharmacol.* **2002**, *135*, 181–187. [CrossRef] [PubMed]
25. Saturnino, C.; Popolo, A.; Ramunno, A.; Adesso, S.; Pecoraro, M.; Plutino, M.R.; Rizzato, S.; Albinati, A.; Marzocco, S.; Sala, M.; et al. Anti-Inflammatory, Antioxidant and Crystallographic Studies of N-Palmitoyl-ethanol Amine (PEA) Derivatives. *Molecules* **2017**, *22*, 616. [CrossRef]
26. Warden, A.; Truitt, J.; Merriman, M.; Ponomareva, O.; Jameson, K.; Ferguson, L.B.; Mayfield, R.D.; Harris, R.A. Localization of PPAR isotypes in the adult mouse and human brain. *Sci. Rep.* **2016**, *6*, 27618. [CrossRef]
27. Shi, Q.X.; Yang, L.K.; Shi, W.L.; Wang, L.; Zhou, S.M.; Guan, S.Y.; Zhao, M.G.; Yang, Q. The novel cannabinoid receptor GPR55 mediates anxiolytic-like effects in the medial orbital cortex of mice with acute stress. *Mol. Brain* **2017**, *10*, 38. [CrossRef]
28. D'Amico, R.; Impellizzeri, D.; Cuzzocrea, S.; Di Paola, R. ALIAmides Update: Palmitoylethanolamide and Its Formulations on Management of Peripheral Neuropathic Pain. *Int. J. Mol. Sci.* **2020**, *21*, 5330. [CrossRef]
29. Beggiato, S.; Tomasini, M.C.; Ferraro, L. Palmitoylethanolamide (PEA) as a Potential Therapeutic Agent in Alzheimer's Disease. *Front. Pharmacol.* **2019**, *10*, 821. [CrossRef]
30. Ardizzone, A.; Fusco, R.; Casili, G.; Lanza, M.; Impellizzeri, D.; Esposito, E.; Cuzzocrea, S. Effect of Ultra-Micronized-Palmitoylethanolamide and Acetyl-l-Carnitine on Experimental Model of Inflammatory Pain. *Int. J. Mol. Sci.* **2021**, *22*, 1967. [CrossRef]

31. Impellizzeri, D.; Bruschetta, G.; Cordaro, M.; Crupi, R.; Siracusa, R.; Esposito, E.; Cuzzocrea, S. Micronized/ultramicronized palmitoylethanolamide displays superior oral efficacy compared to nonmicronized palmitoylethanolamide in a rat model of inflammatory pain. *J. Neuroinflammation* **2014**, *11*, 136. [CrossRef] [PubMed]
32. Noce, A.; Albanese, M.; Marrone, G.; Di Lauro, M.; Pietroboni Zaitseva, A.; Palazzetti, D.; Guerriero, C.; Paolino, A.; Pizzenti, G.; Di Daniele, F.; et al. Ultramicronized Palmitoylethanolamide (um-PEA): A New Possible Adjuvant Treatment in COVID-19 patients. *Pharmaceuticals* **2021**, *14*, 336. [CrossRef] [PubMed]
33. Cordaro, M.; Cuzzocrea, S.; Crupi, R. An Update of Palmitoylethanolamide and Luteolin Effects in Preclinical and Clinical Studies of Neuroinflammatory Events. *Antioxidants* **2020**, *9*, 216. [CrossRef] [PubMed]
34. Ashaari, Z.; Hadjzadeh, M.A.; Hassanzadeh, G.; Alizamir, T.; Yousefi, B.; Keshavarzi, Z.; Mokhtari, T. The flavone luteolin improves central nervous system disorders by different mechanisms: A review. *J. Mol. Neurosci.* **2018**, *65*, 491–506. [CrossRef] [PubMed]
35. Adami, R.; Liparoti, S.; Di Capua, A.; Scogliamiglio, M.; Reverchon, E. Production of PEA composite microparticles with polyvinylpyrrolidone and luteolin using supercritical assisted atomization. *J. Supercrit. Fluids.* **2019**, *143*, 82–89. [CrossRef]
36. Power, R.; Prado-Cabrero, A.; Mulcahy, R.; Howard, A.; Nolan, J.M. The Role of Nutrition for the Aging Population: Implications for Cognition and Alzheimer's Disease. *Annu. Rev. Food Sci. Technol.* **2019**, *10*, 619–639. [CrossRef]
37. Vauzour, D.; Camprubi-Robles, M.; Miquel-Kergoat, S.; Andres-Lacueva, C.; Bánáti, D.; Barberger-Gateau, P.; Bowman, G.L.; Caberlotto, L.; Clarke, R.; Hogervorst, E.; et al. Nutrition for the ageing brain: Towards evidence for an optimal diet. *Ageing Res. Rev.* **2017**, *35*, 222–240. [CrossRef]
38. Kwon, H.S.; Koh, S.H. Neuroinflammation in neurodegenerative disorders: The roles of microglia and astrocytes. *Transl. Neurodegener.* **2020**, *9*, 42. [CrossRef]
39. Azam, S.; Haque, M.E.; Balakrishnan, R.; Kim, I.S.; Choi, D.K. The Ageing Brain: Molecular and Cellular Basis of Neurodegeneration. *Front. Cell. Dev. Biol.* **2021**, *9*, 683459. [CrossRef]
40. Skaper, S.D.; Barbierato, M.; Facci, L.; Borri, M.; Contarini, G.; Zusso, M.; Giusti, P. Co–Ultramicronized Palmitoylethanolamide/Luteolin Facilitates the Development of Differentiating and Undifferentiated Rat Oligodendrocyte Progenitor Cells. *Mol. Neurobiol.* **2018**, *55*, 103–114. [CrossRef]
41. Jakel, S.; Dimou, L. Glial Cells and Their Function in the Adult Brain: A Journey through the History of Their Ablation. *Front. Cell. Neurosci.* **2017**, *11*, 24. [CrossRef]
42. Hussain, R.; Zubair, H.; Pursell, S.; Shahab, M. Neurodegenerative Diseases: Regenerative Mechanisms and Novel Therapeutic Approaches. *Brain Sci.* **2018**, *8*, 177. [CrossRef] [PubMed]
43. Dugger, B.N.; Dickson, D.W. Pathology of Neurodegenerative Diseases. *Cold Spring Harb. Perspect. Biol.* **2017**, *9*, a028035. [CrossRef] [PubMed]
44. Desai, A.K.; Grossberg, G.T. Diagnosis and treatment of Alzheimer's disease. *Neurology* **2005**, *64*, S34–S39. [CrossRef] [PubMed]
45. Chaudhuri, K.R.; Schapira, A.H.V. Non-motor symptoms of Parkinson's disease: Dopaminergic pathophysiology and treatment. *Lancet Neurol.* **2009**, *8*, 464–474. [CrossRef]
46. Rees, K.; Stowe, R.; Patel, S.; Ives, N.; Breen, K.; Clarke, C.E.; Ben-Shlomo, Y. Non-steroidal anti-inflammatory drugs as disease-modifying agents for Parkinson's disease: Evidence from observational studies. *Cochrane Database Syst Rev.* **2011**, Cd008454. [CrossRef] [PubMed]
47. Calabrò, R.S.; Naro, A.; De Luca, R.; Leonardi, S.; Russo, M.; Marra, A.; Bramanti, P. PEALut efficacy in mild cognitive impairment: Evidence from a SPECT case study! *Aging Clin. Exp. Res.* **2016**, *28*, 1279–1282. [CrossRef]
48. Brotini, S.; Schievano, C.; Guidi, L. Ultra–micronized Palmitoylethanolamide: An Efficacious Adjuvant Therapy for Parkinson's Disease. *CNS Neurol. Disord. Drug Targets* **2017**, *16*, 705–713. [CrossRef]
49. Brotini, S. Palmitoylethanolamide/Luteolin as Adjuvant Therapy to Improve an Unusual Case of Camptocormia in a Patient with Parkinson's Disease: A Case Report. *Innov. Clin. Neurosci.* **2021**, *18*, 12–14.
50. Assogna, M.; Casula, E.P.; Borghi, I.; Bonnì, S.; Samà, D.; Motta, C.; Di Lorenzo, F.; D'Acunto, A.; Porrazzini, F.; Minei, M.; et al. Effects of Palmitoylethanolamide Combined with Luteoline on Frontal Lobe Functions, High Frequency Oscillations, and GABAergic Transmission in Patients with Frontotemporal Dementia. *J. Alzheimers Dis.* **2020**, *76*, 1297–1308. [CrossRef]
51. Clemente, S. Amyotrophic lateral sclerosis treatment with ultramicronized palmitoylethanolamide: A case report. *CNS Neurol. Disord. Drug Targets* **2012**, *11*, 933–936. [CrossRef] [PubMed]
52. Palma, E.; Reyes–Ruiz, J.M.; Lopergolo, D.; Roseti, C.; Bertollini, C.; Ruffolo, G.; Cifelli, P.; Onesti, E.; Limatola, C.; Miledi, R.; et al. Acetylcholine receptors from human muscle as pharmacological targets for ALS therapy. *Proc. Natl. Acad. Sci. USA.* **2016**, *113*, 3060–3065. [CrossRef] [PubMed]
53. Jean-Gilles, L.; Feng, S.; Tench, C.R.; Chapman, V.; Kendall, D.A.; Barrett, D.A.; Constantinescu, C.S. Plasma endocannabinoid levels in multiple sclerosis. *J. Neurol. Sci.* **2009**, *287*, 212–215. [CrossRef] [PubMed]
54. Kopsky, D.J.; Hesselink, J.M. Multimodal stepped care approach with acupuncture and PPAR–α agonist palmitoylethanolamide in the treatment of a patient with multiple sclerosis and central neuropathic pain. *Acupunct. Med.* **2012**, *30*, 53–55. [CrossRef]
55. Orefice, N.S.; Alhouayek, M.; Carotenuto, A.; Montella, S.; Barbato, F.; Comelli, A.; Calignano, A.; Muccioli, G.G.; Orefice, G. Oral Palmitoylethanolamide Treatment Is Associated with Reduced Cutaneous Adverse Effects of Interferon–β1a and Circulating Proinflammatory Cytokines in Relapsing–Remitting Multiple Sclerosis. *Neurotherapeutics* **2016**, *13*, 428–438. [CrossRef]

56. Onesti, E.; Frasca, V.; Ceccanti, M.; Tartaglia, G.; Gori, M.C.; Cambieri, C.; Libonati, L.; Palma, E.; Inghilleri, M. Short–Term Ultramicronized Palmitoylethanolamide Therapy in Patients with Myasthenia Gravis: A Pilot Study to Possible Future Implications of Treatment. *CNS Neurol. Disord. Drug Targets* **2019**, *18*, 232–238. [CrossRef]
57. Scuderi, C.; Stecca, C.; Valenza, M.; Ratano, P.; Bronzuoli, M.R.; Bartoli, S.; Steardo, L.; Pompili, E.; Fumagalli, L.; Campolongo, P.; et al. Palmitoylethanolamide controls reactive gliosis and exerts neuroprotective functions in a rat model of Alzheimer's disease. *Cell Death Dis.* **2014**, *5*, e1419. [CrossRef]
58. D'Agostino, G.; Russo, R.; Avagliano, C.; Cristiano, C.; Meli, R.; Calignano, A. Palmitoylethanolamide protects against the amyloid–β25–35–induced learning and memory impairment in mice, an experimental model of Alzheimer disease. *Neuropsychopharmacol. Off. Publ. Am. Coll. Neuropsychopharmacol.* **2012**, *37*, 1784–1792. [CrossRef]
59. Bronzuoli, M.R.; Facchinetti, R.; Steardo, L., Jr.; Romano, A.; Stecca, C.; Passarella, S.; Steardo, L.; Cassano, T.; Scuderi, C. Palmitoylethanolamide Dampens Reactive Astrogliosis and Improves Neuronal Trophic Support in a Triple Transgenic Model of Alzheimer's Disease: In vitro and In vivo Evidence. *Oxid. Med. Cell. Longev.* **2018**, *2018*, 4720532. [CrossRef]
60. Scuderi, C.; Bronzuoli, M.R.; Facchinetti, R.; Pace, L.; Ferraro, L.; Broad, K.D.; Serviddio, G.; Bellanti, F.; Palombelli, G.; Carpinelli, G.; et al. Ultramicronized palmitoylethanolamide rescues learning and memory impairments in a triple transgenic mouse model of Alzheimer's disease by exerting anti–inflammatory and neuroprotective effects. *Transl. Psychiatry* **2018**, *8*, 32. [CrossRef]
61. Beggiato, S.; Tomasini, M.C.; Cassano, T.; Ferraro, L. Chronic Oral Palmitoylethanolamide Administration Rescues Cognitive Deficit and Reduces Neuroinflammation, Oxidative Stress, and Glutamate Levels in A Transgenic Murine Model of Alzheimer's Disease. *J. Clin. Med.* **2020**, *9*, 428. [CrossRef] [PubMed]
62. Esposito, E.; Impellizzeri, D.; Mazzon, E.; Paterniti, I.; Cuzzocrea, S. Neuroprotective activity of palmitoylethanolamide in an animal model of Parkinson's disease. *PLoS ONE* **2012**, *7*, e41880. [CrossRef] [PubMed]
63. Crupi, R.; Impellizzeri, D.; Cordaro, M.; Siracusa, R.; Casili, G.; Evangelista, M.; Cuzzocrea, S. N–palmitoylethanolamide prevents parkinsonian phenotypes in aged mice. *Mol. Neurobiol.* **2018**, *55*, 8455–8472. [CrossRef] [PubMed]
64. Avagliano, C.; Russo, R.; De Caro, C.; Cristiano, C.; La Rana, G.; Piegari, G.; Paciello, O.; Citraro, R.; Russo, E.; De Sarro, G.; et al. Palmitoylethanolamide protects mice against 6–OHDA–induced neurotoxicity and endoplasmic reticulum stress: In vivo and in vitro evidence. *Pharmacol. Res.* **2016**, *113 Pt A*, 276–289. [CrossRef]
65. Siracusa, R.; Paterniti, I.; Impellizzeri, D.; Cordaro, M.; Crupi, R.; Navarra, M.; Cuzzocrea, S.; Esposito, E. The Association of Palmitoylethanolamide with Luteolin Decreases Neuroinflammation and Stimulates Autophagy in Parkinson's Disease Model. *CNS Neurol. Disord. Drug Targets* **2015**, *14*, 1350–1365. [CrossRef]
66. Cordaro, M.; Siracusa, R.; Crupi, R.; Impellizzeri, D.; Peritore, A.F.; D'Amico, R.; Gugliandolo, E.; Di Paola, R.; Cuzzocrea, S. 2–Pentadecyl–2–Oxazoline Reduces Neuroinflammatory Environment in the MPTP Model of Parkinson Disease. *Mol. Neurobiol.* **2018**, *55*, 9251–9266. [CrossRef]
67. Bisogno, T.; Martire, A.; Petrosino, S.; Popoli, P.; Di Marzo, V. Symptom-related changes of endocannabinoid and palmitoylethanolamide levels in brain areas of R6/2 mice, a transgenic model of Huntington's disease. *Neurochem. Int.* **2008**, *52*, 307–313. [CrossRef]
68. Baker, D.; Pryce, G.; Croxford, J.L.; Brown, P.; Pertwee, R.G.; Makriyannis, A.; Khanolkar, A.; Layward, L.; Fezza, F.; Bisogno, T.; et al. Endocannabinoids control spasticity in a multiple sclerosis model. *FASEB J.* **2001**, *15*, 300–302. [CrossRef]
69. Rahimi, A.; Faizi, M.; Talebi, F.; Noorbakhsh, F.; Kahrizi, F.; Naderi, N. Interaction between the protective effects of cannabidiol and palmitoylethanolamide in experimental model of multiple sclerosis in C57BL/6 mice. *Neurosci.* **2015**, *290*, 279–287. [CrossRef]
70. Contarini, G.; Franceschini, D.; Facci, L.; Barbierato, M.; Giusti, P.; Zusso, M. A co–ultramicronized palmitoylethanolamide/luteolin composite mitigates clinical score and disease–relevant molecular markers in a mouse model of experimental autoimmune encephalomyelitis. *J. Neuroinflammation* **2019**, *16*, 126. [CrossRef]
71. Loría, F.; Petrosino, S.; Mestre, L.; Spagnolo, A.; Correa, F.; Hernangómez, M.; Guaza, C.; Di Marzo, V.; Docagne, F. Study of the regulation of the endocannabinoid system in a virus model of multiple sclerosis reveals a therapeutic effect of palmitoylethanolamide. *Eur. J. Neurosci.* **2008**, *28*, 633–641. [CrossRef] [PubMed]
72. Impellizzeri, D.; Siracusa, R.; Cordaro, M.; Crupi, R.; Peritore, A.F.; Gugliandolo, E.; D'Amico, R.; Petrosino, S.; Evangelista, M.; Di Paola, R.; et al. N–Palmitoylethanolamine–oxazoline (PEA–OXA): A new therapeutic strategy to reduce neuroinflammation, oxidative stress associated to vascular dementia in an experimental model of repeated bilateral common carotid arteries occlusion. *Neurobiol. Dis.* **2019**, *125*, 77–91. [CrossRef] [PubMed]
73. Gauthier, S.; Reisberg, B.; Zaudig, M.; Petersen, R.C.; Ritchie, K.; Broich, K.; Belleville, S.; Brodaty, H.; Bennett, D.; Chertkow, H.; et al. International Psychogeriatric Association Expert Conference on mild cognitive impairment. Mild cognitive impairment. *Lancet* **2006**, *367*, 1262–1270. [CrossRef]
74. Ellison, J.M.; Harper, D.G.; Berlow, Y.; Zeranski, L. Beyond the "C" in MCI: Noncognitive symptoms in amnestic and non–amnestic mild cognitive impairment. *CNS Spectr.* **2008**, *13*, 66–72. [CrossRef] [PubMed]
75. Petersen, R.C.; Caracciolo, B.; Brayne, C.; Gauthier, S.; Jelic, V.; Fratiglioni, L. Mild cognitive impairment: A concept in evolution. *Int. J. Med.* **2014**, *275*, 214–228. [CrossRef]
76. Lombardi, G.; Crescioli, G.; Cavedo, E.; Lucenteforte, E.; Casazza, G.; Bellatorre, A.G.; Lista, C.; Costantino, G.; Frisoni, G.; Virgili, G.; et al. Structural magnetic resonance imaging for the early diagnosis of dementia due to Alzheimer's disease in people with mild cognitive impairment. *Cochrane Database Syst. Rev.* **2020**, *3*, CD009628. [CrossRef]

77. Serrano-Pozo, A.; Frosch, M.P.; Masliah, E.; Hyman, B.T. Neuropathological alterations in Alzheimer disease. *Cold Spring Cold Spring Harb. Perspect. Med.* **2011**, *1*, a006189. [CrossRef]
78. Andrews, K.A.; Modat, M.; Macdonald, K.E.; Yeatman, T.; Cardoso, M.J.; Leung, K.K.; Barnes, J.; Villemagne, V.L.; Rowe, C.C.; Fox, N.C.; et al. Acceleration of hippocampal atrophy rates in asymptomatic amyloidosis. *Neurobiol. Aging* **2016**, *39*, 99–107. [CrossRef]
79. Scheltens, P.; Blennow, K.; Breteler, M.M.B.; de Strooper, B.; Frisoni, G.B.; Salloway, S.; Van der Flier, W.M. Alzheimer's disease. *Lancet* **2016**, *388*, 505–517. [CrossRef]
80. Ferreira-Vieira, T.H.; Guimaraes, I.M.; Silva, F.R.; Ribeiro, F.M. Alzheimer's disease: Targeting the cholinergic system. *Curr. Neuropharmacol.* **2016**, *14*, 101–115. [CrossRef]
81. Scuderi, C.; Esposito, G.; Blasio, A.; Valenza, M.; Arietti, P.; Steardo, L., Jr.; Carnuccio, R.; De Filippis, D.; Petrosino, S.; Iuvone, T.; et al. Palmitoylethanolamide counteracts reactive astrogliosis induced by β–amyloid peptide. *J. Cell. Mol. Med.* **2011**, *15*, 2664–2674. [CrossRef] [PubMed]
82. Scuderi, C.; Valenza, M.; Stecca, C.; Esposito, G.; Carratù, M.R.; Steardo, L. Palmitoylethanolamide exerts neuroprotective effects in mixed neuroglial cultures and organotypic hippocampal slices via peroxisome proliferator–activated receptor–α. *J. Neuroinflammation* **2012**, *9*, 49. [CrossRef] [PubMed]
83. Scuderi, C.; Steardo, L. Neuroglial roots of neurodegenerative diseases: Therapeutic potential of palmitoylethanolamide in models of Alzheimer's disease. *CNS Neurol. Disord. Drug Targets* **2013**, *12*, 62–69. [CrossRef]
84. Beggiato, S.; Borelli, A.C.; Ferraro, L.; Tanganelli, S.; Antonelli, T.; Tomasini, M.C. Palmitoylethanolamide blunts amyloid-β42-induced astrocyte activation and improves neuronal survival in primary mouse cortical astrocyte–neuron co-cultures. *J. Alzheimers Dis.* **2018**, *61*, 389–399. [CrossRef] [PubMed]
85. Tomasini, M.C.; Borelli, A.C.; Beggiato, S.; Ferraro, L.; Cassano, T.; Tanganelli, S.; Antonelli, T. Differential Effects of Palmitoylethanolamide against Amyloid–β Induced Toxicity in Cortical Neuronal and Astrocytic Primary Cultures from Wild–Type and 3xTg–AD Mice. *J. Alzheimer's Dis.* **2015**, *46*, 407–421. [CrossRef] [PubMed]
86. Beggiato, S.; Cassano, T.; Ferraro, L.; Tomasini, M.C. Astrocytic palmitoylethanolamide pre–exposure exerts neuroprotective effects in astrocyte–neuron co–cultures from a triple transgenic mouse model of Alzheimer's disease. *Life Sci.* **2020**, *257*, 118037. [CrossRef]
87. Paterniti, I.; Cordaro, M.; Campolo, M.; Siracusa, R.; Cornelius, C.; Navarra, M.; Cuzzocrea, S.; Esposito, E. Neuroprotection by association of palmitoylethanolamide with luteolin in experimental Alzheimer's disease models: The control of neuroinflammation. *CNS Neurol. Disord. Drug Targets* **2014**, *13*, 1530–1541. [CrossRef]
88. Surguchov, A. Biomarkers in Parkinson's Disease. In *Neurodegenerative Diseases Biomarkers*; Peplow, P.V., Martinez, B., Gennarelli, T.A., Eds.; Humana: New York, NY, USA, 2022; Volume 173, pp. 155–180.
89. Alexander, G.E. Biology of Parkinson's disease: Pathogenesis and pathophysiology of a multisystem neurodegenerative disorder. *Dialogues Clin. Neurosci.* **2004**, *6*, 259–280. [CrossRef]
90. DeMaagd, G.; Philip, A. Parkinson's Disease and Its Management: Part 1: Disease Entity, Risk Factors, Pathophysiology, Clinical Presentation, and Diagnosis. *Pharm. Ther.* **2015**, *40*, 504–532.
91. Surmeier, D.J.; Obeso, J.A.; Halliday, G.M. Selective neuronal vulnerability in Parkinson disease. *Nat. Rev. Neurosci.* **2017**, *18*, 101–113. [CrossRef]
92. Michel, P.P.; Hirsch, E.C.; Hunot, S. Understanding Dopaminergic Cell Death Pathways in Parkinson Disease. *Neuron* **2016**, *90*, 675–691. [CrossRef] [PubMed]
93. Wirdefeldt, K.; Adami, H.O.; Cole, P.; Trichopoulos, D.; Mandel, J. Epidemiology and etiology of Parkinson's disease: A review of the evidence. *Eur. J. Epidemiol.* **2011**, *26* (Suppl. 1), S1–S58. [CrossRef] [PubMed]
94. Pajares, M.; Rojo, A.I.; Manda, G.; Boscá, L.; Cuadrado, A. Inflammation in Parkinson's Disease: Mechanisms and Therapeutic Implications. *Cells* **2020**, *9*, 1687. [CrossRef] [PubMed]
95. Erkkinen, M.G.; Kim, M.O.; Geschwind, M.D. Clinical Neurology and Epidemiology of the Major Neurodegenerative Diseases. *Cold Spring Harb. Perspect. Biol.* **2018**, *10*, a033118. [CrossRef] [PubMed]
96. Jenner, P.; Marsden, C.D. The actions of 1–methyl–4–phenyl–1,2,3,6–tetrahydropyridine in animals as a model of Parkinson's disease. *J. Neural Transm. Suppl.* **1986**, *20*, 11–39. [PubMed]
97. McColgan, P.; Tabrizi, S.J. Huntington's disease: A clinical review. *Eur. J. Neurol.* **2018**, *25*, 24–34. [CrossRef]
98. Brinkman, R.R.; Mezei, M.M.; Theilmann, J.; Almqvist, E.; Hayden, M.R. The likelihood of being affected with Huntington disease by a particular age, for a specific CAG size. *Am. J. Hum. Genet.* **1997**, *60*, 1202–1210.
99. Gatto, E.M.; Rojas, N.G.; Persi, G.; Etcheverry, J.L.; Cesarini, M.E.; Perandones, C. Huntington disease: Advances in the understanding of its mechanisms. *Clin. Park. Relat. Disord.* **2020**, *3*, 100056. [CrossRef]
100. Glass, M. The role of cannabinoids in neurodegenerative diseases. *Prog. Neuropsychopharmacol. Biol. Psychiatry* **2001**, *25*, 743–765. [CrossRef]
101. Finger, E.C. Frontotemporal Dementias. *Continuum* **2016**, *22*, 464–489. [CrossRef]
102. Young, J.J.; Lavakumar, M.; Tampi, D.; Balachandran, S.; Tampi, R.R. Frontotemporal dementia: Latest evidence and clinical implications. *Ther. Adv. Psychopharmacol.* **2018**, *8*, 33–48. [CrossRef] [PubMed]
103. Zarei, S.; Carr, K.; Reiley, L.; Diaz, K.; Guerra, O.; Altamirano, P.F.; Pagani, W.; Lodin, D.; Orozco, G.; Chinea, A. A comprehensive review of amyotrophic lateral sclerosis. *Surg. Neurol. Int.* **2015**, *6*, 171. [CrossRef]

104. Hardiman, O.; Al-Chalabi, A.; Chio, A.; Corr, E.M.; Logroscino, G.; Robberecht, W.; Shaw, P.J.; Simmons, Z.; van den Berg, L.H. Amyotrophic lateral sclerosis. *Nat. Rev. Dis. Primers* **2017**, *3*, 17071. [CrossRef] [PubMed]
105. van Es, M.A.; Hardiman, O.; Chio, A.; Al-Chalabi, A.; Pasterkamp, R.J.; Veldink, J.H.; van den Berg, L.H. Amyotrophic lateral sclerosis. *Lancet* **2017**, *390*, 2084–2098. [CrossRef]
106. Chang, J.L.; Lomen-Hoerth, C.; Murphy, J.; Henry, R.G.; Kramer, J.H.; Miller, B.L.; Gorno-Tempini, M.L. A voxel-based morphometry study of patterns of brain atrophy in ALS and ALS/FTLD. *Neurology* **2005**, *65*, 75–80. [CrossRef] [PubMed]
107. Correale, J.; Marrodan, M.; Ysrraelit, M.C. Mechanisms of Neurodegeneration and Axonal Dysfunction in Progressive Multiple Sclerosis. *Biomedicines* **2019**, *7*, 14. [CrossRef]
108. Rahn, K.; Slusher, B.; Kaplin, A. Cognitive impairment in multiple sclerosis: A forgotten disability remembered. *Cerebrum* **2012**, *2012*, 14.
109. Miljković, D.; Spasojević, I. Multiple sclerosis: Molecular mechanisms and therapeutic opportunities. *Antioxid. Redox Signal.* **2013**, *19*, 2286–2334. [CrossRef]
110. Peschl, P.; Bradl, M.; Höftberger, R.; Berger, T.; Reindl, M. Myelin Oligodendrocyte Glycoprotein: Deciphering a Target in Inflammatory Demyelinating Diseases. *Front. Immunol.* **2017**, *8*, 529. [CrossRef]
111. Chang, A.; Tourtellotte, W.W.; Rudick, R.; Trapp, B.D. Premyelinating oligodendrocytes in chronic lesions of multiple sclerosis. *N. Engl. J. Med.* **2002**, *346*, 165–173. [CrossRef]
112. Barbierato, M.; Facci, L.; Marinelli, C.; Zusso, M.; Argentini, C.; Skaper, S.D.; Giusti, P. Co-ultramicronized Palmitoylethanolamide/Luteolin Promotes the Maturation of Oligodendrocyte Precursor Cells. *Sci. Rep.* **2015**, *5*, 16676. [CrossRef] [PubMed]
113. Facci, L.; Barbierato, M.; Fusco, M.; Giusti, P.; Zusso, M. Co-Ultramicronized Palmitoylethanolamide/Luteolin-Induced Oligodendrocyte Precursor Cell Differentiation is Associated with Tyro3 Receptor Upregulation. *Front. Pharmacol.* **2021**, *12*, 698133. [CrossRef] [PubMed]

biomolecules

MDPI

Review

Role of EPA in Inflammation: Mechanisms, Effects, and Clinical Relevance

Rosalia Crupi [1] and Salvatore Cuzzocrea [2,*]

[1] Department of Veterinary Sciences, University of Messina, 98168 Messina, Italy; rcrupi@unime.it
[2] Department of Chemical, Biological, Pharmaceutical and Environmental Sciences, University of Messina, 98166 Messina, Italy
* Correspondence: salvator@unime.it

Abstract: Many chronic inflammatory processes are linked with the continuous release of inflammatory mediators and the activation of harmful signal-transduction pathways that are able to facilitate disease progression. In this context atherosclerosis represents the most common pathological substrate of coronary heart disease, and the characterization of the disease as a chronic low-grade inflammatory condition is now validated. The biomarkers of inflammation associated with clinical cardiovascular risk support the theory that targeted anti-inflammatory treatment appears to be a promising strategy in reducing residual cardiovascular risk. Several literature data highlight cardioprotective effects of the long-chain omega-3 polyunsaturated fatty acids (PUFAs), such as eicosapentaenoic acid (EPA). This PUFA lowers plasma triglyceride levels and has potential beneficial effects on atherosclerotic plaques. Preclinical studies reported that EPA reduces both pro-inflammatory cytokines and chemokines levels. Clinical studies in patients with coronary artery disease that receive pharmacological statin therapy suggest that EPA may decrease plaque vulnerability preventing plaque progression. This review aims to provide an overview of the links between inflammation and cardiovascular risk factors, importantly focusing on the role of diet, in particular examining the proposed role of EPA as well as the success or failure of standard pharmacological therapy for cardiovascular diseases.

Keywords: inflammation; eicosapentaenoic acid; atherosclerosis

Citation: Crupi, R.; Cuzzocrea, S. Role of EPA in Inflammation: Mechanisms, Effects, and Clinical Relevance. *Biomolecules* **2022**, *12*, 242. https://doi.org/10.3390/biom12020242

Academic Editor: Prakash Kulkarni

Received: 22 December 2021
Accepted: 31 January 2022
Published: 1 February 2022

1. Overview of Inflammation

Inflammation is classified as "normal defense mechanism" able to preserve the host from both infections and insults; the involvement of the inflammatory process helps to restore homeostasis at damaged or infected sites. Cardinal signs of inflammation are redness, swelling, heat, pain, and loss of function, and it sees the presence and interaction of different cell types with recruitment of several chemical mediators. If the inflammatory response is well regulated it resolves quickly and without causing damage to the organism, which overall involve the activation of negative feedback mechanisms such as the secretion of anti-inflammatory mediators, the inhibition of pro-inflammatory signaling cascades, shedding of receptors for inflammatory mediators, and activation of regulatory cells [1]; otherwise damage to organs and tissues can occur. Inflammatory mediators normally play a defensive role in the defense of the host; however, if the production of these mediators is not regulated, tissue damage can be caused, inducing the development of different pathologies. Literature data have reported, in fact, that several chronic inflammatory processes are linked with the continuous release of inflammatory mediators and the activation of harmful signal-transduction pathway that facilitate disease progression. In these cases, there are high concentrations of inflammatory markers and cells both at the primary site of damage and at the systemic level, as occurs in cardiovascular disease or rheumatoid arthritis. Recent studies have shown that the condition of low-grade chronic inflammation is decisive in the obesity state, metabolic syndrome, and cardiovascular diseases [2].

For the aforementioned concepts, inflammation is classified as the lowest common denominator of numerous pathologies (Figure 1).

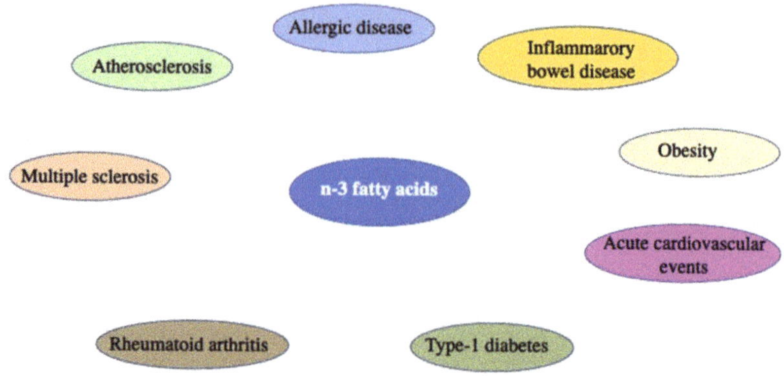

Figure 1. Representative image of inflammatory diseases in which marine n-3 fatty acids might be of benefit.

Classical events that characterized the inflammatory response include the following:

- An increased blood supply to the site of inflammation.
- Elevated capillary permeability due to retraction of endothelial cells.
- Leukocyte migration from the capillaries into the surrounding tissue. This process is facilitated by the release of chemotactic factors from the site of inflammation and from the upregulation of adhesion molecules on the endothelium.
- Release of mediators from leukocytes at the site of inflammation.

Lipid Mediators

There are different lipid mediators (e.g., leukotrienes (LTs), prostaglandins (PGs), peptide mediators (e.g., cytokines), amino acid derivatives (e.g., histamine), reactive oxygen species (e.g., superoxide), and enzymes (e.g., matrix proteases) in relationship to the anatomical site involved, cell type involved, nature of the inflammatory stimulus, and the stage during the inflammatory response. Among the mediators involved in the development of inflammatory processes, the first group is represented by toll-like receptors (TLRs), membrane proteins located on macrophages, and dendritic cells. The receptors have the specificity to recognize the molecular patterns that are associated with pathogens (PAMP) and can also identify the endogenous signals that are activated during tissue or cellular damage related to danger (DAMPS). The second group of mediators is represented by the arachidonic acid (AA) mediators. The phosphorylase enzyme is able to act on membrane phospholipids releasing arachidonic acid. The arachidonic acid can metabolize either through the cyclooxygenase pathway or the 5-lipoxygenase pathway. Through the action of cyclooxygenase, the mediators of PGs are formed, which are prostaglandin D2 (PGD2) and thromboxane, which are bronchoconstrictor prostaglandins, and bronchoprotective or inhibitory prostaglandin E2 (PGE2) and prostacyclin. LT is formed from the 5-lipoxygenase pathway [3]. The third group of mediators is represented by mast cells, cells distributed throughout the body.

Mast cells are immune cells of the myeloid lineage and are present in connective tissues throughout the body. The activation and degranulation of mast cells significantly modulates many aspects of physiological and pathological conditions in various settings. With respect to normal physiological functions, mast cells are known to regulate vasodilation, vascular homeostasis, innate and adaptive immune responses, angiogenesis, and venom detoxification. On the other hand, mast cells have also been implicated in the pathophysiology of many diseases, including allergy, asthma, anaphylaxis, gastrointestinal disorders, many types of malignancies, and cardiovascular diseases. The characteristic of these cells is to undergo a degranulation process when activated by tissue damage; pro-inflammatory molecules are released from the granules such as histamine, tumor necrosis factor (TNF), kinin, and leukotrienes [4].

2. Inflammation in Cardiovascular Diseases: Focus on Atherosclerosis

All cardiovascular diseases are characterized by the presence of an inflammatory environment often resulting from comorbid pathologies such as diabetes mellitus and arterial hypertension. These are then associated with additional risk factors which are represented by the emotional component and lifestyle. Atherosclerotic cardiovascular disease (ASCVD) is the leading cause of morbidity and mortality globally. In the United States, nearly 808,000 people died from ASCVD in 2014, translating to about 1 in every 3 deaths [5], and the death rate from ASCVD increased by 1% in 2015, the first since 1999 [6]. ASCVD is defined as acute coronary syndrome, stable angina, history of myocardial infarction, coronary or other revascularization, ischemic stroke, transient ischemic attack, or peripheral arterial disease presumably of atherosclerotic origin. Atherosclerosis can also be considered a chronic, immuno-inflammatory, fibroproliferative disease of large and medium-sized arteries fueled by lipids. The study carried out by Anichkov, on rabbits who were fed a diet rich in fatty acids, highlighted the importance and involvement of dyslipidemic phenomena in the development of cardiovascular diseases [7]. Arteriosclerosis is a composite disease in which the inflammatory process is found on the one hand, but on the other, it is also a lipid disorder. Indeed, the Framingham Heart Study in the 1950s demonstrated that hypercholesterolemia accelerates atherosclerosis in humans [3]. In the study of cardiovascular diseases, great importance was given to the thrombotic aspect, which still represents the greatest risk in the management of low-grade systemic inflammation; in fact, there are numerous studies that underline the link between the mediators of inflammation and platelet activation [8].

Formation and rupture of atherosclerotic plaque are closely connected with the development of the inflammatory process. In particular, the lipoprotein stimulates both chemokines and endothelial cells that facilitate the recruitment of lymphocytes and adherent monocytes to the forming lesion site; growth factors and cytokines formed by the inflammatory intima stimulate the differentiation of monocytes into macrophages linked by the upregulation of the TLRs, in particular TLR4 and scavenger receptors; the macrophages activated form foam-like cells or stimulate an inflammatory cascade that lead to atherosclerotic plaque formation [9].

3. Fatty Acids: Composition and Role in Inflammatory Processes

Fatty acids are long-chain hydrocarbons and important components of lipids in plants, animals, and microorganisms. Among several role of fatty acids, it was demonstrated that they are able to influence inflammatory processes through several mechanisms by acting, for example, via the cell surface and intracellular receptors/sensors that regulate both gene expression patterns and inflammatory cell signaling. Alterations in fatty acid compositions of cell membranes modulates membrane fluidity, lipid raft formation, and cell signaling leading to altered gene expression. A fatty acid consists of a straight chain of an even number of carbon atoms, with hydrogen atoms along the length of the chain and at one end of the chain and a carboxyl group (—COOH) at the other end. They can be separated into four categories: saturated, mono-unsaturated, polyunsaturated, and *trans* fats. In particular, if the carbon-to-carbon bonds are all single, the acid is saturated; if any of the bonds is double or triple, the acid is unsaturated and is more reactive.

A fatty acid containing two or more double bonds is called a polyunsaturated fatty acid (PUFA). Saturated fatty acids do not contain double bonds in the acyl chain. A few fatty acids have branched chains; others contain ring structures (e.g., prostaglandins). Fatty acids are not found in a free state in nature; commonly, they exist in combination with glycerol in the form of triglycerides. Two principal families of PUFAs are known, namely the n-6 (or omega-6) and the n-3 (or omega-3) (Table 1).

Table 1. PUFAs n-6 (or omega-3) family.

Sources of Dietary n-3 PUFA	Sources of Dietary n-3 PUFA	ALA (α-Linolenic Acid)	EPA (Eicosapentaenoic Acid)	DHA (Docosahexaenoic Acid)	Ref.
Fish oil	Menhaden (oil)	-	13.18	8.56	[10,11]
	Salmon (oil)	-	13.3	18.23	
	Herring (oil)	-	6.28	4.21	
	Sardine (oil)	-	10.15	10.66	
Fish raw	Salmon (raw)	0.09	0.89	1.19	[11]
	Sardine (raw)	-	0.51	1.16	
	Cod (dried)	-	0.02	0.62	
	Trout (raw)	0.1	0.15	0.5	
	Herring (raw)	0.19	1.09	1.01	
Beef	New Zealand, liver (raw)	0.05	0.11	0.04	[10]
	New Zealand, kidney (cooked)	0.08	0.15	0.03	
Oils	Soybean (oil)	7.3	-	-	[10,11]
	Wheat germ (oil)	5.3	-	-	
	Sunflower (oil)	0.33	-	-	
	Flaxseed (oil)	53.37	-	-	
	Safflower (oil)	0.1	-	-	
	Corn (oil)	0.6	-	-	
	Canola (oil)	9.15	-	-	
Seed and nuts	Chia (dried/ground)	17.83	-	-	[10,11]
	Hazelnut (dried/ground)	0.11	-	-	
	Almond (dried/ground)	0.3	-	-	
	Hemp seed (hulled)	8.68	-	-	
	Brazil nuts (dried)	0.02	-	-	
	Walnut (dried/ground)	6.64	-	-	

Linoleic acid (18: 2n-6) and α-linolenic acid (18: 3n-3) belong to these families, which have the particularity of not being able to be synthesized by mammals. Vegetable oils contain high amounts of linoleic acid. Green plants and some vegetable oils, on the other hand, contain α-linolenic acid.

Importantly, although linoleic and α-linolenic acids cannot be synthesized by humans, they can still be metabolized into other fatty acids. In particular, linoleic acid can be converted via γ-linolenic acid (18: 3n-6) and dihomo gamma-linoleic acid (20: 3n-6) into arachidonic acid (20: 4n-6); it is also possible to convert α-linolenic acid in eicosapentaenoic acid (20: 5n-3; EPA). The specific chemical structure of EPA reflects important biological characteristics. EPA can modify the physical properties of cellular membranes by substituting the omega-6 fatty acid arachidonic acid (AA; 20:4, n-6) in membrane phospholipids. Moreover, from the metabolism of EPA, antithrombotic and anti-inflammatory lipid mediators can be generated, which are in stark contrast to the prothrombotic and pro-inflammatory factors that are produced by arachidonic acid (AA) [12,13]. EPA can be further metabolized, giving rise to docosapentaenoic acid (22: 5n-3; DPA) and docosahexaenoic acid (22: 6n-3; DHA).

Fatty Acid Sources

Sources of fatty acids include fruits, vegetable oils, seeds, nuts, animal fats, and fish oils. Essential fatty acids, such as omega-3 fatty acids, serve important cellular functions. They are a necessary part of the human diet because the body has no biochemical pathway to produce these molecules on its own. Particular linoleic and α-linolenic acids are introduced through diet. Arachidonic acid is found in meats, and intake is estimated to be between 50 and 500 mg/day. Fish, especially fatty fish, are rich in EPA, DPA, and DHA. Membrane phospholipids consist of basic units of polyunsaturated fatty acids. Preclinical studies, conducted on laboratory guinea pigs fed according to standard feeding, showed a high content of arachidonic acid (20: 4n-6) and a low content of eicosapentaenoic acid (20: 5n-3; EPA) and docosahexaenoic acid (22: 6n-3; DHA) in the phospholipid composition of tissue lymphocytes [14], peritoneal macrophages [15,16], alveolar macrophages [17], Kupffer cells, and alveolar neutrophils [18]. If the diet of animals was enriched with fish oil, rich in EPA and DHA, there is an accumulation of these in membrane phospholipids both in lymphocytes [14] and in macrophages [19,20], and in Kupffer cells [17]. The even more interesting data that emerges is represented by the decrease in the content of arachidonic acid. Studies conducted on membrane phospholipids of cells such as neutrophils, lymphocytes, and monocytes of patients who consume typical Western diets contain about 10–20% of fatty acids such as arachidonic acid, with about 0.5–1% EPA and about 24% DHA [21–23]. The fatty acid composition of these cells can vary by increasing the intake of marine n-3 fatty acids [23,24]. This occurs in a dose–response mode [23] and over a period of days to weeks, with a new steady-state composition achieved within approximately 4 weeks. There are several mechanisms through which fatty acids can influence inflammatory processes, in particular the following:

- By taking fatty acids we can alter the intracellular concentrations of lipoproteins, metabolites, complex lipids, and hormones, which in turn are modulators of inflammation;
- Fatty acids can undergo oxidation processes and the compounds obtained can act on inflammatory cells by binding to specific receptors;
- Fatty acids can be incorporated into cell membranes where they keep the fluidity of the membranes intact; membrane phospholipids are also substrates for diacylglycerol, and fatty acids can act as transcription factors or precursors for the biosynthesis of lipid mediators.

4. Fatty Acid: Preclinical Studies

Several animal models were used to study the effects of EPA. As reported in a study conducted on a mouse model of hyperlipidemia, represented by ApoE- and LDL-receptor-deficient mice, the administration of EPA reduced the development of atherosclerotic lesions and increased the cell content of omega-3 PUFAs, without changing the total cholesterol or HDL content [25]. The analysis carried out on the atherosclerotic plaques of mice administered with EPA showed a stable morphology associated with a lower deposition of lipids and a reduced accumulation of macrophages accompanied by an increase in smooth muscle cells and collagen content. An anti-inflammatory effect of EPA was also highlighted due to the inhibition of the expression of adhesion molecules and monocyte chemoattractant protein 1 (MCP-1) and inhibition of the production of metalloproteinases by macrophages. The atherosclerotic plaques of mice treated with EPA had a stable morphology, including less lipid deposition, decreased accumulation of macrophages, increased smooth muscle cells, and greater collagen content. In addition, EPA had an anti-inflammatory effect on endothelial cells, inhibiting the expression of adhesion molecules and MCP-1 and by inhibiting production of matrix metalloproteinase by macrophages.

EPA is converted to 18R-hydroxyheicosapentaenoic acid (18R-HEPE) by COX-2-expressing vascular endothelial cells, which is acetylated in the presence of aspirin during inflammation. Endothelial cells expressing COX-2 convert EPA into 18R-HEPE, which is subsequently released by endothelial cells and converted via neutrophil-derived 5-LOX through a common epoxy intermediate in Resolvin E1 (RvE1) and Resolvin E2 (RvE2).

Data from a study in which they were used have recently been acquired. ApoE 3 Leiden transgenic mice transgenic (known as mice able to develop hyperlipidemia and susceptible to diet-induced atherosclerosis) were used to study RvE1 obtained by the EPA on the development of arteriosclerotic lesion. The animals were given a hypercholesterolemic diet for 9 weeks after which they received the administration of RvE1 (low or high dose) for 16 weeks. A group treated with atorvastatin and one with low-dose RvE1 were also added to the study atorvastatin. Compared to the control group, the group treated with RvE1 (low or high dose) and with atorvastatin reduced the arteriosclerotic lesion from 35% to 27%; the combination of RvE1 and atorvastatin reduced the atherosclerotic lesion area by 51% [26]. Data collected from another study conducted on LDL-receptor-deficient mice showed that EPA significantly reduces the size of atherosclerotic plaques as well as total cholesterol levels [27]. Levels of proinflammatory cytokines and chemokines (TNF alpha; interferon gamma) were lower than in the control group. Another target on which EPA acts is represented by dendritic cells whose phenotype is reorganized after administration of EPA; the number of T cells is reduced in the spleen and lymph nodes [28].

5. Fatty Acids: Clinical Studies

The importance of the use of omega-3 fatty acids in general, and of EPA in particular, is also linked to the numerous guidelines on their use; for the National Lipid Association (NLA) omega-3 fatty acids represent the first option for the treatment of patients with high triglyceride (TG) levels (≥500 mg/dL) and as an add-on option to statin therapy for those with high TG levels (200–499 mg/dL) [29]. For the Japan Atherosclerosis Society (JAS) the addition of EPA to a statin therapy represents an added value for the treatment of high-risk patients with LDL-C levels ≥140 mg/dL [30]. Numerous clinical studies have reported data on the importance assumed by omega-3 PUFAs on the prevention of ASCVD; in particular, on the international scene, four studies stand out: Gruppo Italiano per Io Studio della Sopravvivenza nell'Infarto Miocardico (GISSI)-Prevention study [3], the Japan Eicosapentaenoic Acid Lipid Intervention (JELIS trial) [31], the GISSI-Heart Failure study [32], and the Reduction of Cardiovascular Events with Icosapent Ethyl-Intervention Trial (REDUCE-IT) [33] (see Table 2). Moreover, other studies were also formulated in which a low dose of omega-3 PUFA was used (in a range of 840 mg/d of EPA + DHA) but which did not produce positive results. In particular, we can count the ORIGIN trial [34], the Risk and Prevention Study [35], the ALPHA OMEGA trial [36], A Study of Cardiovascular Events in Diabetes (ASCEND) trial [37], and the Vitamin D and Omega-3 Trial (VITAL) [38].

Table 2. Clinical studies showed data on the relevance of omega-3 PUFAs on the prevention of ASCVD.

Clinical Trial	Patient Characteristics	Dose PUFA	Outcomes	Ref
(GISSI)-Prevention study	Men and women (15%) after myocardial infarction	850 mg EPA/DHA	The group treated with omega-3 PUFAs were shown to have a 20% reduction in major CV events, a 30% reduction of CV death, and a 45% reduction in SCD	[3]
JELIS trial	Hypercholesterolemic men and women (69%), with and without CHD, already receiving statin therapy	1800 mg EPA	Treatment was associated with a 19% reduction in major CV events	[29]

Table 2. *Cont.*

Clinical Trial	Patient Characteristics	Dose PUFA	Outcomes	Ref
GISSI-Heart Failure study	Men and women (22%) with congestive heart failure	850 mg EPA/DHA	Treatment was associated with a 6% reduction in CV death or hospitalization	[30]
REDUCE-IT	Middle-aged, history of CVD or DM; TG 135–499 mg/dL; LDL-C 40–100 mg/dL with statin	4000 mg EPA	Treatment was associated with a reduction risk of ischemic events	[31]

EPA: Outcomes Studies

In the Outcomes to Assess Statin Residual Risk Reduction with Epanova in High CV Risk Patients with Hypertriglyceridemia (STRENGTH) study, the effects of a 4 g/day administration of EPA and DHA in a carboxylic acid formulation were analyzed [39]. A meta-analysis conducted on 20 studies that enrolled 68,680 patients showed that the administration of omega-3 PUFAs led to a reduction in the risk of cardiac death (RR 0.91, 95% CI: 0.85–0.98) in secondary prevention [40]. Data obtained from this survey led the American Heart Association (AHA) of 2017 to express a scientific opinion in which it is reported that the low-dose supplementation of omega-3 PUFA was conceivable to prevent the secondary manifestations of coronary diseases in patients with overt coronary heart disease. [41] A recent meta-analysis formulated by expanding the data with those obtained from three recent large-scale RCTs of omega-3 PUFAs (REDUCE-IT, ASCEND, and VITAL) clearly highlighted the beneficial properties of omega-3 PUFAs in limiting the risk of developing cardiovascular events. These reductions appear to be associated with the supplemental omega-3 PUFA dose [42]. In the analysis carried out by the REDUCE-IT study, the effects obtained from the use of the ethyl ester EPA (icosapent ethyl 4 g/day) on the development of high-risk cardiovascular events in patients on drug treatment with statins were examined [33]. The clinical condition of the patients was subsequently assessed with 4.9-year follow-up, which demonstrated a positive effect of using EPA [33]. Data obtained from REDUCE-IT led the National Lipid Association to recommend the administration of icosapent ethyl for patients aged ≥45 years with clinical ASCVD, or aged ≥50 years with diabetes mellitus requiring drugs plus ≥1 factor of additional risk, with fasting triglycerides 135 to 499 mg/dL on high-intensity or maximally tolerated statin therapy to reduce the risk of ASCVD [43]. The FDA has approved the administration of icosapent ethyl to both treat patients with triglyceride levels ≥500 mg/dL, and to reduce the risk of ASCVD in people with diabetes mellitus and two or more additional risk factors for cardiovascular disease in drug therapy with statins [43]. The difference between the positive effect of icosapent ethyl reported in the REDUCE-IT [33] study and the lack of an effect of the mixed carboxylic acids EPA and DHA in the STRENGTH [39] study is still questionable. The studies cited used omega-3 PUFAs at high doses but in different formulations, so comparisons should be made with caution. The studies were designed differently in at least three aspects. First, the omega-3 PUFAs were administered as an ethyl ester formula in the REDUCE-IT study and as non-esterified fatty acids, rapidly ionized transforming into molecules with detergent properties (soaps), in the STRENGTH study. Second, unlike the corn oil used in the STRENGTH study, the placebo mineral oil used in the REDUCE-IT study may have influenced expectations. Third, the DHA component of omega-3 PUFAs may be ineffective or even harmful, although there are currently no studies on the ASCVD outcomes of DHA monotherapy. The significant therapeutic efficacy of EPA in combination with statin on ASCVD was not found in other triglyceride-lowering agents, including fenofibrate and niacin, which failed to reduce cardiovascular events as compared to statin treatment alone [44,45]. With the withdrawal of recommendations by the FDA on the combination of statins with fibrates or niacin in the prevention or treatment of ASCVD, icosapent ethyl remains a viable non-LDL target therapy for patients with increased ASCVD risk and hyper-triglyceridemia. Pemafibrate to Reduce Cardiovascular Outcomes by Reducing Triglycerides in Patients with Diabetes (PROMINENT; NCT03071692), an ongoing trial of pemafibrate (a selective PPAR α modulator that significantly lowers triglyceride) in patients with type 2 diabetes mellitus, mild-to-moderate hypertriglyceridemia, and low HDL-cholesterol might further shed light on the mechanism of triglyceride-lowering agents on ASCVD [46].

6. Role of EPA on Athero-Inflammatory-Thrombotic Processes

The scientific literature demonstrates how EPA plays a beneficial function in the regulation of endothelial tone. The endothelial cells, in fact, release nitric oxide (NO), which has the ability to modulate the vasomotor tone in response to acetylcholine and other vasoactive agonists [47]. Under physiological conditions, there is, in the body, an endothelium-dependent vasodilation due to the release of NO. If one has endothelial dysfunction, then NO release is reduced or absent. In this case, we are witnessing the appearance of toxic effects due to reactive oxygen costs, including peroxynitrite (ONOO_1) [47,48]. EPA is able to significantly reduce the formation of reactive oxygen species, as well as the expression of adhesion molecules, the release of pro-inflammatory cytokines, and the apoptotic cascade, as demonstrated by the Din in vitro studies conducted on HUVEC cells [49]. EPA was also found to be able to reduce or inhibit lipid peroxidation in membrane vesicles with even high cholesterol levels. The highlighted effect is also easy due to the presence of statins [44]. It is known that glucose contributes to the development of lipid peroxidation, the direct consequence of which is the appearance of high cholesterol levels. In this process, EPA also plays a key role, since it inhibits the formation of lipids [50]. The antioxidant effect of EPA seems to be due to its ability to intercalate in the lipid bilayer of the membrane while preserving its structural component (Figure 2) [50].

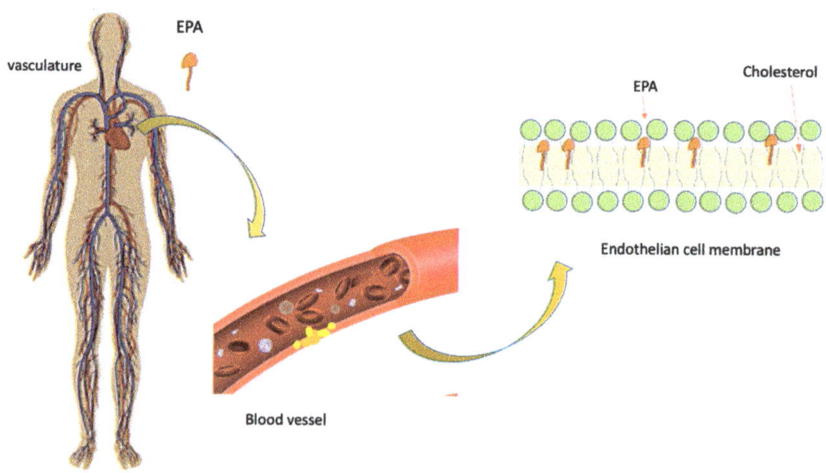

Figure 2. Possible effect of EPA on the endothelial membrane: inhibition of the propagation of free radicals. According to this model, it is hypothesized that EPA is able to intercalate between membrane phospholipids, in the central region, inhibiting the propagation of free radicals and thus preserving a more homogeneous distribution of cholesterol.

EPA and Atherosclerotic Plaque

Recent studies show how EPA is able to reduce the neo vasculogenic process in human endothelial cells by acting on the modulation of the expression of the c-lit protein on the PI3-K/Akt/eNOS pathway, thus preventing ischemic injury [51]. Two studies in particular demonstrate these EPA actions; in the ANCHOR study, the administration of EPA at a dose of 4 g/day for 12 weeks reduced the oxidation of LDL by 13.3% compared to the placebo group in patients treated with statins and high triglycerides (from 200 to <500 mg/dL). In the MARINE study, there was a reduction of 6.6% ($P = 0.055$) in patients with higher levels of triglycerides (500 to <2000 mg/dL) [52]. In patients with type 2 diabetes mellitus and on drug therapy with statins, the addition of EPA at a dose of 1.8 g/day for 6 months) improved the endothelial function, as well as the conditions of CHD patients [53]. Similarly, in hyperlipidemic patients, the administration of EPA at a dose of 1.8 g/day for 3 months restored endothelium-dependent vasodilation at a superimposable level compared to that found in the group of normolipidemic patients [54]. Co-administration of EPA at a dose of 1.8 g/day for 48 weeks) with statins compared to statin monotherapy significantly inhibited the progression of arterial stiffness as observed in the analysis of the b-index stiffness parameter of the carotid artery in CHD patients ($p = 0.02$) [55]. EPA at a dose of 4 g/day for 12 weeks was shown to be effective in significantly reducing RLP-C by 25.8% in statin-treated patients who exhibited high levels of triglycerides ($p = 0.0001$) and 29.8%, respectively, in patients with even higher triglyceride levels ($p = 0.0041$) than in the placebo group [56]. Analyzing the ANCHOR study even more selectively, it was found that EPA significantly reduces RLP-C by 25.0% ($p < 0.0001$) and VLDL triglycerides by 28.9% ($p < 0.01$) compared to the placebo group in the subgroup of patients with type 2 diabetes mellitus and on drug treatment with statins and elevated triglycerides [57]. In another study, in patients with type 2 diabetes mellitus and metabolic syndrome, EPA administered at a dose of 1.8 g/day for 3 months significantly reduced the volume of dense LDL particles ($p < 0.01$) and RLP triglycerides ($p < 0.05$) compared to baseline [58]. Numerous studies have shown that EPA is able to significantly reduce the volume of lipid plaque. By adding EPA at a dose of 1.8 g/day to background drug therapy with high-intensity statins, it significantly reduces the volume of lipid plaque and that of fibrotic plaque after 6 months (both $p < 0.05$) as measured by ultrasound intravascular [59]. Similarly, EPA at a dose of 1.8 g/day in co-administration with pitavastatin significantly reduced coronary plaque volume after 8 months of treatment compared to pitavastatin alone (−24% vs. −2%, $p < 0.01$) in patients with impaired glucose tolerance and angina pectoris [60]. In patients treated with EPA for 1 year, the tomography reported an important reduction in volume of soft plaque, contrary to what was seen in patients undergoing pharmacological treatment with ezetimibe [61]. The aforementioned studies have shown that the EPA, in general, exerts protective effects on the development of arteriosclerosis processes; in particular, it is clear in its involvement in the protection of endothelial function, in the reduction of oxidative stress levels, and in the maintenance of low inflammatory levels, which see the release of cytokines, platelet aggregation, and thrombus formation. Furthermore, EPA is also able to act on the reduction of atherogenic dyslipidemia levels with numerous benefits given by the ability to intercalate in the lipid bilayer of the plasma membrane. Of considerable pharmacological interest is the data that demonstrate how the beneficial effects of EPA are still maintained if administered together with statins. A clinical trial, randomized and controlled, called REDUCE-IT (NCT01492361), is now available, developed to evaluate the administration of EPA in a highly purified formulation [62]. The drug currently under study is called icosapent ethyl and represents a high purity formulation containing the ethyl ester of EPA. The recommended therapeutic dose in the trial is 4 g/day to be taken as two 1 g capsules twice daily with food. Therapeutic results were also achieved in the MARINE and ANCHOR studies. In particular, in the MARINE study, conducted in patients with elevated triglyceride levels, treatment with icosapent ethyl at a dose of 4 g/day for 12 weeks significantly reduced their levels by 33.1% ($p < 0.0001$), VLDL triglycerides by 25.8% ($p = 0.0023$), VLDL cholesterol (VLDL-C) by 28.6% ($p = 0.0002$), non-high density lipoprotein cholesterol (not HDL-C) by 17.7% ($p < 0.0001$), and Apo B by 8.5% ($p = 0.0019$) compared to the placebo values [63]. In the ANCHOR study conducted in patients on drug treatment with statins and elevated triglycerides, treatment with icosapent ethyl at a dose of 4 g/day for 12 weeks significantly reduced triglyceride levels by 21.5% ($p < 0.0001$), LDL-C by 6.2% ($p = 0.0067$), VLDL triglycerides by 26.5% ($p < 0.0001$), C-VLDL by 24.4% ($p < 0.0001$), C-non-HDL by 13.6% ($p < 0.0001$), and Apo B by 9.3% ($p < 0.0001$) compared to placebo [63]. On the basis of the data obtained from the analysis of the different clinical trials, a new pharmacoeconomic model was prepared that highlighted how the combination of EPA and statin for the secondary prevention of cardiovascular diseases is now associated with important economic savings and utility therapeutics when compared with single statin treatment [64].

7. Conclusions

Numerous studies have now shown that the inflammatory process is the basis of various diseases; in this context, it is now recognized that the role of fatty acids is to control inflammation by changing their composition in cell membranes, for example making it more fluid or altering its gene expression. The key cells of the inflammatory process are rich in arachidonic acid n-6 fatty acid, but their content can vary through oral administration of EPA and DHA. The increase in the membrane content of EPA and DHA causes a change in the production pattern of eicosanoids and probably also of resolvins. Given the involvement that n-3 marine PUFAs have in modulating inflammatory responses, it is understood how these can be decisive in inflammatory process and resolution. The clinical data obtained from the anti-inflammatory evaluations obtained thanks to the role of EPA have raised the awareness that an increase in the diet could bring a clinical benefit. There have been numerous studies that have provided encouraging data in patients with rheumatoid arthritis [65], though fewer in patients suffering from inflammatory bowel diseases [2]. With regard to these diseases, the data between adults and children are conflicting; in the latter, there are greater therapeutic successes [66]. A very interesting study was conducted in pregnant women where the administration of EPA determined beneficial effects on both the maternal and fetal immune systems [67,68], so as to reduce the risk of onset of allergic diseases during early childhood [69]. These first evidences opened the field to study the effects of EPA on the immune system [70]. In the plethora of inflammatory diseases, an exception is represented by cardiovascular disease, of which, to date, studies on EPA have been most concentrated [71,72], whose effects are particularly evident in the stabilization of atherosclerotic plaques [73]. A univocal and clear dose of EPA to be used has not yet been identified, although it is clear that the therapeutic effects are strictly dose-dependent [23]. Studies conducted on volunteers have highlighted that an intake of >2 g EPA + DHA/day is required to modulate inflammatory processes. Studies in patients with rheumatoid arthritis used 1.5 to 7 g EPA + DHA/day (average about 3.5 g/day) and were long-lasting (3 to 12 months), with effects which appear after several months [74]. One study compared two different doses [75], showing that the greatest benefits were obtained with the higher dose. Reading data clearly demonstrate that fatty acids modulate inflammatory phenomena through various motive mechanisms related to changes in the fatty acid composition of cell membranes. Inflammatory cells are rich in n-6 arachidonic acid, but the content of arachidonic acid and n-3 fatty acids EPA and DHA may vary after oral administration of EPA and DHA. The EPA in itself can be a source of eicosanoids with characteristics different from those deriving from arachidonic acid. In fact, both EPA and DHA are sources of recently discovered resolvins with anti-inflammatory properties. The altered fatty acid composition of inflammatory cells also alters the synthesis of peptide mediators of inflammation such as adhesion molecules or cytokines.

Summarizing, data related to clinical investigation reported the following:

- EPA lowers both triglyceride and cholesterol levels [7,8];
- Unlike DHA, EPA does not cause the increase of low-density lipoprotein cholesterol (LDL-C) levels [63,76];
- EPA protects against oxidative damage and improves endothelium formation [77];
- EPA inhibits monocyte movement into early lesions and subsequent conversion to macrophages and foam cells [78];
- EPA supports anti-oxidant and anti-inflammatory functions of high- density lipoprotein (HDL) [79];
- EPA promotes HDL-mediated cholesterol efflux from macrophages [79];

- EPA reduces atherosclerotic plaque formation, progression, and vulnerability to rupture [80];
- EPA decreases platelet-mediated thrombus formation [81];
- EPA reduces blood pressure, likely attributable to improvement of endothelial function [82]; importantly, many of these effects have been observed with EPA alone or are additive to those of statins.

Author Contributions: Conceptualization, R.C. and S.C.; investigation (literature review), R.C.; writing—original draft preparation, R.C.; writing—review and editing, S.C. All authors have read and agreed to the published version of the manuscript.

Funding: This research received no external funding.

Institutional Review Board Statement: Not applicable.

Informed Consent Statement: Not applicable.

Data Availability Statement: Not applicable.

Conflicts of Interest: The authors declare no conflict of interest.

References

1. Liu, C.H.; Abrams, N.; Carrick, D.M.; Chander, P.; Dwyer, J.; Hamlet, M.R.J.; Macchiarini, F.; Prabhudas, M.; Shen, G.L.; Tandon, P.; et al. Biomarkers of chronic inflammation in disease development and prevention: Challenges and opportunities. *Nat. Immunol.* **2017**, *18*, 1175–1180. [CrossRef] [PubMed]
2. Calder, P.C. Fatty acids and immune function: Relevance to inflammatory bowel diseases. *Int. Rev. Immunol.* **2009**, *28*, 506–534. [CrossRef] [PubMed]
3. Liu, Q.K. Triglyceride-lowering and anti-inflammatory mechanisms of omega-3 polyunsaturated fatty acids for atherosclerotic cardiovascular risk reduction. *J. Clin. Lipidol.* **2021**, *15*, 556–568. [CrossRef] [PubMed]
4. de Winther, M.P.J.; Lutgens, E. The link between hematopoiesis and atherosclerosis. *N. Engl. J. Med.* **2019**, *380*, 1869–1871. [CrossRef]
5. Liu, A.G.; Ford, N.A.; Hu, F.B.; Zelman, K.M.; Mozaffarian, D.; Kris-Etherton, P.M. A healthy approach to dietary fats: Understanding the science and taking action to reduce consumer confusion. *Nutr. J.* **2017**, *16*, 53. [CrossRef]
6. Mehta, N.K.; Abrams, L.R.; Myrskyla, M. US life expectancy stalls due to cardiovascular disease, not drug deaths. *Proc. Natl. Acad. Sci. USA* **2020**, *117*, 6998–7000. [CrossRef]
7. Sergienko, I.V.; Ansheles, A.A.; Drapkina, O.M.; Gornyakova, N.B.; Zubareva, M.Y.; Shepel, R.N.; Kuharchuk, V.V.; Boytsov, S.A. ANICHKOV study: The effect of combined hypotensive and lipid-lowering therapy on cardiovascular complications in patients of high and very high risk. *Ter. Arkh.* **2019**, *91*, 90–98. [CrossRef]
8. Siasos, G.; Skotsimara, G.; Oikonomou, E.; Sagris, M.; Vasiliki-Chara, M.; Bletsa, E.; Stampouloglou, P.; Theofilis, P.; Charalampous, G.; Tousoulis, D. Antithrombotic treatment in diabetes mellitus: A review of the literature about antiplatelet and anticoagulation strategies used for diabetic patients in primary and secondary prevention. *Curr. Pharm. Des.* **2020**, *26*, 2780–2788. [CrossRef]
9. Bäck, M.; Yurdagul, A., Jr.; Tabas, I.; Öörni, K.; Kovanen, P.T. Inflammation and its resolution in atherosclerosis: Mediators and therapeutic opportunities. *Nat. Rev. Cardiol.* **2019**, *16*, 389–406. [CrossRef]
10. Saini, R.K.; Keum, Y.S. Omega-3 and omega-6 polyunsaturated fatty acids: Dietary sources, metabolism, and significance—A review. *Life Sci.* **2018**, *203*, 255–267. [CrossRef]
11. Balić, A.; Vlašić, D.; Žužul, K.; Marinović, B.; Bukvić Mokos, Z. Omega-3 versus omega-6 polyunsaturated fatty acids in the prevention and treatment of inflammatory skin diseases. *Int. J. Mol. Sci.* **2020**, *21*, 741. [CrossRef]
12. Weintraub, H.S. Overview of prescription omega-3 fatty acid products for hypertriglyceridemia. *Postgrad. Med.* **2014**, *126*, 7–18. [CrossRef] [PubMed]
13. Larsson, S.C.; Kumlin, M.; Ingelman-Sundberg, M.; Wolk, A. Dietary long-chain n-3 fatty acids for the prevention of cancer: A review of potential mechanisms. *Am. J. Clin. Nutr.* **2004**, *79*, 935–945. [CrossRef] [PubMed]
14. Yaqoob, P.; Newsholme, E.A.; Calder, P.C. Influence of cell culture conditions on diet-induced changes in lymphocyte fatty acid composition. *Biochim. Biophys. Acta* **1995**, *1255*, 333–340. [CrossRef]
15. Surette, M.E.; Whelan, J.; Lu, G.; Hardard'Ottir, I.; Kinsella, J.E. Dietary n-3 polyunsaturated fatty acids modify Syrian hamster platelet and macrophage phospholipid fatty acyl composition and eicosanoid synthesis: A controlled study. *Biochim. Biophys. Acta.* **1995**, *1255*, 185–191. [CrossRef]
16. Calder, P.C.; Bond, J.A.; Harvey, D.J.; Gordon, S.; Newsholme, E.A. Uptake and incorporation of saturated and unsaturated fatty acids into macrophage lipids and their effect upon macrophage adhesion and phagocytosis. *Biochem. J.* **1990**, *269*, 807–814. [CrossRef]

17. Palombo, J.D.; Demichele, S.J.; Lydon, E.; Bistrian, B.R. Cyclic vs continuous enteral feeding with omega-3 and gamma-linolenic fatty acids: Effects on modulation of phospholipid fatty acids in rat lung and liver immune cells. *JPEN J. Parenter. Enter. Nutr.* **1997**, *21*, 123–132. [CrossRef]
18. Careaga-Houck, M.; Sprecher, H. Effect of a fish oil diet on the composition of rat neutrophil lipids and the molecular species of choline and ethanolamine glycerophospholipids. *J. Lipid Res.* **1989**, *30*, 77–87. [CrossRef]
19. Lokesh, B.R.; Hsieh, H.L.; Kinsella, J.E. Peritoneal macrophages from mice fed dietary (n-3) polyunsaturated fatty acids secrete low levels of prostaglandins. *J. Nutr.* **1986**, *116*, 2547–2552. [CrossRef]
20. Chapkin, R.; Hubbard, N.E.; Erickson, K.L. 5-series peptido-leukotriene synthesis in mouse peritoneal macrophages: Modulation by dietary n-3 fatty acids. *Biochem. Biophys. Res. Commun.* **1990**, *171*, 764–769. [CrossRef]
21. Endres, S.; Ghorbani, R.; Kelley, V.E.; Georgilis, K.; Lonnemann, G.; Van Der Meer, J.W.M.; Cannon, J.G.; Rogers, T.S.; Klempner, M.S.; Weber, P.C.; et al. The effect of dietary supplementation with n-3 polyunsaturated fatty acids on the synthesis of interleukin-1 and tumor necrosis factor by mononuclear cells. *N. Engl. J. Med.* **1989**, *320*, 265–271. [CrossRef]
22. Thies, F.; Nebe-von-Caron, G.; Powell, J.R.; Yaqoob, P.; Newsholme, E.A.; Calder, P.C. Dietary supplementation with gamma-linolenic acid or fish oil decreases T lymphocyte proliferation in healthy older humans. *J. Nutr.* **2001**, *131*, 1918–1927. [CrossRef]
23. Rees, D.; Miles, E.A.; Banerjee, T.; Wells, S.J.; Roynette, C.E.; Wahle, K.W.; Calder, P.C. Dose-related effects of eicosapentaenoic acid on innate immune function in healthy humans: A comparison of young and older men. *Am. J. Clin. Nutr.* **2006**, *83*, 331–342. [CrossRef] [PubMed]
24. Miles, E.A.; Banerjee, T.; Calder, P.C. The influence of different combinations of gamma-linolenic, stearidonic and eicosapentaenoic acids on the fatty acid composition of blood lipids and mononuclear cells in human volunteers. *Prostaglandins Leukot. Essent. Fat. Acids* **2004**, *70*, 529–538. [CrossRef]
25. Matsumoto, M.; Sata, M.; Fukuda, D.; Tanaka, K.; Soma, M.; Hirata, Y.; Nagai, R. Orally administered eicosapentaenoic acid reduces and stabilizes atherosclerotic lesions in ApoE-deficient mice. *Atherosclerosis* **2008**, *197*, 524–533. [CrossRef] [PubMed]
26. Salic, K.; Morrison, M.C.; Verschuren, L.; Wielinga, P.Y.; Wu, L.; Kleemann, R.; Gjorstrup, P.; Kooistra, T. Resolvin E1 attenuates atherosclerosis in absence of cholesterol-lowering effects and on top of atorvastatin. *Atherosclerosis* **2016**, *250*, 158–165. [CrossRef] [PubMed]
27. Nakajima, K.; Yamashita, T.; Kita, T.; Takeda, M.; Sasaki, N.; Kasahara, K.; Shinohara, M.; Rikitake, Y.; Ishida, T.; Yokoyama, M.; et al. Orally administered eicosapentaenoic acid induces rapid regression of atherosclerosis via modulating the phenotype of dendritic cells in LDL receptor-deficient mice. *Arter. Thromb. Vasc. Biol.* **2011**, *31*, 1963–1972. [CrossRef]
28. Lau, A.C.; Jongstra-Bilen, J.; Cybulsky, M.I. Eicosapentaenoic acid and regression of atherosclerotic lesions: A role for dendritic cells. *Arter. Thromb. Vasc. Biol.* **2011**, *31*, 1943–1945. [CrossRef]
29. Jacobson, T.A.; Ito, M.K.; Maki, K.C.; Orringer, C.E.; Bays, H.E.; Jones, P.H.; McKenney, J.M.; Grundy, S.M.; Gill, E.A.; Wild, R.A.; et al. National Lipid Association recommendations for patient-centered management of dyslipidemia: Part 1—Executive summary. *J. Clin. Lipidol.* **2014**, *8*, 473–488. [CrossRef]
30. Teramoto, T.; Sasaki, J.; Ishibashi, S.; Birou, S.; Daida, H.; Dohi, S.; Egusa, G.; Hiro, T.; Hirobe, K.; Iida, M.; et al. Executive summary of the Japan atherosclerosis society (JAS) guidelines for the diagnosis and prevention of atherosclerotic cardiovascular diseases in Japan 2012 version. *J. Atheroscler. Thromb.* **2013**, *20*, 517–523. [CrossRef] [PubMed]
31. Yokoyama, M.; Origasa, H.; Matsuzaki, M.; Matsuzawa, Y.; Saito, Y.; Ishikawa, Y.; Oikawa, S.; Sasaki, J.; Hishida, H.; Itakura, H.; et al. Effects of eicosapentaenoic acid on major coronary events in hypercholesterolaemic patients (JELIS): A randomised open-label, blinded endpoint analysis. *Lancet* **2007**, *369*, 1090–1098. [CrossRef]
32. Tavazzi, L.; Maggioni, A.P.; Marchioli, R.; Barlera, S.; Franzosi, M.G.; Latini, R.; Lucci, D.; Nicolosi, G.L.; Porcu, M.; Tognoni, G. Effect of n-3 polyunsaturated fatty acids in patients with chronic heart failure (the GISSI-HF trial): A randomised, double-blind, placebo-controlled trial. *Lancet* **2008**, *372*, 1223–1230. [PubMed]
33. Bhatt, D.L.; Steg, P.G.; Miller, M.; Brinton, E.A.; Jacobson, T.A.; Ketchum, S.B.; Doyle, R.T., Jr.; Juliano, R.A.; Jiao, L.; Granowitz, C.; et al. Cardiovascular Risk Reduction with Icosapent Ethyl for Hypertriglyceridemia. *N. Engl. J. Med.* **2019**, *380*, 11–22. [CrossRef] [PubMed]
34. ORIGIN Trial Investigators. n-3 Fatty acids and cardiovascular outcomes in patients with dysglycemia. *N. Engl. J. Med.* **2012**, *367*, 309–318. [CrossRef]
35. Risk and Prevention Study Collaborative Group. n-3 Fatty acids in patients with multiple cardiovascular risk factors. *N. Engl. J. Med.* **2013**, *368*, 1800–1808. [CrossRef] [PubMed]
36. Kromhout, D.; Giltay, E.J.; Geleijnse, J.M. n-3 Fatty acids and cardiovascular events after myocardial infarction. *N. Engl. J. Med.* **2010**, *363*, 2015–2026. [CrossRef]
37. ASCEND Study Collaborative Group. Effects of n-3 fatty acid supplements in diabetes mellitus. *N. Engl. J. Med.* **2018**, *379*, 1540–1550. [CrossRef]
38. Manson, J.E.; Cook, N.R.; Lee, I.-M.; Christen, W.; Bassuk, S.S.; Mora, S.; Gibson, H.; Albert, C.M.; Gordon, D.; Copeland, T.; et al. Marine n-3 fatty acids and prevention of cardiovascular disease and cancer. *N. Engl. J. Med.* **2019**, *380*, 23–32. [CrossRef]
39. Nicholls, S.J.; Lincoff, A.M.; Garcia, M.; Bash, D.; Ballantyne, C.M.; Barter, P.J.; Davidson, M.H.; Kastelein, J.J.; Koenig, W.; McGuire, D.K.; et al. Effect of high-dose omega-3 fatty acids vs. corn oil on major adverse cardiovascular events in patients at high cardiovascular risk: The STRENGTH randomized clinical trial. *JAMA* **2020**, *324*, 2268–2280. [CrossRef]

40. Rizos, E.C.; Ntzani, E.E.; Bika, E.; Kostapanos, M.S.; Elisaf, M.S. Association between omega-3 fatty acid supplementation and risk of major cardiovascular disease events: A systematic review and meta-analysis. *JAMA* **2012**, *308*, 1024–1033. [CrossRef]
41. Elbadawi, A.; Olorunfemi, O.; Ogunbayo, G.O.; Saad, M.; Elgendy, I.Y.; Arif, Z.; Badran, H.; Saheed, D.; Ahmed, H.M.A.; Rao, M. Cardiovascular outcomes with surgical left atrial appendage exclusion in patients with atrial fibrillation who underwent valvular heart surgery (from the national inpatient sample database). *Am. J. Cardiol.* **2017**, *119*, 2056–2060. [CrossRef] [PubMed]
42. Hu, Y.; Hu, F.B.; Manson, J.E. Marine omega-3 supplementation and cardiovascular disease: An updated meta-analysis of 13 randomized controlled trials involving 127 477 participants. *J. Am. Heart Assoc.* **2019**, *8*, e013543. [CrossRef] [PubMed]
43. Orringer, C.E.; Jacobson, T.A.; Maki, K.C. National Lipid Association Scientific Statement on the use of icosapent ethyl in statin-treated patients with elevated triglycerides and high or very-high ASCVD risk. *J. Clin. Lipidol.* **2019**, *13*, 860–872. [CrossRef] [PubMed]
44. Aim-High Investigators. Niacin in patients with low HDL cholesterol levels receiving intensive statin therapy. *N. Engl. J. Med.* **2011**, *365*, 2255–2267. [CrossRef] [PubMed]
45. Keech, A.C.; Simes, R.J.; Barter, P.J.; Best, J.; Scott, R.A.P.; Taskinen, M.-R.; Forder, P.M.; Pillai, A.; Davis, T.M.; Glasziou, P.; et al. Effects of long-term fenofibrate therapy on cardiovascular events in 9795 people with type 2 diabetes mellitus (the FIELD study): Randomised controlled trial. *Lancet* **2005**, *366*, 1849–1861. [CrossRef]
46. Pradhan, A.D.; Paynter, N.P.; Everett, B.M.; Glynn, R.J.; Amarenco, P.; Elam, M.; Ginsberg, H.; Hiatt, W.R.; Ishibashi, S.; Koenig, W.; et al. Rationale and design of the pemafibrate to reduce cardiovascular outcomes by reducing triglycerides in patients with diabetes (PROMINENT) study. *Am. Hear. J.* **2018**, *206*, 80–93. [CrossRef] [PubMed]
47. Tousoulis, D.; Kampoli, A.-M.; Tentolouris, C.; Papageorgiou, N.; Stefanadis, C. The role of nitric oxide on endothelial function. *Curr. Vasc. Pharmacol.* **2012**, *10*, 4–18. [CrossRef]
48. Montezano, A.; Touyz, R.M. Reactive oxygen species and endothelial function—Role of nitric oxide synthase uncoupling and nox family nicotinamide adenine dinucleotide phosphate oxidases. *Basic Clin. Pharmacol. Toxicol.* **2012**, *110*, 87–94. [CrossRef]
49. Lee, C.-H.; Lee, S.-D.; Ou, H.-C.; Lai, S.-C.; Cheng, Y.-J. Eicosapentaenoic acid protects against palmitic acid-induced endothelial dysfunction via activation of the AMPK/eNOS pathway. *Int. J. Mol. Sci.* **2014**, *15*, 10334–10349. [CrossRef]
50. Mason, R.P.; Jacob, R.F. Eicosapentaenoic acid inhibits glucose-induced membrane cholesterol crystalline domain formation through a potent antioxidant mechanism. *Biochim. Biophys. Acta* **2015**, *1848*, 502–509. [CrossRef]
51. Chiu, S.-C.; Chiang, E.-P.I.; Tsai, S.-Y.; Wang, F.-Y.; Pai, M.-H.; Syu, J.-N.; Cheng, C.-C.; Rodriguez, R.L.; Tang, F.-Y. Eicosapentaenoic acid induces neovasculogenesis in human endothelial progenitor cells by modulating c-kit protein and PI3-K/Akt/eNOS signaling pathways. *J. Nutr. Biochem.* **2014**, *25*, 934–945. [CrossRef] [PubMed]
52. Bays, H.E.; Ballantyne, C.M.; Braeckman, R.A.; Stirtan, W.G.; Soni, P.N. Icosapent ethyl, a pure ethyl ester of eicosapentaenoic acid: Effects on circulating markers of inflammation from the MARINE and ANCHOR studies. *Am. J. Cardiovasc. Drugs* **2013**, *13*, 37–46. [CrossRef]
53. Toyama, K.; Nishioka, T.; Isshiki, A.; Ando, T.; Inoue, Y.; Kirimura, M.; Kamiyama, T.; Sasaki, O.; Ito, H.; Maruyama, Y.; et al. Eicosapentaenoic acid combined with optimal statin therapy improves endothelial dysfunction in patients with coronary artery disease. *Cardiovasc. Drugs Ther.* **2014**, *28*, 53–59. [CrossRef] [PubMed]
54. Yamakawa, K.; Shimabukuro, M.; Higa, N.; Asahi, T.; Ohba, K.; Arasaki, O.; Higa, M.; Oshiro, Y.; Yoshida, H.; Higa, T.; et al. Eicosapentaenoic acid supplementation changes fatty acid composition and corrects endothelial dysfunction in hyperlipidemic patients. *Cardiol. Res. Pract.* **2012**, *2012*, 754181. [CrossRef]
55. Takaki, A.; Umemoto, S.; Ono, K.; Seki, K.; Ryoke, T.; Fujii, A.; Itagaki, T.; Harada, M.; Tanaka, M.; Yonezawa, T.; et al. Add-on therapy of EPA reduces oxidative stress and inhibits the progression of aortic stiffness in patients with coronary artery disease and statin therapy: A randomized controlled study. *J. Atheroscler. Thromb.* **2011**, *18*, 857–866. [CrossRef]
56. Ballantyne, C.M.; Bays, H.E.; Philip, S.; Doyle, R.T.; Braeckman, R.A.; Stirtan, W.G.; Soni, P.N.; Juliano, R.A. Icosapent ethyl (eicosapentaenoic acid ethyl ester): Effects on remnant-like particle cholesterol from the MARINE and ANCHOR studies. *Atherosclerosis* **2016**, *253*, 81–87. [CrossRef]
57. Brinton, E.A.; Ballantyne, C.M.; Bays, H.E.; Kastelein, J.J.; Braeckman, R.A.; Soni, P.N. Effects of icosapent ethyl on lipid and inflammatory parameters in patients with diabetes mellitus-2, residual elevated triglycerides (200–500 mg/dL), and on statin therapy at LDL-C goal: The ANCHOR study. *Cardiovasc. Diabetol.* **2013**, *12*, 100. [CrossRef]
58. Satoh, N.; Shimatsu, A.; Kotani, K.; Sakane, N.; Yamada, K.; Suganami, T.; Kuzuya, H.; Ogawa, Y. Purified eicosapentaenoic acid reduces small dense LDL, remnant lipoprotein particles, and C-reactive protein in metabolic syndrome. *Diabetes Care* **2007**, *30*, 144–146. [CrossRef]
59. Karyagina, A.S.; Grunina, T.; Lyaschuk, A.M.; Voronina, E.V.; Marigin, R.A.; Cherepushkin, S.; Trusova, I.N.; Grishin, A.V.; Poponova, M.S.; Orlova, P.A.; et al. Recombinant human erythropoietin proteins synthesized in escherichia coli cells: Effects of additional domains on the in vitro and in vivo activities. *Biochemistry* **2019**, *84*, 20–32. [CrossRef] [PubMed]
60. Takashima, H.; Ozaki, Y.; Yasukawa, T.; Waseda, K.; Asai, K.; Wakita, Y.; Kuroda, Y.; Kosaka, T.; Kuhara, Y.; Ito, T. Impact of lipid-lowering therapy with pitavastatin, a new HMG-CoA reductase inhibitor, on regression of coronary atherosclerotic plaque. *Circ. J.* **2007**, *71*, 1678–1684. [CrossRef]
61. Goto, Y.Y.; Kawasaki, T.; Koga, N.; Tanaka, H.; Koga, H.; Orita, Y.; Ikeda, S.; Shintani, Y.; Kajiwara, M.; Fukuyama, T. Plaque distribution patterns in left main trunk bifurcations: Prediction of branch vessel compromise by multidetector row computed topography after percutaneous coronary intervention. *EuroIntervention* **2012**, *8*, 708–716. [CrossRef] [PubMed]

62. Nichols, G.A.; Philip, S.; Reynolds, K.; Granowitz, C.B.; Fazio, S. Increased residual cardiovascular risk in patients with diabetes and high versus normal triglycerides despite statin-controlled LDL cholesterol. *Diabetes Obes. Metab.* **2019**, *21*, 366–371. [CrossRef]
63. Ballantyne, C.M.; Bays, H.E.; Kastelein, J.J.; Stein, E.; Isaacsohn, J.L.; Braeckman, R.A.; Soni, P.N. Efficacy and safety of eicosapentaenoic acid ethyl ester (AMR101) therapy in statin-treated patients with persistent high triglycerides (from the ANCHOR study). *Am. J. Cardiol.* **2012**, *110*, 984–992. [CrossRef] [PubMed]
64. Philip, S.; Chowdhury, S.; Nelson, J.R.; Everett, P.B.; Hulme-Lowe, C.K.; Schmier, J.K. A novel cost-effectiveness model of prescription eicosapentaenoic acid extrapolated to secondary prevention of cardiovascular diseases in the United States. *J. Med. Econ.* **2016**, *19*, 1003–1010. [CrossRef] [PubMed]
65. Calder, P.C. Polyunsaturated fatty acids, inflammatory processes and inflammatory bowel diseases. *Mol. Nutr. Food Res.* **2008**, *52*, 885–897. [CrossRef]
66. Kremmyda, L.-S.; Vlachava, M.; Noakes, P.S.; Diaper, N.D.; Miles, E.A.; Calder, P.C. Atopy risk in infants and children in relation to early exposure to fish, oily fish, or long-chain omega-3 fatty acids: A systematic review. *Clin. Rev. Allergy Immunol.* **2011**, *41*, 36–66. [CrossRef]
67. Dunstan, J.A.; Mori, T.A.; Barden, A.; Beilin, L.J.; Taylor, A.L.; Holt, P.G.; Prescott, S.L. Fish oil supplementation in pregnancy modifies neonatal allergen-specific immune responses and clinical outcomes in infants at high risk of atopy: A randomized, controlled trial. *J. Allergy Clin. Immunol.* **2003**, *112*, 1178–1184. [CrossRef]
68. Krauss-Etschmann, S.; Hartl, D.; Rzehak, P.; Heinrich, J.; Shadid, R.; del Carmen Ramírez-Tortosa, M.; Campoy, C.; Pardillo, S.; Schendel, D.J.; Decsi, T.; et al. Decreased cord blood IL-4, IL-13, and CCR4 and increased TGF-beta levels after fish oil supplementation of pregnant women. *J. Allergy Clin. Immunol.* **2008**, *121*, 464–470 e6. [CrossRef]
69. Olsen, S.F.; Østerdal, M.L.; Salvig, J.D.; Mortensen, L.M.; Rytter, D.; Secher, N.J.; Henriksen, T.B. Fish oil intake compared with olive oil intake in late pregnancy and asthma in the offspring: 16 y of registry-based follow-up from a randomized controlled trial. *Am. J. Clin. Nutr.* **2008**, *88*, 167–175. [CrossRef]
70. Sala-Vila, A.; Miles, E.A.; Calder, P. Fatty acid composition abnormalities in atopic disease: Evidence explored and role in the disease process examined. *Clin. Exp. Allergy* **2008**, *38*, 1432–1450. [CrossRef]
71. Calder, P.C.; Yaqoob, P. Omega-3 (n-3) fatty acids, cardiovascular disease and stability of atherosclerotic plaques. *Cell. Mol. Biol.* **2010**, *56*, 28–37.
72. Calder, P.C. n-3 Fatty acids and cardiovascular disease: Evidence explained and mechanisms explored. *Clin. Sci.* **2004**, *107*, 1–11. [CrossRef] [PubMed]
73. Thies, F.; Garry, J.M.C.; Yaqoob, P.; Rerkasem, K.; Williams, J.; Shearman, C.P.; Gallagher, P.J.; Calder, P.; Grimble, R.F. Association of n-3 polyunsaturated fatty acids with stability of atherosclerotic plaques: A randomised controlled trial. *Lancet* **2003**, *361*, 477–485. [CrossRef]
74. Calder, P.C.; Campoy, C.; Eilander, A.; Fleith, M.; Forsyth, S.; Larsson, P.-O.; Schelkle, B.; Lohner, S.; Szommer, A.; Van De Heijning, B.J.M.; et al. A systematic review of the effects of increasing arachidonic acid intake on PUFA status, metabolism and health-related outcomes in humans. *Br. J. Nutr.* **2019**, *121*, 1201–1214. [CrossRef] [PubMed]
75. Kremer, J.M.; Lawrence, D.A.; Jubiz, W.; Digiacomo, R.; Rynes, R.; Bartholomew, L.E.; Sherman, M. Dietary fish oil and olive oil supplementation in patients with Rheumatoid Arthritis clinical and immunologic effects. *Arthritis Care Res.* **1990**, *33*, 810–820. [CrossRef]
76. Bays, H.E.; Ballantyne, C.M.; Kastelein, J.J.; Isaacsohn, J.L.; Braeckman, R.A.; Soni, P.N. Eicosapentaenoic acid ethyl ester (AMR101) therapy in patients with very high triglyceride levels (from the multi-center, plAcebo-controlled, randomized, double-blINd, 12-week study with an open-label extension [MARINE] trial). *Am. J. Cardiol.* **2011**, *108*, 682–690. [CrossRef] [PubMed]
77. Mason, R.P.; Walter, M.F.; Jacob, R.F. Effects of HMG-CoA reductase inhibitors on endothelial function: Role of microdomains and oxidative stress. *Circulation* **2004**, *109*, II34–II41. [CrossRef] [PubMed]
78. Cawood, A.L.; Ding, R.; Napper, F.L.; Young, R.H.; Williams, J.A.; Ward, M.J.; Gudmundsen, O.; Vige, R.; Payne, S.P.; Ye, S.; et al. Eicosapentaenoic acid (EPA) from highly concentrated n-3 fatty acid ethyl esters is incorporated into advanced atherosclerotic plaques and higher plaque EPA is associated with decreased plaque inflammation and increased stability. *Atherosclerosis* **2010**, *212*, 252–259. [CrossRef]
79. Tanaka, N.; Ishida, T.; Nagao, M.; Mori, T.; Monguchi, T.; Sasaki, M.; Mori, K.; Kondo, K.; Nakajima, H.; Honjo, T.; et al. Administration of high dose eicosapentaenoic acid enhances anti-inflammatory properties of high-density lipoprotein in Japanese patients with dyslipidemia. *Atherosclerosis* **2014**, *237*, 577–583. [CrossRef]
80. Liu, L.; Yang, W.; Nagahara, Y.; Li, Y.; Lamooki, S.R.; Muramatsu, T.; Kitslaar, P.; Sarai, M.; Ozaki, Y.; Barlis, P.; et al. The impact of image resolution on computation of fractional flow reserve: Coronary computed tomography angiography versus 3-dimensional quantitative coronary angiography. *Int. J. Cardiovasc. Imaging* **2016**, *32*, 513–523. [CrossRef]
81. Gajos, G.; Rostoff, P.; Undas, A.; Piwowarska, W. Effects of polyunsaturated omega-3 fatty acids on responsiveness to dual antiplatelet therapy in patients undergoing percutaneous coronary intervention: The OMEGA-PCI (OMEGA-3 fatty acids after pci to modify responsiveness to dual antiplatelet therapy) study. *J. Am. Coll. Cardiol.* **2010**, *55*, 1671–1678. [PubMed]
82. Mozaffarian, D.; Wu, J.H.Y. Omega-3 fatty acids and cardiovascular disease: Effects on risk factors, molecular pathways, and clinical events. *J. Am. Coll. Cardiol.* **2011**, *58*, 2047–2067. [CrossRef] [PubMed]

MDPI
St. Alban-Anlage 66
4052 Basel
Switzerland
www.mdpi.com

Biomolecules Editorial Office
E-mail: biomolecules@mdpi.com
www.mdpi.com/journal/biomolecules

www.ingramcontent.com/pod-product-compliance
Lightning Source LLC
Chambersburg PA
CBHW061459210926
51850CB00008B/1759
9783036594743